Birkhäuser

Progress in Probability
Volume 69

More information about this series at
http://www.springer.com/series/4839

Ramsés H. Mena • Juan Carlos Pardo •
Víctor Rivero • Gerónimo Uribe Bravo
Editors

XI Symposium on Probability and Stochastic Processes

CIMAT, Mexico, November 18-22, 2013

Birkhäuser

Editors
Ramsés H. Mena
Probabilidad y Estadística
IIMAS UNAM
Mexico

Juan Carlos Pardo
CIMAT
Guanajuato, Mexico

Víctor Rivero
CIMAT
Guanajuato, Mexico

Gerónimo Uribe Bravo
Instituto de Matemáticas
UNAM
Mexico

ISSN 1050-6977 ISSN 2297-0428 (electronic)
Progress in Probability
ISBN 978-3-319-13983-8 ISBN 978-3-319-13984-5 (eBook)
DOI 10.1007/978-3-319-13984-5

Library of Congress Control Number: 2015945576

Mathematics Subject Classification (2010): 60-06

Springer Cham Heidelberg New York Dordrecht London

Printed on acid-free paper

Springer International Publishing AG Switzerland is part of Springer Science+Business Media (www.birkhauser-science.com)

Introduction

The present volume contains contributions and lecture notes of the XI Symposium on Probability and Stochastic Processes, held at Centro de Investigación en Matemáticas, México (CIMAT), November 18–22 in 2013.

This Symposium traces its roots back to December 1988 at CIMAT, when it was held for the first time, and constitutes one of the main events in the area happening biannually in various academic institutions in Mexico. During these more than 25 years, this series of symposia continuously accomplishes its main goal of exchanging ideas and discussing the current developments in the field by gathering national and international researchers as well as graduate students.

On this occasion, the Symposium was part of many activities organised in Mexico to celebrate the International Year of Statistics 2013. It gathered scholars from over 10 countries and included a wide set of topics that highlighted the interaction between statistics and stochastic processes. The scientific programme included two courses: 'Probabilistic aspects of minimal spanning trees' by Louigi Addario-Berry and 'Spatial point pattern analysis' by Carlos Díaz-Avalos. The event also benefited from six plenary conferences delivered by Loïc Chaumont, Janos Englander, Enrique Figueroa, Daniel Hernández, Andreas Kyprianou and Mark Podolskij; eight thematic sessions; eight contributed talks; and several poster presentations.

The volume begins with the lecture notes by Addario-Berry, providing with an accessible description of some features of the multiplicative coalescent and its connection with random graphs and minimum spanning tree. The tutorial is then followed by the illustrative article by Arizmendi and Gaxiola where they show that the large N-fold limit of the spectral distribution of a connected graph converges to a certain centred Bernoulli distribution. Baudoin studies stochastic differential equations driven by Brownian loops in a free Carnot group providing with sufficient conditions to ensure that solutions admit smooth densities. By considering a simple age distribution, Blath, Eldon, González Casanova and Kurt investigate the behaviour of the genealogy of a Wright-Fisher population model under the influence of a strong seed-bank effect. A Vervaat-like pathwise construction of a process with cyclically exchangeable increments with a predetermined minimum is introduced and analysed by Chaumont and Uribe; it consists of inverting the paths of the process at an adequately chosen random time τ, which surprisingly happens to be independent and to follow a uniform distribution. Motivated by different extensions of the so-called Bercovici-Pata bijection, Dominguez-Molina and Rocha Arteaga obtain an interesting sample path representation by covariation processes of matrix Lévy processes of rank-one jumps. Using the thermodynamic interpretation of the α parameter of a Poisson-Dirichlet distribution, i.e. as the ratio between the temperature T and a critical temperature Tc, Feng and Zhou study the asymptotic behaviour of such distribution as the temperature approaches the critical value. Gordienko, Martinez and Ruiz de Chavez provide estimates of the stability index with respect to the total variation metric and Prokhorov distance for a total-reward Markov decision chains with an absorbing set. The problem of determining a price for a contingent claim in an incomplete market is analysed by Hernández-Hernández and Sheu; they propose a pricing based on the utility, which is assumed to be exponential, and show that there exists a unique solution to optimal control problems. Relying heavily on the use of the so-called Dynkin-Kuznetsov N-measures, Murillo and Pérez Garmendia provide a pathwise backbone decomposition for supercritical superprocesses with nonlocal branching. Their result complements a related result obtained for supercritical superprocesses without nonlocal branching. Pedersen and Sauri present a detailed study of the stationary distribution of Lévy semi-stationary processes with a gamma kernel; they establish conditions for absolute continuity, infinite divisibility and self-decomposability, together with descriptions of its characteristics. The paper by Podolskij constitutes a thorough review of the theory of ambit fields, which is a flexible model for dynamical structures in time and/or space; the interesting list of open problems included will challenge any reader and generate deep research in the field.

In summary, the high quality and variety of these contributions give a broad panorama of the rich academic programme of the Symposium and of its impact. It is worth noting that all papers, including the invited course lecture notes, were subject to a strict refereeing process with high international standards. We are very grateful to the referees, many of whom are leading experts in their fields, for their careful and useful reports. Their comments were implemented by the authors allowing to improve the material here presented.

We would also like to extend our gratitude to all the authors whose original contributions appear published here as well as to all the speakers and session organisers in the Symposium for their stimulating talks and support. Their valuable contributions encourage the interest and activity in the area of probability and stochastic processes in Mexico.

We hold in high regard the editors of the series Progress in Probability: Davar Khoshnevisan, Andreas E. Kyprianou and Sidney I. Resnick for giving us the possibility to publish the Symposium Volume in this prestigious series.

Special thanks go to the Symposium venue CIMAT and its staff for its great hospitality and for providing excellent conference facilities. We are indebted to Rosy Davalos whose outstanding organisational work permitted us to concentrate mainly in the academic aspects of the conference.

The Symposium as well as the edition of this volume would not have been possible without the generous support of our sponsors: Centro de Investigación en Matemáticas, International Year of Statistics 2013 (Mexico), Laboratorio Internacional Solomon Lefschetz CNRS-CONACYT, Instituto de Investigaciones en Matemáticas Aplicadas y Sistemas and Instituto de Matemáticas, UNAM.

Finally, we hope the reader of this volume enjoys learning about the various topics treated as much as we did editing it.

Ramsés H. Mena Chavez
Juan Carlos Pardo
Víctor Manuel Rivero
Gerónimo Uribe

Previous Volumes from the Symposium on Probability and Stochastic Processes

- M. E. Caballero and L. G. Gorostiza, editors. *Simposio de Probabilidad y Procesos Estocásticos*, volume 4 of *Aportaciones Matemáticas: Notas de Investigación [Mathematical Contributions: Research Notes]*. Sociedad Matemática Mexicana, México, 1989.
  Held at the Universidad de Guanajuato, Guanajuato, December 12–14, 1988.
- M. E. Caballero and L. G. Gorostiza, editors. *II Simposio de Probabilidad y Procesos Estocásticos. I Encuentro México-Chile de Análisis Estocástico*, volume 7 of *Aportaciones Matemáticas: Notas de Investigación [Mathematical Contributions: Research Notes]*. Sociedad Matemática Mexicana, México, 1992.
  Held at the Universidad de Guanajuato, Guanajuato, January 27–31, 1992.
- M. E. Caballero and L. G. Gorostiza, editors. *III Simposio de Probabilidad y Procesos Estocásticos*, volume 11 of *Aportaciones Matemáticas: Notas de Investigación [Mathematical Contributions: Research Notes]*. Sociedad Matemática Mexicana, México, 1994.
  Held at the Universidad de Sonora, Hermosillo, March 7–11, 1994.
- L. G. Gorostiza, J. A. León, and J. A. López-Mimbela, editors. *IV Simposio de Probabilidad y Procesos Estocásticos*, volume 12 of *Aportaciones Matemáticas: Notas de Investigación [Mathematical Contributions: Research Notes]*. Sociedad Matemática Mexicana, México, 1996.
  Held at the Centro de Investigación en Matemáticas, Guanajuato, April 1–5, 1996.
- J. M. González Barrios and L. G. Gorostiza, editors. *Modelos estocásticos*, volume 14 of *Aportaciones Matemáticas: Investigación [Mathematical Contributions: Research]*. Sociedad Matemática Mexicana, México, 1998. Papers from the 5th Symposium on Probability and Stochastic Processes.
  Held at the Centro de Investigación en Matemáticas, Guanajuato, March 29–April 3, 1998.
- D. Hernández, J. A. López-Mimbela, and R. Quezada, editors. *Modelos estocásticos II*, volume 16 of *Aportaciones Matemáticas: Investigación [Mathematical Contributions: Research]*. Sociedad Matemática Mexicana, México, 2001. Papers from the 6th Symposium on Probability and Stochastic Processes.
  Held at the Centro de Investigación en Matemáticas, Guanajuato, May 23–27, 2000.
- J. M. González-Barrios, J. A. León, and A. Meda, editors. *Stochastic models*, volume 336 of *Contemporary Mathematics*. American Mathematical Society, Providence, RI; Sociedad Matemática Mexicana, México, 2003. Aportaciones Matemáticas. [Mathematical Contributions]. Proceedings of the 7th Symposium on Probability and Stochastic Processes.
  Held at the National University of Mexico, Mexico City, June 23–28, 2002

- M. Bladt, J. A. Lopez Mimbela, J. Ruiz de Chavez, and R. Navarro, editors. *Special issue: Proceedings of the 8th Symposium on Probability and Stochastic Processes*, volume 22. Stochastic Models, 2006.

  Held at the Universidad de las Americas in Puebla, June 20–25, 2004.
- M. E. Caballero, V. Rivero, and J. Ruiz de Chávez, editors. *Ninth Symposium on Probability and Stochastic Processes*, volume 24. *Stochastic Models*, 2008.

  Held at the Centro de Investigación en Matemáticas, Guanajuato, November 20–24, 2006.
- M. E. Caballero, L. Chaumont, D. Hernández-Hernández, and V. Rivero, editors. *X Symposium on Probability and Stochastic Processes and the First Joint Meeting France-Mexico of Probability*, volume 31 of *ESAIM Proceedings*. EDP Sciences, Les Ulis, 2011.

  Held at the Centro de Investigación en Matemáticas, Guanajuato, November 3–7, 2008.

Contents

Partition Functions of Discrete Coalescents: From Cayley's Formula to Frieze's $\zeta(3)$ Limit Theorem 1
Louigi Addario-Berry

Asymptotic Spectral Distributions of Distance-k Graphs of Star Product Graphs 47
Octavio Arizmendi and Tulio Gaxiola

Stochastic Differential Equations Driven by Loops 59
Fabrice Baudoin

Genealogy of a Wright-Fisher Model with Strong SeedBank Component 81
Jochen Blath, Bjarki Eldon, Adrián González Casanova, and Noemi Kurt

Shifting Processes with Cyclically Exchangeable Increments at Random 101
Loïc Chaumont and Gerónimo Uribe Bravo

Stochastic Integral and Covariation Representations for Rectangular Lévy Process Ensembles 119
J. Armando Domínguez-Molina and Alfonso Rocha-Arteaga

Asymptotic Behaviour of Poisson-Dirichlet Distribution and Random Energy Model 141
Shui Feng and Youzhou Zhou

Stability Estimation of Transient Markov Decision Processes 157
Evgueni Gordienko, Jaime Martinez, and Juan Ruiz de Chávez

Solution of the HJB Equations Involved in Utility-Based Pricing 177
Daniel Hernández–Hernández and Shuenn-Jyi Sheu

The Backbone Decomposition for Superprocesses with Non-local Branching 199
Antonio Murillo-Salas and José Luis Pérez

On Lévy Semistationary Processes with a Gamma Kernel 217
Jan Pedersen and Orimar Sauri

Ambit Fields: Survey and New Challenges 241
Mark Podolskij

Partition Functions of Discrete Coalescents: From Cayley's Formula to Frieze's $\zeta(3)$ Limit Theorem

Louigi Addario-Berry

Abstract In these expository notes, we describe some features of the multiplicative coalescent and its connection with random graphs and minimum spanning trees. We use Pitman's proof (Pitman, J Combin Theory Ser A 85:165–193, 1999) of Cayley's formula, which proceeds via a calculation of the partition function of the additive coalescent, as motivation and as a launchpad. We define a random variable which may reasonably be called the empirical partition function of the multiplicative coalescent, and show that its typical value is exponentially smaller than its expected value. Our arguments lead us to an analysis of the susceptibility of the Erdős-Rényi random graph process, and thence to a novel proof of Frieze's $\zeta(3)$-limit theorem for the weight of a random minimum spanning tree.

1 Introduction

Consider a discrete time process $(P_i, 1 \leq i \leq n)$ of coalescing blocks, with the following dynamics. The process starts from the partition of $[n] = \{1, \ldots, n\}$ into singletons: $P_1 = \{\{1\}, \ldots, \{n\}\}$. To form P_{i+1} from P_i choose two parts P, P' from P_i and merge them. We assume there is a function κ such that the probability of choosing parts P, P' is proportional to $\kappa(|P|, |P'|)$; call κ a *rate kernel*.

Different rate kernels lead to different dynamics. Three kernels whose dynamics have been studied in detail are $\kappa(x, y) = 1$, $\kappa(x, y) = x + y$, and $\kappa(x, y) = xy$; these are often called Kingman's coalescent, the additive coalescent, and the multiplicative coalescent, respectively. In these cases there is a natural way to enrich the process and obtain a *forest-valued* coalescent.

These notes are primarily focussed on the properties of the forest-valued multiplicative coalescent. We proceed from a statistical physics perspective, and begin by analyzing the partition functions of the three coalescents. Here is what we mean by this. Say that a sequence $(P_1, \ldots, P_n)$ of partitions of $[n]$ is an *n-chain* if

L. Addario-Berry (✉)
Department of Mathematics and Statistics, McGill University, 805 Sherbrooke Street West, Montréal, QC H3A 2K6, Canada
e-mail: louigi@math.mcgill.ca

R.H. Mena et al. (eds.), *XI Symposium on Probability and Stochastic Processes*,
Progress in Probability 69, DOI 10.1007/978-3-319-13984-5_1

$P_1 = \{\{1\}, \ldots, \{n\}\}$ is the partition of n into singletons, and for $1 \le i < n$, P_{i+1} can be formed from P_i by merging two parts of P_i. Think of $\kappa(x, y)$ as the number of possible ways to merge a block of size x with one of size y. Then corresponding to an n-chain $P = (P_1, \ldots, P_n)$ there are

$$\prod_{i=1}^{n-1} \kappa(|A_i(P)|, |B_i(P)|)$$

possible ways that the coalescent may have unfolded; here we write $A_i(P)$ and $B_i(P)$ for the blocks of P_i that are merged in P_{i+1}. Writing $\mathcal{P}_n$ for the set of n-chains, it follows that the total number of possibilities for the coalescent with rate kernel κ is

$$\sum_{P=(P_1,\ldots,P_n)\in\mathcal{P}_n} \prod_{i=1}^{n-1} \kappa\left(|A_i(P)|, |B_i(P)|\right) ,$$

and we view this quantity as the partition function of the coalescent with kernel κ.

The partition functions of Kingman's coalescent and the additive and multiplicative coalescents have particularly simple forms: they are

$$\begin{aligned} Z_{\mathrm{KC}}(n) &= n!(n-1)! , \\ Z_{\mathrm{AC}}(n) &= n^{n-1}(n-1)! , \text{ and} \\ Z_{\mathrm{MC}}(n) &= n^{n-2}(n-1)! . \end{aligned}$$

These formulae are proved in Sect. 2. A corollary of the formula for $Z_{\mathrm{KC}}(n)$ is that the number of increasing trees with n vertices is $(n-1)!$; this easy fact is well-known. The formula for $Z_{\mathrm{AC}}(n)$ is due to Pitman [13], who used it to give a beautiful proof of Cayley's formula; this is further detailed in Sect. 2.1.

It may seem surprising that the partition function of the multiplicative coalescent is so similar to that of the additive coalescent: near the start of the process, when most blocks have size 1, the additive coalescent has twice as many choices as the multiplicative coalescent. Later in the process, blocks should be larger, and one would guess that usually $xy > x + y$. Why these two effects should almost exactly cancel each other out is something of a mystery. On the other hand, the similarity of the partition functions may suggest that the additive and multiplicative coalescents have similar behaviour.

A more detailed investigation will reveal interesting behaviour whose subtleties are not captured by the above formulae. We will see in Sect. 2.3 that there is a naturally defined "empirical partition function" $\hat{Z}_{\mathrm{MC}}(n)$ such that $Z_{\mathrm{MC}}(n) = \mathbf{E}\left[\hat{Z}_{\mathrm{MC}}(n)\right]$. However, $\hat{Z}_{\mathrm{MC}}(n)$ is typically *exponentially smaller* than $Z_{\mathrm{MC}}(n)$ (see Corollary 4.3), so in a quantifiable sense, the partition function $Z_{\mathrm{MC}}(n)$ takes the value it does due to extremely rare events. Correspondingly, it turns out that the behaviour of the additive and multiplicative coalescents are typically quite different.

To analyze the typical value of $\hat{Z}_{\mathrm{MC}}(n)$, we are led to develop the connection between the multiplicative coalescent and the classical Erdős-Rényi random graph process $(G(n,p), 0 \leq p \leq 1)$. The most technical part of the notes is the proof of a concentration result for the susceptibility of $G(n,p)$; this is Theorem 4.4, below. Using a well-known coupling between the multiplicative coalescent and Kruskal's algorithm for the minimum weight spanning tree problem, our susceptibility bound leads easily to a novel proof of the $\zeta(3)$ limit for the total weight of the minimum spanning tree of the complete graph (this is stated in Theorem 5.1, below).[1]

1.1 Stylistic Remarks

The primary purpose of these notes is expository (though there are some new results, notably Theorems 4.2 and 4.4). Accordingly, we have often opted for repetition over concision. We have also included plenty of exercises and open problems (the open problems are mostly listed in Sect. 7). Some exercises state facts which are required later in the text; these are distinguished by a ⊛.

2 A Tale of Three Coalescents

2.1 Cayley's Formula and Pitman's Coalescent

We begin by describing the beautiful proof of Cayley's formula found by Jim Pitman, and its link with uniform spanning trees. Cayley's formula states that the number of trees with vertices $\{1, 2, \ldots, n\}$ is n^{n-2}, or equivalently that the number of *rooted* trees with vertices labeled by $[n] := \{1, 2, \ldots, n\}$ is n^{n-1}. To prove this formula, Pitman [13] analyzes a process we call Pitman's coalescent. To explain the process, we need some basic definitions. A *forest* is a graph with no cycles; its connected components are its trees. A *rooted forest* is a forest in which each tree t has a distinguished root vertex $r(t)$.

Pitman's Coalescent, Version 1. The process has n steps, and at step i consists of a rooted forest $F_i = \{T_1^{(i)}, \ldots, T_{n+1-i}^{(i)}\}$ with $n+1-i$ trees. (At step

(continued)

[1]We find this proof of the $\zeta(3)$ limit for the MST weight pleasing, as it avoids lemmas which involve estimating the number of unicyclic and complex components in $G(n,p)$; morally, the cycle structure of components of $G(n,p)$ should be unimportant, since cycles are never created in Kruskal's algorithm!

1, these trees are simply isolated vertices with labels $1, \ldots, n$.) To obtain F_{i+1} from F_i, choose a pair (U_i, V_i), where $U_i \in [n]$ and V_i is the root of some tree of F_i not containing U_i, uniformly at random from among all such pairs. Add an edge from U_i to V_i, and root the resulting tree at the root of U_i's old tree. The forest F_{i+1} consists of this new tree together with the $n-i-1$ unaltered trees from F_i.

The coalescents we consider all have the general form of Pitman's coalescent: they are forest-valued stochastic processes $(F_i, 1 \le i \le n)$, where $F_i = \{T_1^{(i)}, \ldots, T_{n+1-i}^{(i)}\}$ is a forest with vertices labeled by $[n]$.

Pitman's Coalescent, Version 2. Consider the directed graph $K_n^{\rightarrow}$ with vertices $\{1, \ldots, n\}$ and an oriented edge from k to ℓ for each $1 \le k \ne \ell \le n$. Let $\mathbf{W} = \{W_{(k,\ell)} : 1 \le k \ne \ell \le n\}$ be independent copies of a continuous random variable W, that weight the edges of $K_n^{\rightarrow}$. Let F_1 be as in Version 1. For $i \in \{1, \ldots, n-1\}$, form F_{i+1} from F_i by adding the smallest weight edge (k, ℓ) whose head ℓ is the root of one of the trees in F_i. (Each tree of F_i is rooted at its unique vertex having indegree zero in F_i.)

Note that in Version 2, for each $i \in \{1, \ldots, n\}$ and each tree T of F_i, all edges of T are oriented away from a single vertex of T; so, viewing this vertex as the root of T, the orientation of edges in T is fully specified by the location of its root.

Exercise 1 View the trees of Version 2 as rooted rather than oriented. Then the sequences of forests $(F_1, \ldots, F_n)$ described in Version 1 and Version 2 have the same distribution.

Say that a finite set $\{X_i, i \in I\}$ of random variables is *exchangeable* if for any two deterministic orderings of I as, say, $i_1, \ldots, i_k$ and $i'_1, \ldots, i'_i$, the vectors $(X_{i_1}, \ldots, X_{i_k})$ and $(X_{i'_1}, \ldots, X_{i'_k})$ are identically distributed. In particular, if the elements of $\{X_i, i \in I\}$ are iid then the set is exchangeable.

Exercise 2 Suppose that the edge weights $\mathbf{W}$ are only assumed to be exchangeable and a.s. pairwise distinct. Show that the sequences of forests $(F_1, \ldots, F_n)$ described in Version 1 and Version 2 still have the same distribution.

To prove Cayley's formula, we compute the *partition function* of Pitman's coalescent: this is the total number of possibilities for its execution. (To do so, it's easiest to think about Version 1 of the procedure.) For example, when $n = 3$, there are 6 possibilities for the first step of the process: 3 choices for the first vertex, then 2 choices of a tree not containing the first vertex. For the second step, there are 3 choices for the first vertex; there is only one component not containing the chosen vertex, and we must choose it. Thus, for $n = 3$, the partition function has value $Z_{AC}(3) = 6 \cdot 3 = 18$. More generally, for the n-vertex process, when adding the i'th edge we have n choices for the first vertex and $n - i$ choices of tree not containing the first vertex, so a total of $n(n - i)$ possibilities. Thus the partition function is

$$Z_{AC}(n) = \prod_{i=1}^{n-1} n \cdot (n-i) = n^{n-1}(n-1)! \tag{2.1}$$

It is not possible to recover the entire execution path of the additive coalescent from the final tree, since there is no way to tell in which order the edges were added. If we wish to retain this information, we may label each edge of $T_1^{(n)}$ with the step at which it was added. More precisely, $L(e)$ is the unique integer $i \in \{1, \ldots, n-1\}$ such that e is not an edge of F_i but is an edge of F_{i+1}. It follows from the definition of the process that the edge labels are distinct, so $L : E(T_1^{(n)}) \to \{1, \ldots, n-1\}$ is a bijective map.

Now fix a rooted tree t with vertices $\{1, \ldots, n\}$, and consider the *restricted* partition function $Z_{AC,t}(n)$; this is simply the number of possibilities for the execution of the process for which the end result is the tree t. We claim that $Z_{AC,t}(n) = (n-1)!$. This is easy to see: for any labelling ℓ of the edges of t with integers $\{1, \ldots, n-1\}$, there is a unique execution path for which $(T_1^{(n)}, L) = (t, \ell)$, and there are $(n-1)!$ possible labellings ℓ. Thus, the probability of ending with the tree t is $Z_{AC,t}(n)/Z_{AC}(n) = 1/n^{n-1}$. Since this number doesn't depend on t, only on n, it follows that every rooted labelled tree with n vertices is equally likely, and so there must be n^{n-1} such trees.

Note *The preceding argument is correct, but treads lightly around an important point. When performing the process, the number of possibilities for the i'th edge does not depend on the first $i-1$ choices, so the probability of building a particular tree t by adding its edges in a particular order is $[n^{n-1}(n-1)!]^{-1}$ regardless of the order. Of course, the* set *of possible choices at a given step must depend on the history of the process – for example, we must not add a single edge twice. More generally, thinking of Version 2, applying the procedure to a graph other than $K_n^{\rightarrow}$ need not yield a uniform spanning tree of the graph, and indeed may not even build a tree. (Consider, for example, applying the procedure to a two-edge path.)*

By stopping Pitman's coalescent before the end, one can use a similar analysis to obtain counting formulae for forests. Write $Z_{AC}(n, k)$ for the total number of possibilities for Pitman's coalescent stopped at step k (so ending with $n + 1 - k$ forests). We write $(m)_\ell$ to denote the *falling factorial* $\prod_{i=0}^{\ell-1}(m - i)$.

Exercise 3

(a) Show that $Z_{AC}(n, k) = n^{k-1}(n-1)_{k-1}$ for each for $1 \le k \le n$.
(b) An *ordered labeled forest* is a sequence $(t_1, \ldots, t_\ell)$ where each t_i is a rooted labelled tree and all labels of vertices in the forest are distinct. Show that

for each $1 \leq \ell \leq n$ the number of ordered labeled forests $(t_1, \ldots, t_\ell)$ with $\bigcup_{i=1}^{\ell} V(t_i) = [n]$, is $\ell \cdot n^{n-\ell-1} \cdot (n)_\ell$.

We briefly discuss a special case of Version 2. Suppose that $W_{(k,l)}$ is exponential with rate $X_{(k,\ell)}$, where $\mathbf{X} = \{X_{(k,\ell)} : 1 \leq k \neq \ell \leq n\}$ are independent copies of any non-negative random variable X. By standard properties of exponentials and the symmetry of the process, the dynamics in this case may be described as follows.

Pitman's Coalescent, Version 3. Let F_1 be as in Version 1. For $i \in \{1, \ldots, n-1\}$, choose an edge whose head is the root of any one of the trees in F_i, each such edge (k, l) chosen with probability proportional to its weight $X_{(k,l)}$; add the chosen edge to create the forest F_{i+1}.

Consider Version 3 of the procedure after $i - 1$ edges have been added. Conditional on $\mathbf{X}$ and on the forest $(T_1^{(i)}, \ldots, T_{n-i+1}^{(i)})$, the probability of adding a particular edge (k, ℓ) whose head is a root, is proportional to $X_{(k,\ell)}$, so is equal to

$$\frac{X_{(k,\ell)}}{\sum_{m=1}^{n-i+1} \sum_{j\in\{1,\ldots,n\}\setminus V(T_m^{(i)})} X_{(j,r(T_m^{(i)}))}}.$$

Now fix any sequence $f_1, \ldots, f_n$ of forests that can arise in the process. Write $f_i = (t_k^{(i)}, 1 \leq k \leq n+1-i)$ and for $i = 1, \ldots, n-1$ write (k_i, ℓ_i) for the unique edge of f_{i+1} not in f_i. Then by the above,

$$\mathbf{P}\{F_i = f_i, 1 \leq i \leq n \mid \mathbf{X}\} = \prod_{i=1}^{n-1} \frac{X_{(k_i,\ell_i)}}{\sum_{m=1}^{n-i+1} \sum_{j\in\{1,\ldots,n\}\setminus V(t_m^{(i)})} X_{(j,r(t_m^{(i)}))}}.$$

By Exercise 1 and the above analysis, it follows that for any such sequence $f_1, \ldots, f_n$,

$$\mathbf{E}\left[\prod_{i=1}^{n-1} \frac{X_{(k_i,\ell_i)}}{\sum_{m=1}^{n-i+1} \sum_{j\in\{1,\ldots,n\}\setminus V(t_m^{(i)})} X_{(j,r(t_m^{(i)}))}}\right] = \frac{1}{n^{n-1}(n-1)!}.$$

It is by no means obvious at first glance that this expectation should not depend on law of X, let alone that it should have such a simple form.

2.2 *Kingman's Coalescent and Random Recursive Trees*

Pitman's coalescent starts from isolated vertices labeled from $\{1, \ldots, n\}$, and builds a rooted tree by successive edge addition. At each step, an edge is added *to* some vertex, *from* some root (of a component not containing the chosen vertex). When we calculated $Z_{AC}(n)$, it was important that the number of possibilities at each step depended only on the *number* of trees in the current forest and not, say, their sizes, or some other feature.

Pitman's merging rule (*to* any vertex, *from* a root) yielded a beautiful proof of Cayley's formula. It is natural to ask what other rules exist, and what information may be gleaned from them. Of course, *from* any vertex, *to* a root just yields the additive coalescent, with edges of the resulting tree oriented towards the root rather than towards the leaves. What about *from* any root, *to* any (other) root, as in the following procedure? In a very slight abuse of terminology, we call this rule *Kingman's coalescent.* We again start from a rooted forest F_1 of n isolated vertices $\{1, \ldots, n\}$. Recall that we write $F_i = \{T_1^{(i)}, \ldots, T_{n+1-i}^{(i)}\}$.

Kingman's Coalescent. At step i, choose an ordered pair (U_i, V_i) of distinct roots from $\{r(T_1^{(i)}), \ldots, r(T_{n+1-i}^{(i)})\}$, uniformly at random from among the $(n+1-i)(n-i)$ such pairs. Add an edge from U_i to V_i, and root the resulting tree at U_i. The forest F_{i+1} consists of this new tree together with the $n-i-1$ unaltered trees from F_i.

Our convention is that when an edge is added from u to v, the root of the resulting tree is u; this maintains that edges are always oriented towards the leaves. For Kingman's coalescent, when i trees remain there are $i(i-1)$ possibilities for which oriented edge to add. Like for Pitman's coalescent, this number depends only on the number of trees, and it follows that the total number of possible execution paths for the process is

$$Z_{KC}(n) = \prod_{i=2}^{n} i(i-1) = n!(n-1)!\,. \tag{2.2}$$

What does this number count?

To answer the preceding question, as in the additive coalescent let $L : E(T_1^{(n)}) \to \{1, \ldots, n-1\}$ label the edges of $T_1^{(n)}$ in their order of addition. It is easily seen that for Kingman's coalescent, the edge labels decrease along any root-to-leaf path; we call such a labelling a *decreasing edge labelling.*[2] Furthermore, any decreasing edge labelling of $T_1^{(n)}$ can arise. Once again, the full behaviour of the coalescent is described by pair $(T_1^{(n)}, L)$, and conversely, the coalescent determines $T_1^{(n)}$ and L. These observations yield that the number of rooted trees with vertices labelled

[2]It is more common to order by *reverse* order of addition, so that labels increase along root-to-leaf paths; this change of perspective may help with Exercise 4.

$\{1, \ldots, n\}$, additionally equipped with a decreasing edge labelling, is $n!(n-1)!$. The factor $n!$ simply counts the number of ways to assign the labels $\{1, \ldots, n\}$ to the vertices. By symmetry, each vertex labelling of a given tree is equally likely to arise, and so we have the following.

Proposition 2.1 *The number of pairs (T, L), where T is a rooted tree with n vertices and L is a decreasing edge labelling of T, is $(n-1)!$.*

Exercise 4 (Random recursive trees) Prove Proposition 2.1 by introducing and analyzing an n-step procedure that at step i constructs a rooted tree with i vertices.

Before the next exercise, we state a few definitions. For a graph G, write $|G|$ for the number of vertices of G. If T is a rooted tree and u is a vertex of T, write T_u for the subtree of T consisting of u together with its descendants in T (we call T_u the subtree of T rooted at u). Also, if u is not the root, write $p(u)$ for the parent of u in T. Finally, write $\mathrm{aut}(T)$ for the number of rooted automorphisms of T.

Exercise 5 Show that for a fixed rooted tree T, the number of decreasing edge labellings of T with labels $1, 2, \ldots, |V(T)| - 1$ is

$$\frac{1}{\mathrm{aut}(T)} \cdot \prod_{v \in V(T)} \frac{(|T_v| - 1)!}{\prod_{\{u \in V(T): p(u) = v\}} |T_u|!} .$$

Our convention is that an empty product equals 1; a special case is that $0! = 1$. It follows from the preceding exercise that, writing $\mathcal{T}_n$ for the set of rooted trees with n vertices,

$$\sum_{T \in \mathcal{T}_n} \frac{1}{\mathrm{aut}(T)} \cdot \prod_{v \in V(T)} \frac{|E(T_u)|!}{\prod_{\{u \in V(T): p(u) = v\}} |V(T_u)|!} = (n-1)!\,;$$

another formula that one may not find obvious at first glance.

To finish the section, note that just like for Pitman's coalescent, we might well consider a version of this procedure that is "driven by" iid non-negative weights $\mathbf{X} = \{X_{(k,\ell)} : 1 \le k \ne \ell \le n\}$. (Recall that we viewed these weights as exponential *rates*, then used the resulting exponential clocks at each step to determine which edge to add.) At each step, add an oriented edge whose tail and head are both the roots of some tree of the current forest, each such edge chosen with probability proportional to its weight. For this procedure, conditional on $\mathbf{X}$, after adding the first $i-1$ edges, the conditional probability of adding a particular edge (k, ℓ) is

$$\frac{X_{(k,\ell)}}{\sum_{1 \le j \ne m \le n} X_{(r(T_j^{(i)}), r(T_m^{(i)}))}} .$$

Now fix any sequence $f_1, \ldots, f_n$ of forests that can arise in the process, write $f_i = (t_k^{(i)}, 1 \le k \le n+1-i)$, and for $i = 1, \ldots, n-1$ write (k_i, ℓ_i) for the unique edge of f_{i+1} not in f_i. Then we have

$$\mathbf{P}\{F_i = f_i, 1 \le i \le n \mid \mathbf{X}\} = \prod_{i=1}^{n-1} \frac{X_{(k_i,\ell_i)}}{\sum_{1 \le m \ne j \le n} X_{(r(t_m^{(i)}), r(t_j^{(i)}))}} .$$

It follows from the above analysis that for any such sequence $f_1, \ldots, f_n$,

$$\mathbf{E}\left[\prod_{i=1}^{n-1} \frac{X_{(k_i,\ell_i)}}{\sum_{1 \le m \ne j \le n} X_{(r(t_m^{(i)}), r(t_j^{(i)}))}}\right] = \frac{1}{n!(n-1)!} .$$

Once again, it is not even a priori clear that this expectation should not depend on the law of X.

Exercise 6 (First-passage percolation) Develop and analyze a "Version 3" variant of the tree growth procedure from Exercise 4, using exponential edge weights.

2.3 The Multiplicative Coalescent and Minimum Spanning Trees

The previous two sections considered merging rules of the form any-to-root and root-to-root, and obtained Pitman's coalescent and Kingman's coalescent, respectively. We now take up the "any-to-any" merging rule. This is arguably the most basic of the three rules, but its behaviour is arguably the hardest to analyze. We begin as usual from a forest F_1 of n isolated vertices $\{1, \ldots, n\}$, and write $F_i = \{T_1^{(i)}, \ldots, T_{n+1-i}^{(i)}\}$. In the multiplicative coalescent there is no natural way to maintain the property that edges are oriented toward some root vertex, so we view the trees of the forests as unrooted, and their edges as unoriented. Given a set S, write $\binom{S}{k}$ for the set of k-element subsets of S.

The multiplicative coalescent. To obtain F_{i+1} from F_i, choose an pair $\{U_i, V_i\}$ uniformly at random from the set of pairs $\{u, v\} \in \binom{[n]}{2}$ for which u and v are different trees of F_i. Add an edge from U_i to V_i to form the forest F_{i+1}.

This is known as the *multiplicative* coalescent, because the number of possible choices of an edge joining trees $T_j^{(i)}$ and $T_k^{(i)}$ is $|T_j^{(i)}||T_k^{(i)}|$. It follows that the number of possible edges that may be added to the forest F_i is

$$\sum_{1\le j\neq k\le n+1-i} |T_j^{(i)}||T_k^{(i)}| = \frac{1}{2}\left(n^2 - \sum_{T\in F_i} |T|^2\right).$$

The above expression is more complicated than for the additive coalescent or Kingman's coalescent: it depends on the forest F_i, for one.

In much of the remainder of these notes, we investigate an expression for the partition function $Z_{\mathrm{MC}}(n)$ of the multiplicative coalescent that arises from the preceding formula. To obtain this expression, recall the definition of an n-chain from Sect. 1, and that $\mathcal{P}_n$ is the set of n-chains.

Exercise 7 Show that $|\mathcal{P}_n| = \frac{(n!)^2}{n\cdot 2^{n-1}}$.

The multiplicative coalescent determines an n-chain in which the i'th partition is simply $P(F_i) := \{V(T_j^{(i)}), 1 \le j \le n+1-i\}$. It is straightforward to see that the number of possibilities for the multiplicative coalescent that give rise to a particular n-chain $P = (P_1, \ldots, P_n)$ is simply

$$\prod_{i=1}^{n-1} |A_i(P)||B_i(P)|,$$

where $A_i(P)$ and $B_i(P)$ are the parts of P_i that are combined in P_{i+1}. It follows that

$$Z_{\mathrm{MC}}(n) = \sum_{P=(P_1,\ldots,P_n)\in\mathcal{P}_n} \prod_{i=1}^{n-1} (|A_i(P)||B_i(P)|).$$

This certainly looks more complicated than in the previous two cases. However, there is an exact formula for $Z_{\mathrm{MC}}(n)$ whose derivation is perhaps easier than for either $Z_{\mathrm{AC}}(n)$ or $Z_{\mathrm{KC}}(n)$ (though it does rely on Cayley's formula).

Proposition 2.2 $Z_{\mathrm{MC}}(n) = n^{n-2}(n-1)!$

Proof Let $\mathcal{S}$ be the set of pairs (t, ℓ) where t is an unrooted tree with $V(t) = [n]$ and $\ell : E(t) \to [n-1]$ is a bijection. By Cayley's formula, the number of trees t with $V(t) = [n]$ is n^{n-2}, so $\mathcal{S} = n^{n-2}(n-1)!$.

For $e \in E(T_n^{(1)})$, let $L(e) = \sup\{i : e \notin E(F_i)\}$. Then $L : E(T_n^{(1)}) \to [n-1]$ is a bijection. Thus the pair $(T_n^{(1)}, L)$ is an element of $\mathcal{S}$. To see this map is bijective, note that if $(T_n^{(1)}, L) = (t, \ell)$ then for each $1 \le i \le n$, F_i is the forest on $[n]$ with edges $\{\ell^{-1}(j), 1 \le j < i\}$. The result follows. □

The above proposition yields that $Z_{\mathrm{MC}}(n) = Z_{\mathrm{AC}}(n)/n$. If we were to additionally choose a root for $T_n^{(1)}$, we would obtain identical partition functions. This suggests that perhaps the additive and multiplicative coalescents have similar structures. One might even be tempted to believe that the trees built by the two coalescents are identically distributed; the following exercise (an observation of Aldous [3]), will disabuse you of that notion.

Exercise 8 Let T be built by the multiplicative coalescent, and let T' be obtained from the additive coalescent by unrooting the final tree. Show that if $n \geq 4$ then T and T' are not identically distributed.

Despite the preceding exercise, it is tempting to guess that the two trees are still similar in structure; this was conjectured by Aldous [3], and only recently disproved [2]. In the remainder of the section, we begin to argue for the difference between the two coalescents, from the perspective of their partition functions. For $1 \leq k \leq n$, write $Z_{\mathrm{MC}}(n, k)$ for the partition function of the first k steps of the multiplicative coalescent,

$$Z_{\mathrm{MC}}(n, k) = \sum_{P=(P_1,\ldots,P_k)\in\mathcal{P}_{n,k}} \prod_{i=1}^{k-1} (|A_i(P)||B_i(P)|),$$

where $\mathcal{P}_{n,k}$ is the set of length-k initial segments of n-chains. We have, e.g., $Z_{\mathrm{MC}}(n, 1) = 1$, $Z_{\mathrm{MC}}(n, 2) = \binom{n}{2}$, and $Z_{\mathrm{MC}}(n, n) = Z_{\mathrm{MC}}(n)$.

The argument of Proposition 2.2 shows that $Z_{\mathrm{MC}}(n, k) = u_{n,k} \cdot (k-1)!$, where $u_{n,k}$ is the number of unrooted forests with vertices $[n]$ and $k-1$ total edges. The identity

$$u_{n,k} = \binom{n}{n+1-k} n^{k-2} \sum_{i=0}^{n+1-k} \left(\frac{-1}{2n}\right)^i \binom{n+1-k}{i} (n+1-k+i)\cdot(k-1)_i,$$

was derived by Rényi [14], and I do not know of an exact formula that simplifies the above expression. We begin to see that there is more to the multiplicative coalescent than first meets the eye.

If we can't have a nice, simple identity, what about bounds? Of course, there is the trivial upper bound $Z_{\mathrm{MC}}(n, k) \leq (n(n-1)/2)^{k-1}$, since at each step there are at most $\binom{n}{2}$ pairs to choose from; similar bounds hold for the other two coalescents. To improve this bound, and more generally to develop a deeper understanding of the dynamics of the multiplicative coalescent, our starting point is the following observation.

Given an n-chain $P = (P_1, \ldots, P_n)$, for the multiplicative coalescent we have

$$\mathbf{P}\{(P(F_i), 1 \leq i \leq n) = P\} = \prod_{i=1}^{n-1} \frac{2|A_i(P)||B_i(P)|}{n^2 - \sum_{\pi\in P_i} |\pi|^2}.$$

This holds since for $1 \le i \le n-1$, given that $P(F_j) = P_j$ for $1 \le j \le i$, there are $(n^2 - \sum_{\pi \in P_i} |\pi|^2)/2$ choices for which oriented edge to add to form F_{i+1}, and $P(F_{i+1}) = P_{i+1}$ for precisely $|A_i(P)||B_i(P)|$ of these. It follows that

$$\begin{aligned}
Z_{\mathrm{MC}}(n) &= \sum_{P=(P_1,\ldots,P_n)\in\mathcal{P}_n} \mathbf{P}\{(P(F_i), 1 \le i \le n) = P\} \cdot \prod_{i=1}^{n-1} \frac{n^2 - \sum_{\pi \in P_i} |\pi|^2}{2} \\
&= \sum_{P=(P_1,\ldots,P_n)\in\mathcal{P}_n} \mathbf{P}\{(P(F_i), 1 \le i \le n) = P\} \cdot 2^{-(n-1)} \\
&\qquad \cdot \mathbf{E}\left\{ \prod_{i=1}^{n-1} \left(n^2 - \sum_{T \in F_i} |T|^2 \right) \,\middle|\, (P(F_i), 1 \le i \le n) = P \right\} \\
&= 2^{-(n-1)} \cdot \mathbf{E}\left[\prod_{i=1}^{n-1} \left(n^2 - \sum_{T \in F_i} |T|^2 \right) \right]. \qquad (2.3)
\end{aligned}$$

A mechanical modification of the logic leading to (2.3) yields the following expression, valid for each $1 \le k \le n$:

$$Z_{\mathrm{MC}}(n,k) = 2^{-(k-1)} \mathbf{E}\left[\prod_{i=1}^{k-1} \left(n^2 - \sum_{T \in F_i} |T|^2 \right) \right]. \qquad (2.4)$$

Write

$$\hat{Z}^{\rightarrow}_{\mathrm{MC}}(n,k) = \prod_{i=1}^{k-1} \left(n^2 - \sum_{T \in F_i} |T|^2 \right),$$

let $\hat{Z}^{\rightarrow}_{\mathrm{MC}}(n) = \hat{Z}^{\rightarrow}_{\mathrm{MC}}(n,1)$, and let $\hat{Z}_{\mathrm{MC}}(n,k) = 2^{-(k-1)} \hat{Z}^{\rightarrow}_{\mathrm{MC}}(n,k)$ and $\hat{Z}_{\mathrm{MC}}(n) = \hat{Z}_{\mathrm{MC}}(n,n)$. With this notation, (2.3) and the subsequent equation state that

$$\mathbf{E}\left[\hat{Z}_{\mathrm{MC}}(n,k) \right] = Z_{\mathrm{MC}}(n,k) = \frac{1}{2^{k-1}} \mathbf{E}\left[\hat{Z}^{\rightarrow}_{\mathrm{MC}}(n,k) \right]. \qquad (2.5)$$

The random variable $\hat{Z}_{\mathrm{MC}}(n)$ is a sort of *empirical partition function* of the multiplicative coalescent. The superscript arrow on $\hat{Z}^{\rightarrow}_{\mathrm{MC}}(n,k)$ is because the factor 2^{k-1} may be viewed as corresponding to a choice of orientation for each edge of F_k. The random variable $\hat{Z}_{\mathrm{MC}}(n)$ of course contains more information than simply its expected value, so by studying it we might hope to gain a greater insight into the behaviour of the coalescent. Much of the remainder of these notes is devoted to showing that $\mathbf{E}\left[\hat{Z}_{\mathrm{MC}}(n) \right] = Z_{\mathrm{MC}}(n)$ is a *terrible* predictor of the typical value of $\hat{Z}_{\mathrm{MC}}(n)$. More precisely, there are unlikely execution paths along which the multiplicative coalescent has many more possibilities than along a typical path; such paths swell the *expected* value of $\hat{Z}_{\mathrm{MC}}(n)$ to exponentially larger than its *typical* size.

The logic leading to (2.3) and (2.4) may also be applied to the additive coalescent; the result is boring but instructive. First note that

$$Z_{\mathrm{AC}}(n,k) = \sum_{P=(P_1,\ldots,P_k)\in\mathcal{P}_{n,k}} \prod_{i=1}^{k-1} (|A_i(P)| + |B_i(P)|)\,.$$

For the additive coalescent, the total number of choices at step i is $n(n-i)$, and given that $P(F_i) = P_i$, the number of choices which yield $P(F_{i+1}) = P_{i+1}$ is $A_i(P)+B_i(P)$. Writing $\mathbf{P}_{\mathrm{AC}}$ for probabilities under the additive coalescent, we thus have

$$\mathbf{P}_{\mathrm{AC}}\left\{(P(F_i), 1 \le i \le k) = (P_1,\ldots,P_k)\right\} = \prod_{i=1}^{k-1} \frac{|A_i(P)| + |B_i(P)|}{n(n-i)}$$

Following the logic through yields

$$Z_{\mathrm{AC}}(n,k) = \mathbf{E}_{\mathrm{AC}}\left[\prod_{i=1}^{k-1} n(n-i)\right] = \mathbf{E}_{\mathrm{AC}}\left[n^{k-1}(n-1)_{k-1}\right].$$

Thus, the "empirical partition function" of the additive coalescent is a constant, so contains no information beyond its expected value. (This fact is essentially the key to Pitman's proof of Cayley's formula.)

The terms of the products (2.3) and (2.4), though random, turn out to behave in a very regular manner (but proving this will take some work). Through a study of these terms, we will obtain control of $\mathbf{E}\left[\log \hat{Z}_{\mathrm{MC}}(n)\right]$, and thereby justify the above assertion that $\hat{Z}_{\mathrm{MC}}(n)$ is typically very different from its mean.

2.3.1 The Growth Rate of $Z_{\mathrm{MC}}(n, \lfloor n/2 \rfloor)$

As a warmup, and to introduce a key tool, we approximate the value of $Z_{\mathrm{MC}}(n, \lfloor n/2\rfloor)$ using a connection between the multiplicative coalescent and a process we call (once again with a very slight abuse of terminology) the Erdős-Rényi coalescent. Write K_n for the *complete graph*, i.e. the graph with vertices $[n]$ and edges $(\{i,j\}, 1 \le i < j \le n)$.

The Erdős-Rényi coalescent. Choose a uniformly random permutation $e_1,\ldots,e_{\binom{n}{2}}$ of $E(K_n)$. For $0 \le i \le \binom{n}{2}$, let $G_i^{(n)}$ have vertices $[n]$ and edges $\{e_1,\ldots,e_i\}$.

Our indexing here starts at zero, unlike in the multiplicative coalescent; this is slightly unfortunate, but it is standard for the Erdős-Rényi graph process to index so that $G_i^{(n)}$ has i edges. This process is different from the previous coalescent processes, most notably because it creates graphs with cycles.

Note that we can recover the multiplicative coalescent from the Erdős-Rényi coalescent in the following way. Informally, simply ignore any edges added by the Erdős-Rényi coalescent that fail to join distinct components. More precisely, for each $0 \le m \le \binom{n}{2}$, let τ_m be the number of edges $\{U_i, V_i\}$, $0 < i \le m$ such that U_i and V_i lie in different components of $G_{i-1}^{(n)}$. (See Fig. 1 for an example.)

Observe that

$$\tau_m + 1 = \begin{cases} \tau_m & \text{if } G_{m+1}^{(n)} \text{ and } G_m^{(n)} \text{ have the same number of components} \\ \tau_m + 1 & \text{if } G_{m+1}^{(n)} \text{ has one fewer component than } G_m^{(n)}\,. \end{cases}$$

In other words, τ_m increases precisely when the endpoints of the edge added to $G_m^{(n)}$ are in different components. Further, the set

$$\{e_m : m \ge 1, \tau_m > \tau_{m-1}\}$$

contains $n-1$ edges, since $G_0^{(n)}$ has n components and $G_{\binom{n}{2}}^{(n)}$ almost surely has only one component.

Set $I_1 = 0$ and for $1 < k \le n$ let

$$I_k = \inf\{m \ge 1 : \tau_m = k-1\}\,.$$

Then for $1 < k \le n$, the edge e_{I_k} joins distinct components of $G_{I_k-1}^{(n)}$, and by symmetry is equally likely to be any such edge. Thus, letting F_k be the graph with edges $\{e_{I_j} : 1 \le j \le k\}$ for $1 \le k \le n$, the process $\{F_k, 1 \le k \le n\}$ is precisely distributed as the multiplicative coalescent. This is a *coupling* between the Erdős-Rényi graph process and the multiplicative coalescent; its key property is that for all

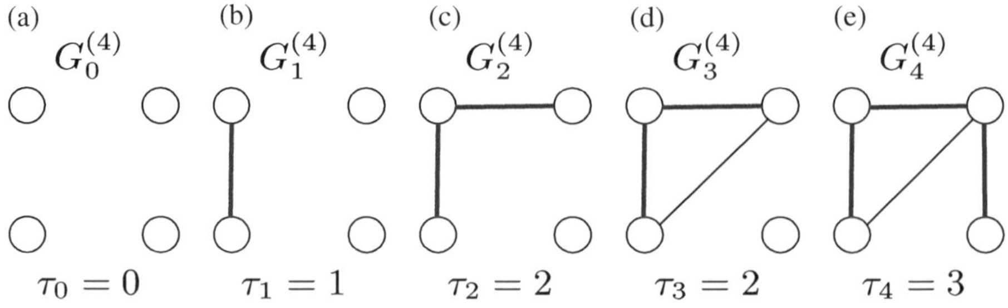

Fig. 1 An example of the first steps of the Erdős-Rényi coalescent. The multiplicative coalescent is obtained by keeping only the *thicker, blue edges*

$1 \le k \le n$, the vertex sets of the trees of F_k are the same as those of the components of $G^{(n)}_{I_k}$.

Having found the multiplicative coalescent within the Erdős-Rényi coalescent, we can now use known results about the latter process to study the former. For a graph G, and $v \in V(G)$, we write $N(v) = N_G(v)$ for the set of nodes adjacent to v (the *neighbours* of v), and write $C(v) = C_G(v)$ for the connected component of G containing v. We will use the results of the following exercise.[3]

Exercise 9 ⊛

(a) Show that in the Erdős-Rényi coalescent, if all components have size at most s then the probability a uniformly random edge from among the remaining edges has both endpoints in the same component is at most $(s-1)/(n-1)$.

(b) Show that for all $0 \le m \le n/2$, in $G^{(n)}_m$, $\mathbf{E}|N(v)| \le 2m/n$.

(c) Prove by induction that for all $0 \le m < n/2$, in $G^{(n)}_m$, $\mathbf{E}\left[|C(1)|\right] \le n/(n-2m)$. (**Hint.** First condition on $N(1)$, then average.)

(d) Prove that for all $\epsilon > 0$,

$$\limsup_{n\to\infty} \mathbb{P}\left(G^{(n)}_{(1-\epsilon)n/2} \text{ has a component of size } > \epsilon n\right) = 0\,.$$

(**Hint.** Given that the largest component of $G^{(n)}_m$ has size s, with probability at least s/n vertex 1 is in such a component.)

Using the above exercise, we now fairly easily prove a lower bound on the partition function of the first half of the multiplicative coalescent.

Proposition 2.3 *For all $\beta > 0$,*

$$\mathbf{P}\left\{\hat{Z}^{\rightarrow}_{\mathrm{MC}}(n, \lfloor n/2 \rfloor) \ge n^{(1-\beta)n}\right\} \to 1 \text{ as } n \to \infty\,.$$

We begin by showing that typically $I_t = (1 + o(1))t$ until $t \ge n/2$.

Lemma 2.4 *For all $\epsilon > 0$, $\limsup_{n\to\infty} \mathbb{P}\left(I_{(1-\epsilon)n/2} \ge n/2\right) = 0$.*

Proof Fix $\epsilon > 0$, let $\delta = \epsilon/3$, and let E be the event that all components of $G_{(1-\delta)n/2}$ have size at most δn. For $m \ge 0$, conditional on $G^{(n)}_m$, by Exercise 9(a), $\tau_{m+1} - \tau_m$ stochastically dominates a Bernoulli$(1 - (s-1)/(n-1))$ random variable, where s is the size of the largest component of $G^{(n)}_m$.

For n large and $s \le \delta n$ we have $1 - (s-1)/(n-1) \ge 1 - \epsilon/2$. Therefore, on E and for large n the sequence $(\tau_{m+1} - \tau_m, 0 \le m < (1-\delta)n/2)$ stochastically dominates a sequence $(B_m, 0 \le m < (1-\delta)n/2)$ of iid Bernoulli$(1 - \epsilon/2)$ random

[3] Until further notice, we omit ceilings and floors for readability.

variables. It follows that

$$\begin{aligned}\mathbf{P}\left\{\tau_{(1-\delta)n/2} \le (1-\epsilon)n/2\right\} &\le \mathbf{P}\{E^c\} + \mathbf{P}\left\{\tau_{(1-\delta)n/2} \le (1-\epsilon)n/2 \mid E\right\} \\ &\le \mathbf{P}\{E^c\} + \mathbf{P}\{\mathrm{Bin}((1-\delta)n/2, 1-\epsilon/2) < (1-\epsilon)n/2\} \\ &= o(1)\,,\end{aligned}$$

the last line Exercise 9(d) and Chebyshev's inequality (note that $(1-\delta)(1-\epsilon/2)n/2 > (1-5\epsilon/6)n/2$). On the other hand, if $\tau_{(1-\delta)n/2} > (1-\epsilon)n/2$ then $I_{(1-\epsilon)n/2} \le (1-\delta)n/2 < n/2$. □

Proof of Proposition 2.3 View $(F_1, \ldots, F_n)$ as coupled with the Erdős-Rényi coalescent as above, so that F_k and $G^{(n)}_{I_k}$ have the same components. Fix $\delta \in (0, 1/4)$ and let $k = k(n) = n/2 - 2\delta n$. Let E_1 be the event that $I_{n/2-\delta n} < n/2$.[4] Since $I_{m+1} \ge I_m + 1$ for all m, we have

$$I_k \le I_{n/2-\delta n} - ((n/2 - \delta n) - k) = I_{n/2-\delta n} - \delta n\,.$$

Thus, on E_1 we have $I_k \le (1-2\delta)n/2$.

Next let E_2 be the event that all component sizes in $G^{(n)}_{(1-2\delta)n/2}$ are at most δn. The components of F_k are precisely the components of $G^{(n)}_{I_k}$, so if $E_1 \cap E_2$ occurs then since on E_1 we have $I_k \le (1-2\delta)n/2$, all components of F_k have size at most δn. In this case, for all $i \le k$ the components of F_i clearly also have size at most δn.

It follows[5] that on $E_1 \cap E_2$, for all $i \le k$,

$$\sum_{T \in F_i} |T|^2 \le \delta n^2$$

so on $E_1 \cap E_2$,

$$\begin{aligned}\hat{Z}_{\overrightarrow{\mathrm{MC}}}(n, k+1) &= \prod_{i=1}^{k} \left(n^2 - \sum_{T \in F_i} |T|^2 \right) \\ &\ge n^{2k}(1-\delta)^k \qquad (2.6) \\ &= n^{n(1-4\delta)} \cdot (1-\delta)^{n/2} \qquad > n^{n(1-5\delta)}\,,\end{aligned}$$

the last inequality holding for n large. By Exercise 9(d) and Lemma 2.4, $\mathbb{P}(E_1 \cap E_2) \to 1$ as $n \to \infty$. Since $\hat{Z}_{\overrightarrow{\mathrm{MC}}}(n, \lfloor n/2 \rfloor) \ge \hat{Z}_{\overrightarrow{\mathrm{MC}}}(n, k+1)$ for n large, the result follows. □

The following exercise is to test whether you are awake.

---

[4]We omit the dependence on n in the notation for E_1; similar infractions occur later in the proof.

[5]To maximize $\sum_j x_j^2$ subject to the conditions that $\sum_j x_j = 1$ and that $\max_j x_j \le \delta$, take $x_j = \delta$ for $1 \le j \le \delta^{-1}$.

Exercise 10 Prove that

$$\frac{\log Z_{\mathrm{MC}}(n, \lfloor n/2 \rfloor)}{n \log n} \to 1\,,$$

as $n \to \infty$.

We next use Proposition 2.3 (more precisely, the inequality (2.6) obtained in the course of its proof) to obtain a first lower bound on $Z_{\mathrm{MC}}(n)$.

Corollary 2.5 *It holds that*

$$\frac{Z_{\mathrm{MC}}(n, \lfloor n/2 \rfloor)}{Z_{\mathrm{AC}}(n, \lfloor n/2 \rfloor)} = \left(\frac{e}{4}\right)^{(1+o(1))n/2}.$$

Proof By Proposition 2.3 and (2.5), we have

$$Z_{\mathrm{MC}}(n, \lfloor n/2 \rfloor) \geq 2^{-(\lfloor n/2 \rfloor - 1)} n^{(1+o(1))n},$$

so by Exercise 3,

$$\frac{Z_{\mathrm{MC}}(n, \lfloor n/2 \rfloor)}{Z_{\mathrm{AC}}(n, \lfloor n/2 \rfloor)} = \frac{n^{(1+o(1))n}}{2^{\lfloor n/2 \rfloor - 1} n^{\lfloor n/2 \rfloor - 1} (n-1)_{\lfloor n/2 \rfloor - 1}} = \frac{n^{(1+o(1))n} (n/2)!}{2^{n/2} n^{n/2} n!}.$$

Using Stirling's approximation,[6] it follows easily that

$$\frac{Z_{\mathrm{MC}}(n, \lfloor n/2 \rfloor)}{Z_{\mathrm{AC}}(n, \lfloor n/2 \rfloor)} \geq \left(\frac{e}{4}\right)^{(1+o(1))n/2}.$$

The corresponding upper bound follows similarly, using that $Z_{\mathrm{MC}}(n, \lfloor n/2 \rfloor) \leq (n(n-1)/2)^{\lfloor n/2 \rfloor - 1} = n^{n(1+o(1))}/2^{n/2}$. □

Exercise 11 Perform the omitted calculation using Stirling's formula from the proof of Corollary 2.5.

The preceding corollary is evidence that despite the similarity of the partition functions $Z_{\mathrm{MC}}(n)$ and $Z_{\mathrm{AC}}(n)$, the fine structure of the multiplicative coalescent is may be interestingly different from that of the additive coalescent.

[6] Stirling's approximation says that $m!/(\sqrt{2\pi m}(m/e)^m) \to 1$ as $m \to \infty$; in fact the (much less precise) fact that $\log(m!) = m \log m - m + o(m)$ is enough for the current situation.

2.3.2 The Multiplicative Coalescent and Kruskal's Algorithm

There is a pleasing interpretation of "Version 2" of the multiplicative coalescent, which is driven by exchangeable distinct edge weights $\mathbf{W} = \{W_{\{j,k\}}, 1 \le j < k \le n\} = \{W_e, e \in E(K_n)\}$. (A special case is that the elements of $\mathbf{W}$ are iid continuous random variables.). The symmetry of the model makes it straightforward to verify that this results in a sequence $(F_1, \ldots, F_n)$ with the same distribution as the multiplicative coalescent.

Multiplicative Coalescent Version 2: Kruskal's algorithm. Let F_1 be a forest of n isolated vertices $1, \ldots, n$. For $1 \le i < n$: $\star$ Let $\{j,k\} \in E(K_n)$ minimize $\{W_{\{j,k\}} : j,k$ in distinct trees of $F_i\}$. $\star$ Form F_{i+1} from F_i by adding $\{j,k\}$.

Exercise 12 ⊛ Prove that any exchangeable, distinct edge weights $\mathbf{W} = \{W_e, e \in E(K_n)\}$ again yield a process with the law of the multiplicative coalescent.

At step i, the edge-weight driven multiplicative coalescent simply adds the smallest weight edge whose endpoints lie in distinct components of F_i. In other words, it adds the smallest weight edge whose addition will not create a *cycle* in the growing graph. This is simply *Kruskal's algorithm* for building the minimum weight spanning tree. When the weights $W_{\{j,k\}}$ are all non-negative, the tree obtained at the end of the Version 2 multiplicative coalescent, $T_1^{(n)}$, is the minimum weight spanning tree of K_n with weights $\mathbf{W}$. We denote it $\mathrm{MST}(K_n, \mathbf{W})$, and refer to it as the *random MST of* K_n.

Order $E(K_n)$ by increasing order of $\mathbf{W}$-weight as $e_1, \ldots, e_{\binom{n}{2}}$. The exchangeability of $\mathbf{W}$ implies this is a uniformly random permutation of $E(K_n)$. Letting $G_k^{(n)}$ have edges $e_1, \ldots, e_k$ thus yields an important instantiation of our coupling of the Erdős-Rényi coalescent and the multiplicative coalescent; we return to this in Sect. 5.

2.3.3 Other Features of the Multiplicative Coalescent

The remainder of the section is not essential to the main development. The following exercise was inspired by a discussion with Remco van der Hofstad.

Exercise 13 (First-passage percolation) Consider the multiplicative coalescent driven by exchangeable, distinct edge weights $\mathbf{W}$ and for $1 \le i < j \le n$, let $d(i,j) = \min \sum_{e \in \gamma} W_e$, the minimum taken over paths from i to j in K_n. Show that the minimum is attained by a unique path $\gamma_{i,j}$. Find exchangeable edge weights $\{W_e, e \in E(K_n)\}$ for which, for each for each $1 \le i < j \le n$, $\gamma_{i,j}$ is a path of $T_1^{(n)}$.

Finally, we turn to Version 3 of the process, in which we view arbitrary iid non-negative weights $\mathbf{X} = \{X_{i,j}, 1 \le i < j \le n\}$ as *rates* for edge addition. In view of the preceding paragraph, this gives a process that results in a tree with the same *distribution* as the random MST of K_n, but which is not necessarily equal to the MST. In particular, the tree is *not* a deterministic function of the edge weights; for example, we may take $\mathbf{X}$ to be a deterministic vector such as the all-ones vector, whereas the resulting tree always is random.

Exercise 14 Find (iid random) *rates* $\mathbf{X}$ for which, in version 3 of the process, the resulting tree $T_1^{(n)}$ is equal to the random MST of K_n with *weights* $\mathbf{X}$, with probability tending to one as $n \to \infty$.

3 Intermezzo: The Heights of the Three Coalescent Trees

To date we have been primarily studying the partition functions of the coalescent processes. The processes have many other interesting features, however. In this section we discuss differences between the structures of the trees formed by the three coalescents.

Write $T_{\mathrm{KC}}^{(n)}, T_{\mathrm{AC}}^{(n)}$, and $T_{\mathrm{MC}}^{(n)}$, respectively, for the trees formed by Kingman's coalescent, the additive coalescent, and the multiplicative coalescent. In each case the coalescent starts from n isolated vertices $\{1, \ldots, n\}$, so each of these trees has vertices $\{1, \ldots, n\}$. If T is any of these trees and e is an edge of T, we write $L(e) = i$ if e was the i'th edge added during the execution of the coalescent. Above, we established the following facts about the distributions of these random trees.

1. Ignoring vertex labels, $(T_{\mathrm{KC}}^{(n)}, L)$ is uniformly distributed over pairs (t, ℓ), where t is a rooted tree with n vertices and ℓ is a decreasing edge labelling of t. (We simply refer to such pairs as *decreasing trees with n vertices*, for short.)
2. $T_{\mathrm{AC}}^{(n)}$ is uniformly distributed over the set of rooted trees with vertices $\{1, \ldots, n\}$. (We refer to such trees as *rooted labeled trees with n vertices*.)
3. $T_{\mathrm{MC}}^{(n)}$ is distributed as the minimum weight spanning tree of the complete graph K_n, with iid continuous edge weights $\mathbf{W} = \{W_{i,j}, 1 \le i < j \le n\}$.

What is known about these three distributions? To illustrate the difference between them, we consider a fundamental tree parameter, the *height*: this is simply the greatest number of edges in any path starting from the root.[7] The third tree, $T_{\mathrm{MC}}^{(n)}$ is not naturally rooted, but one may check that any choice of root will yield the same height up to a multiplicative factor of two; we root $T_{\mathrm{MC}}^{(n)}$ at vertex 1 by convention. Given a rooted tree t, we write $r(t)$ for its root and $h(t)$ for its height. The following exercise develops a fairly straightforward route to upper bounds on $h(T_{\mathrm{KC}}^{(n)})$ that are tight, at least to first order.

[7]A glance back at Figs. 2, 3 and 4 gives a hint as to the relative heights of the three trees.

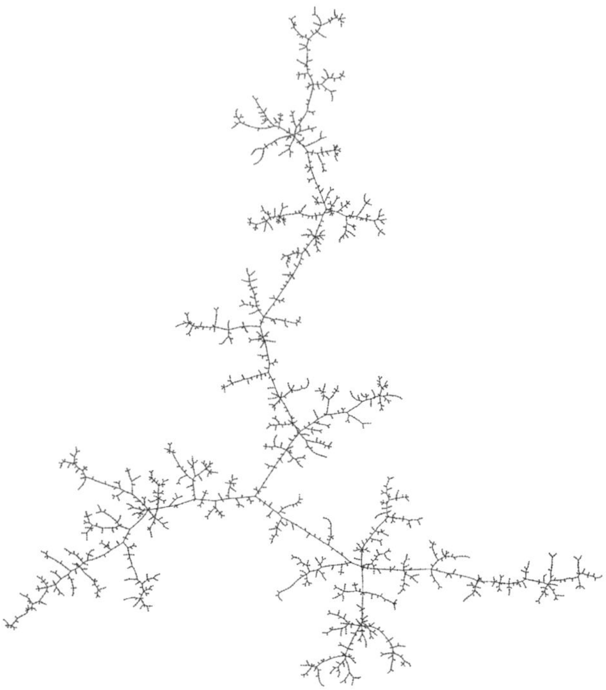

Fig. 2 One of the $3{,}000^{2{,}998}$ labeled trees with 3,000 vertices, selected uniformly at random

Exercise 15 Let D_i be the number of edges on the path from vertex i to $r(T_{\mathrm{KC}}^{(n)})$.

(a) Show that $\{D_1, \ldots, D_n\}$ are exchangeable random variables.
(b) (i) Show that D_1 can be written as a sum of independent Bernoulli random variables.
  (ii) For $\alpha > 0$, find λ such that Bernoulli$(\alpha) \preceq_{\mathrm{st}}$ Poisson(λ).
  (iii) Show that D_1 is stochastically dominated by a Poisson$(\log n)$ random variable.
(c) Show that for X a Poisson(μ) random variable, for integer $x > \mu$,

$$\mathbf{P}\{X > x\} \leq \frac{e^{-\mu}\mu^x}{(x-\mu)(x-1)!}.$$

(d) Show that $\mathbf{P}\{\max_{1\leq i\leq n} D_i \geq e\log n\} \to 0$ as $n \to \infty$.
(e) Show that $\limsup_{n\to\infty}(\max_{1\leq i\leq n} D_i - e\log n) \to -\infty$ in probability.

We next turn to $T_{\mathrm{AC}}^{(n)}$. I am not aware of an easy way to directly use the additive coalescent to analyze the height of $T_{\mathrm{AC}}^{(n)}$. However, one can use the additive coalescent to derive combinatorial results which, together with exchangeability, yields lower bounds of the correct order of magnitude, and upper bounds that

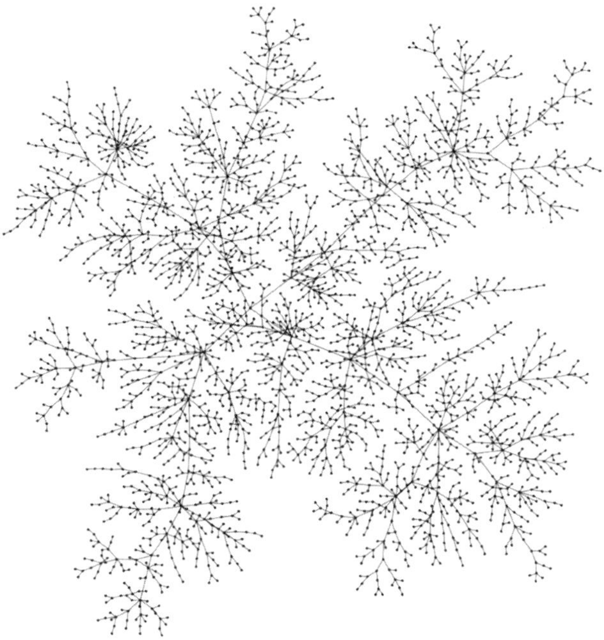

Fig. 3 One of the 2,999! rooted trees on 3,000 vertices with a decreasing edge labelling (labels suppressed)

are tight up to poly-logarithmic corrections; such bounds are the content of the following exercise. A non-negative random variable R has the *standard Rayleigh* distribution if it has density $f(x) = xe^{-x^2/2}$ on $[0, \infty)$.

Exercise 16 Let D_i be the number of edges on the path from vertex i to $r(T_{\mathrm{AC}}^{(n)})$.

(a) Show that $\{D_1, \ldots, D_n\}$ are exchangeable random variables.
(b) Show that the number of pairs (t, i), where t is a rooted labeled tree with $V(t) = [n]$ and $i \in V(t)$ has $d(r(t), i) = k - 1$, is $k \cdot (n)_k \cdot n^{n-k-1}$.
(c) Show that for $1 \leq k \leq n$, $\mathbf{P}\{D_1 = k - 1\} = \frac{k}{n}\prod_{i=1}^{k-1}\left(1 - \frac{i}{n}\right)$. Conclude that $D_1/\sqrt{n}$ converges in distribution to a standard Rayleigh.
(d) Using (c) and a union bound, show that if $(c_n, n \geq 1)$ are constants with $c_n \to \infty$ then $\mathbf{P}\left\{\max_{1\leq i\leq n} D_i > c_n\sqrt{n\log n}\right\} \to 0$.
(e) Use the exchangeability of the trees in a uniformly random ordered labeled forest to prove that $\mathbf{P}\{|\{i : D_i \geq k/2\}| \geq n/2 \mid D_1 = k\} \geq 1/2$ for all $1 \leq k \leq n$.
(f) Use (c) and (e) to show that if $(c_n, n \geq 1)$ are constants with $c_n \to \infty$ then $\mathbf{P}\left\{\max_{1\leq i\leq n} D_i > c_n\sqrt{n}\right\} \to 0$, strengthening the result from (d).

From the preceding exercise, we see immediately that $T_{\mathrm{AC}}^{(n)}$ has a very different structure from $T_{\mathrm{KC}}^{(n)}$, which had logarithmic height. Moreover, the heights of the two trees are *qualitatively* different. The height of $T_{\mathrm{KC}}^{(n)}$ is concentrated: $h(T_{\mathrm{KC}}^{(n)})/\log n \to e$

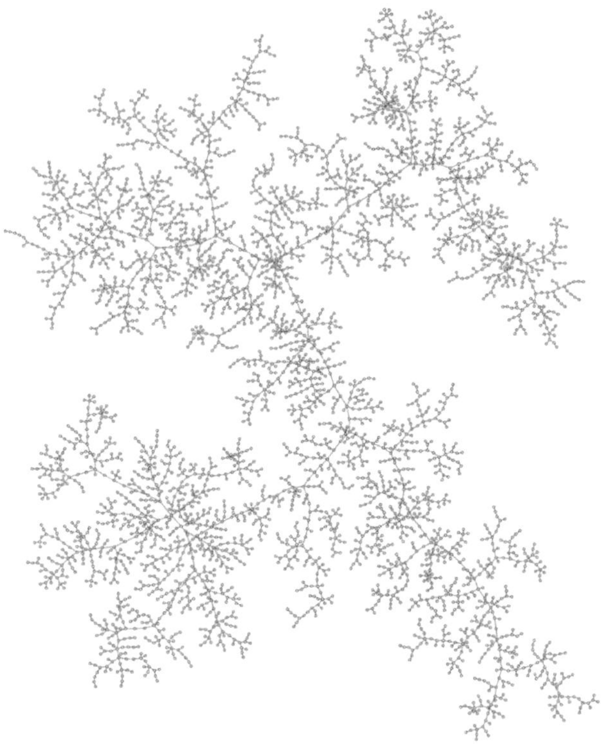

Fig. 4 The tree resulting from the multiplicative coalescent on 3,000 points

in probability. On the other hand, $h(T_{\mathrm{AC}}^{(n)})$ is diffuse: $h(T_{\mathrm{AC}}^{(n)})/n^{1/2}$ converges in distribution to a non-negative random variable with a density.[8,9]

What about the tree $T_{\mathrm{MC}}^{(n)}$ built by the multiplicative coalescent? Probabilistically, this is the most challenging of the three to study. For $T_{\mathrm{KC}}^{(n)}$ and $T_{\mathrm{AC}}^{(n)}$, Exercises 15 and 16 yielded exact or nearly exact expressions for the distance between the root and a fixed vertex (by exchangeability, this is equivalent to the distance between the root and a *uniformly random* vertex). The partition function $Z_{\mathrm{MC}}(n)$ seems too complex for such a direct argument to be feasible.

The coalescent procedure can be used to obtain lower bounds on the height, but with greater effort than in the two preceding cases. Our approach is elucidated by the following somewhat challenging exercise. Let K_n have iid Exponential[0, 1] edge

---

[8]Neither of these convergence statements follows from the exercises, and both require some work to prove. The fact that $h(T_{\mathrm{KC}}^{(n)})/\log n \to e$ in probability was first shown by Devroye [8]. The distributional convergence of $h(T_{\mathrm{AC}}^{(n)})/n^{1/2}$ is a result of Rényi and Szekeres [15].

[9]In fact, if edge lengths in $T_{\mathrm{AC}}^{(n)}$ are multiplied by $n^{-1/2}$ then the resulting object converges in distribution to a random compact metric space called the *Brownian continuum random tree* (or *CRT*), and $h(T_{\mathrm{AC}})/n^{1/2}$ converges in distribution to the height of the CRT. For more on this important result, we refer the reader to [4, 10].

weights, and let H be the subgraph of K_n with the same vertices, but containing only edges of weight at most $1/n$. A *tree component* of H is a connected component of H that is a tree.

Exercise 17

(a) Let N be the number of vertices in tree components of H of size at most $\lfloor n^{1/4} \rfloor$. Using Chebyshev's inequality, show that $\mathbf{P}\{N = 0\} \to 0$ as $n \to \infty$.
(b) Fix $S \subset \{1, \ldots, n\}$. Show that, given that H contains a tree component whose vertices are precisely S, then such a component is uniformly distributed over labeled trees with vertices S.
(c) Use Kruskal's algorithm to show that any tree component of H is a subtree of the minimum weight spanning tree of $T_{\mathrm{MC}}^{(n)}$.
(d) Use Exercise 16(c) to conclude that, as $n \to \infty$,

$$\mathbf{P}\left\{ T_{\mathrm{MC}}^{(n)} \text{ has height at most } \frac{n^{1/8}}{\log^2 n} \right\} \to 0\,.$$

This shows that $T_{\mathrm{MC}}^{(n)}$ is quite different from $T_{\mathrm{KC}}^{(n)}$.[10] It is not as straightforward to bound the height of $T_{\mathrm{MC}}^{(n)}$ away from $n^{1/2}$ using the tools currently at our disposal. It turns out that $T_{\mathrm{MC}}^{(n)}$ has height of order $n^{1/3}$ (and has non-trivial fluctuations on this scale), but proving this takes a fair amount of work [1] and is beyond the scope of these notes.

Exercise 18 (Open problem – two point function of random MSTs) Let D_n be the distance from vertex n to vertex 1 in $T_{\mathrm{MC}}^{(n)}$. Obtain an explicit expression for the distributional limit of $D_n/n^{1/3}$.

4 The Susceptibility Process

The remainder of the paper focusses exclusively on the multiplicative coalescent, which we continue to denote $(F_1, \ldots, F_n)$. Recall that $\hat{Z}_{\overrightarrow{\mathrm{MC}}}(n,k) = \prod_{i=1}^{k-1} \left(n^2 - \sum_{T \in F_i} |T|^2\right)$. The terms in the preceding product are not independent; linearity of expectation makes the "empirical entropy" $\log \hat{Z}_{\overrightarrow{\mathrm{MC}}}(n)$ easier to study.

$$\mathbf{E}\left[\log \hat{Z}_{\overrightarrow{\mathrm{MC}}}(n)\right] = \sum_{k=1}^{n-1} \mathbf{E}\left[\log(n^2 - \sum_{T \in F_k} |T|^2) \right]. \tag{4.1}$$

[10] With more care, one can show that with high probability H contains tree components containing around $n^{2/3}$ vertices and with height around $n^{1/3}$, which yields that with high probability $T_{\mathrm{MC}}^{(n)}$ has height of order at least $n^{1/3}$.

The expectation in the latter sum is closely related to the *susceptibility* of the forest F_i. More precisely, given a finite graph G, write $\mathcal{C}(G)$ for the set of connected components of G. The susceptibility of G is the quantity

$$\chi(G) = \frac{1}{|G|} \sum_{C \in \mathcal{C}(G)} |C|^2.$$

Recalling that $C(v) = C_G(v)$ is the component of G containing v, we may also write $\chi(G) = |G|^{-1} \sum_{v \in V(G)} |C(v)|$, so $\chi(G)$ is the expected size of the component containing a uniformly random vertex from G.

Exercise 19 Let G be any graph, write L and S for the number of vertices in the largest and second-largest components of G, respectively. Then

$$\frac{L^2}{|G|} \leq \chi(G) \leq \frac{L^2}{|G|} + S.$$

Viewing F_i as a graph with vertices $\{1, \ldots, n\}$, (4.1) becomes

$$\mathbf{E}\left[\log \hat{Z}_{\mathrm{MC}}^{\rightarrow}(n)\right] = 2(n-1)\log n + \mathbf{E}\left[\sum_{k=1}^{n-1} \log\left(1 - \frac{\chi(F_k)}{n}\right)\right]. \tag{4.2}$$

In order to analyze this expression, we use the connection with the Erdős-Rényi coalescent $(G_m^{(n)}, 0 \leq m \leq \binom{n}{2})$, which we described in Sect. 2.3.1; in brief, we coupled to $(F_k, 1 \leq k \leq n)$ by letting F_k have edges $\{e_{I_j}, 1 \leq j \leq k\}$, where I_k was the first time m that $G_m^{(n)}$ had $n + 1 - k$ components. In the next proposition and throughout the remainder of the paper, we interpret $x \log x$ as 0 when $x = 1$.

Proposition 4.1

$$\mathbf{E}\left[\log \hat{Z}_{\mathrm{MC}}^{\rightarrow}(n)\right] = 2(n-1)\log n +$$
$$\sum_{m=0}^{\binom{n}{2}-1} \left(1 - \frac{n+2m}{n^2}\right)^{-1} \mathbf{E}\left[\log\left(1 - \frac{\chi(G_m^{(n)})}{n}\right) \cdot \left(1 - \frac{\chi(G_m^{(n)})}{n}\right)\right].$$

Proof In the coupling with the Erdős-Rényi coalescent, F_k and $G_{I_k}^{(n)}$ have the same connected components, so $\chi(F_k) = \chi(G_{I_k}^{(n)})$. We obtain the identity

$$\sum_{k=1}^{n-1} \log\left(1 - \frac{\chi(F_k)}{n}\right) = \sum_{k=1}^{n-1} \log\left(1 - \frac{\chi(G_{I_k}^{(n)})}{n}\right)$$
$$= \sum_{m=0}^{\binom{n}{2}-1} \log\left(1 - \frac{\chi(G_m^{(n)})}{n}\right) \mathbf{1}_{[\chi(G_{m+1}^{(n)}) > \chi(G_m^{(n)})]}$$

Using the tower law for conditional expectations, we thus have

$$
\begin{aligned}
&\mathbf{E}\left[\sum_{k=1}^{n-1} \log\left(1-\frac{\chi(F_k)}{n}\right)\right] \\
&= \sum_{m=0}^{\binom{n}{2}-1} \mathbf{E}\left[\log\left(1-\frac{\chi(G_m^{(n)})}{n}\right)\mathbf{1}_{[\chi(G_{m+1}^{(n)})>\chi(G_m^{(n)})]}\right]. \\
&= \sum_{m=0}^{\binom{n}{2}-1} \mathbf{E}\left[\mathbf{E}\left[\log\left(1-\frac{\chi(G_m^{(n)})}{n}\right)\mathbf{1}_{[\chi(G_{m+1}^{(n)})>\chi(G_m^{(n)})]} \mid G_m^{(n)}\right]\right] \\
&= \sum_{m=0}^{\binom{n}{2}-1} \mathbf{E}\left[\log\left(1-\frac{\chi(G_m^{(n)})}{n}\right)\cdot \mathbf{P}\left\{\chi(G_{m+1}^{(n)}) > \chi(G_m^{(n)}) \mid G_m^{(n)}\right\}\right]
\end{aligned}
$$

For any finite graph G, the quantity $\chi(G)/|G| = \sum_{C\in\mathcal{C}(G)} |C|^2/|G|^2$ is simply the probability that a pair (U, V) of independent, uniformly random vertices of G lie in the same component of G. Let (U, V) be independent, uniformly random elements of $[n] = V(G_{m-1}^{(n)})$. Then

$$
\mathbf{P}\left\{U \neq V, \{U, V\} \notin E(G_m^{(n)}) \mid G_m^{(n)}\right\} = 1 - \frac{1}{n} - \frac{2|E(G_m^{(n)})|}{n^2} = 1 - \frac{n+2m}{n^2}
$$

Conditionally given that $U \neq V$ and $\{U, V\} \notin E(G_m^{(n)})$, the pair $\{U, V\}$ has the same law as e_{m+1}. It follows that

$$
\begin{aligned}
&\mathbf{P}\left\{\chi(G_{m+1}^{(n)}) > \chi(G_m^{(n)}) \mid G_m^{(n)}\right\} \\
=&\mathbf{P}\left\{C_{G_m^{(n)}}(U) \neq C_{G_m^{(n)}}(V) \mid G_m^{(n)}, U \neq V, \{U, V\} \notin E(G_m^{(n)})\right\} \\
=&\left(1-\frac{\chi(G_{m-1}^{(n)})}{n}\right)\left(1-\frac{n+2m}{n^2}\right)^{-1},
\end{aligned}
\tag{4.3}
$$

so

$$
\begin{aligned}
&\mathbf{E}\left[\sum_{k=1}^{n-1} \log\left(1-\frac{\chi(F_k)}{n}\right)\right] \\
&= \sum_{m=0}^{\binom{n}{2}-1} \left(1-\frac{n+2m}{n^2}\right)^{-1} \mathbf{E}\left[\log\left(1-\frac{\chi(G_m^{(n)})}{n}\right)\cdot\left(1-\frac{\chi(G_m^{(n)})}{n}\right)\right].
\end{aligned}
\tag{4.4}
$$

The proposition now follows from (4.2). □

It turns out that there is a deterministic, increasing function $f : [0, \infty) \to [0, 1]$ such that $\sup_{0 \le m < \binom{n}{2}} |\chi(G_m^{(n)})/n - f(m/n)| \to 0$ in probability, as $n \to \infty$. Much of the rest of the paper is devoted to explaining this fact in more detail. However, imagine for the moment that such a function f exists and, moreover, that terms in the sum with $m \gg n$ have an insignificant total contribution. With these assumptions, the sum in (4.4) looks like a Riemann approximation for $\int_0^\infty (1 - f(x)) \log(1 - f(x)) \mathrm{d}x$ with mesh $1/n$. We should then expect that

$$\mathbf{E}\left[\log \hat{Z}_{\overrightarrow{\mathrm{MC}}}(n)\right] = 2(n-1)\log n - (1 + o(1))n \cdot \int_0^\infty (1 - f(x)) \log(1 - f(x)) \mathrm{d}x .$$

This is indeed the case. Furthermore, enough is known about f that explicit evaluation of the integral is possible, and we obtain the following theorem.

Theorem 4.2 *Let*

$$\zeta_{\mathrm{MC}} = \zeta(2) - 3 + \log 2 - \log^2 2 . \tag{4.5}$$

Then

$$\mathbf{E}\left[\log \hat{Z}_{\mathrm{MC}}(n)\right] = n \cdot (2 \log n + \zeta_{\mathrm{MC}} + o(1)).$$

Numerically, ζ_{MC} is around -1.14237.

Corollary 4.3 *There is $c > 0$ such that $\mathbf{P}\left\{\hat{Z}_{\mathrm{MC}}(n)/\mathbf{E}\hat{Z}_{\mathrm{MC}}(n) < e^{-cn}\right\} \to 1$ as $n \to \infty$.*

Proof Fix $c \in \mathbb{R}$ and suppose that $\mathbf{P}\left\{\hat{Z}_{\mathrm{MC}}(n) \ge n^{2n} e^{cn}\right\} > \epsilon > 0$. Then

$$\begin{aligned} &\mathbf{E}\left[\log \hat{Z}_{\mathrm{MC}}(n) - 2n \log n - cn + 1\right] \\ \ge &\epsilon \mathbf{E}\left[\log \hat{Z}_{\mathrm{MC}}(n) - 2n \log n + cn + 1 \mid \hat{Z}_{\mathrm{MC}}(n) \ge n^{2n} e^{-cn}\right] \\ \ge &\epsilon . \end{aligned}$$

Thus, if $\liminf_{n\to\infty} \mathbf{P}\left\{\hat{Z}_{\mathrm{MC}}(n) \ge n^{2n} e^{cn}\right\} > 0$ then for all n large enough,

$$\mathbf{E}\left[\log \hat{Z}_{\mathrm{MC}}(n)\right] > 2n \log n + cn - 1 .$$

It thus follows from Theorem 4.2 that for all $\epsilon > 0$,

$$\mathbf{P}\left\{\hat{Z}_{\overrightarrow{\mathrm{MC}}}(n) \ge n^{2n} e^{(\zeta_{\overrightarrow{\mathrm{MC}}} + \epsilon)n}\right\} \to 0$$

as $n \to \infty$. On the other hand,

$$\mathbf{E}\hat{Z}_{\overrightarrow{\mathrm{MC}}}(n) = n^{n-2}(n-1)! = n^{2n}e^{-n(1+o(1))} .$$

Since $\zeta_{\mathrm{MC}} < -1$, the result follows. □

The form of the constant ζ_{MC} is unimportant, though intriguing. What is clear from the above is that information about the susceptibility process of the multiplicative coalescent yields bounds for $\hat{Z}_{\mathrm{MC}}(n)$. The aim of the next section is thus to understand the susceptibility process in more detail.

4.1 Bounding χ Using a Graph Exploration

The coupling between the "Version 2" multiplicative coalescent (Kruskal's algorithm) and the Erdős-Rényi coalescent from Sect. 2.3.2 applied to arbitrary exchangeable, distinct edge weights $\mathbf{W}$. In this coupling, for $m \in [\binom{n}{2}]$, we took $G_m^{(n)}$ to be the subgraph of K_n consisting of the m edges of smallest $\mathbf{W}$-weight.

In the current section, it is useful to be more specific. We suppose the entries of $\mathbf{W}$ are iid Uniform[0, 1] random variables. Write $G(n,p)$ for the graph with vertices $[n]$ and edges $\{e_j : W_{e_j} \le p\}$. In $G(n,p)$, each edge of K_n is independently present with probability p. Furthermore, we have $G(n,p) = G_{m_p}^{(n)}$, where $m_p = \max\{i : W_{e_i} \le p\}$, so this also couples the Erdős-Rényi coalescent with the process $(G(n,p), 0 \le p \le 1)$. The next exercise is standard, but important.

Exercise 20 ⊛ Show that for any $p \in (0,1)$ and $m \in \binom{n}{2}$, given that $|E(G(n,p))| = m$, the conditional distribution of $G(n,p)$ is the same as that of $G_m^{(n)}$.

For $c > 0$, let $\alpha = \alpha(c)$ be the largest real solution of $e^{-cx} = 1 - x$. The aim of this section is to prove the following result.

Theorem 4.4 *For all $n \ge 1$ and $0 \le p \le n^{-19/20}$,*

$$\mathbf{P}\left\{|\chi(G(n,p)) - \alpha(np)^2 n| > 22n^{4/5}\right\} < 6ne^{-12n^{1/10}}$$

The coupling with the Erdős-Rényi coalescent will allow us to derive corresponding results for $G_m^{(n)}$. While the ingredients for the proof are all in the literature, and closely related results have certainly appeared in many places, we were unable to find a reference for the form we require. Some of the basic calculations required for the proof appear as exercises; the first such exercise relates to properties of the function α.

Exercise 21 ⊛

(a) Show that α is continuous and that α is concave and strictly positive on $(1,\infty)$.
(b) Show that for $0 < c \le 1$, $\alpha(c) = 0$, and for $c \ge 2$, $1 - 2e^{-c} \le \alpha(c) \le 1 - e^{-c}$.

(c) Show that $\alpha(c)$ is increasing and $c(1-\alpha(c))$ is decreasing.
(d) Show that $\alpha'(c) < 2$ for all $c > 1$ and that $\frac{d}{dc}\alpha(c) \uparrow 2$ as $c \downarrow 1$. Conclude that $2\epsilon(1-o(1)) \le \alpha(1+\epsilon) \le 2\epsilon$, the first inequality holding as $\epsilon \downarrow 0$.
(e) Show that $\alpha(c)$ is the survival probability of a Poisson(c) branching process. *(This exercise is not used directly.)*

Our proof of Theorem 4.4 hinges on a variant of the well-known and well-used depth-first search exploration procedure. In depth-first search, at each step one vertex is "explored": its neighbours are revealed, and those neighbours lying in the undiscovered region of the graph are added to the "depth-first search queue" for later exploration. In our variant, if the queue is ever empty, in the next step we add each undiscovered vertex to the queue independently with probability p. (It is more standard to add a *single* undiscovered vertex, but adding randomness turns out to simplify the formula for the expected number of unexplored vertices.)

We now formally state our search procedure for $G(n,p)$. At step i the vertex set $[n]$ is partitioned into sets E_i, D_i and U_i, respectively containing *explored, discovered*, and *undiscovered* vertices. We always begin with $E_0 = \emptyset$, $D_0 = \{1\}$, and $U_0 = [n] \setminus \{1\}$. For a set S, we write $\text{Bin}(S,p)$ to denote a random subset of S which contains each element of S independently with probability p. For $v \in [n]$ we write $N(v)$ for the neighbours of v in $G(n,p)$. Finally, we define the priority of a vertex $v \in [n]$ to be its time of discovery $\inf\{j : v \in D_j\}$, so vertices that are discovered later have higher priority.

Search process for $G(n,p)$**.**

STEP i:

⋆ If $D_i \neq \emptyset$ then choose $v \in D_i$ with highest priority (if there is a tie, pick the vertex with smallest label among highest-priority vertices). Let $E_{i+1} = E_i \cup \{v\}$, let $D_{i+1} = (D_i \cup (N(v) \cap U_i)) \setminus \{v\}$ and let $U_{i+1} = U_i \setminus (N(v) \cap U_i)$.

⋆ If $D_i = \emptyset$ then let $D_{i+1} = \text{Bin}(U_i, p)$, independently of all previous steps. Let $E_{i+1} = E_i$ and let $U_{i+1} = U_i \setminus D_{i+1}$.

Observe that the sequence $((D_i, E_i, U_i), i \ge 0)$ describing the process may be recovered from either $(D_i, i \ge 0)$ or $(U_i, i \ge 0)$. The order of exploration yields the following property of the search process. Suppose $D_i = \emptyset$ for a given i. Then D_{i+1} may contain several nodes, all of which have priority $(i+1)$. Starting at step $(i+1)$, the search process will fully explore the component containing the smallest labelled vertex of D_{i+1} before exploring any vertex in any other component. More strongly, the search process will explore the components that intersect D_{i+1} in order of their smallest labeled vertices.

For $i > 0$ such that $E_i \neq E_{i-1}$, write v_i for the unique element of $E_i \setminus E_{i-1}$. Say that a component exploration concludes at time t if v_{t+1} and v_t are in distinct components of $G(n,p)$. The observation of the preceding paragraph implies the following fact about the search process. Set $D_0 = \emptyset$ for convenience.

Fact 4.5 Fix $t > 0$ and let $i = i(t) = \max\{j < t : D_j = \emptyset\}$. If a component exploration concludes at time t then $|D_{i+1}| \geq n - t - |U_t|$.

Proof Since a component exploration concludes at time t we have $D_t \subset D_{i+1}$. Furthermore, $|E_t| \leq t$ because $|E_0| = \emptyset$ and $|E_{j+1} \setminus E_j| \leq 1$ for all $j \geq 0$. As D_t, E_t and U_t partition $[n]$, we thus have

$$|U_t| = n - |E_t| - |D_t| \geq n - t - |D_{i+1}|.\square$$

In proving Theorem 4.4 we use a concentration inequality due to McDiarmid [11]. Let $\mathrm{X} = (X_i, 1 \leq i \leq m)$ be independent Bernoulli(q) random variables. Suppose that $f : \{0,1\}^m$ is such that for all $1 \leq k \leq m$, for all $(x_1, \ldots, x_k) \in \{0,1\}^k$,

$$|\mathbf{E}\left[f(x_1, \ldots, x_k, X_{k+1}, \ldots, X_m)\right] - \mathbf{E}\left[f(x_1, \ldots, 1 - x_k, X_{k+1}, \ldots, X_m)\right]| \leq 1.$$

In other words, given the values of the first $k - 1$ variables, knowledge of the k'th variable changes the conditional expectation by at most one.

Theorem 4.6 (McDiarmid's inequality) *Let* X *and f be as above. Write $\mu = \mathbf{E}\left[f(\mathrm{X})\right]$. Then for $x > 0$,*

$$\mathbf{P}\{f(X) \geq \mu + x\} \leq e^{-x^2/(2mq+2x/3)}, \quad \mathbf{P}\{f(X) \leq \mu - x\} \leq e^{-x^2/(2mq+2x/3)}.$$

Our probabilistic analysis of the search process begins with the following observation. For each $i \geq 0$, the set $U_i \setminus U_{i+1} = D_{i+1} \setminus D_i$ of vertices discovered at step i has law Bin(U_i, p). This observation also allows us to couple the search process with a family $\mathrm{B} = (B_{i,j}, i \geq 1, j \geq 1)$ of iid Bernoulli(p) random variables, by inductively letting $U_i \setminus U_{i+1} = D_{i+1} \setminus D_i$ equal $\{j \in U_i : B_{i,j} = 1\}$, for each $i \geq 1$. The coupling shows that for all $i \geq 1$, $|U_i|$ satisfies the hypotheses of Theorem 4.6, with $m = ni$ and $q = p$. Also, using the preceding coupling, the next exercise is an easy calculation.

Exercise 22 ⊛ Show that for $i \geq 0$, $\mathbf{E}\left[|U_{i+1}| \mid (U_j, j \leq i)\right] = |U_i|(1 - p)$; conclude that $\mathbf{E}|U_i| = (n - 1)(1 - p)^i$ for all $i \geq 0$.

The exploration of the component $C(1)$ is completed precisely at the first time j that $D_j = \emptyset$; this is also the first time j that $|U_j| = n - j$, and for earlier times k we have $|U_k| < n - k$. If we had $|U_i| = \mathbf{E}|U_i|$ for all i then the above exercise would imply that $|C(1)| = \min\{t \in \mathbb{N} : (n - 1)(1 - p)^t \geq n - t\}$. Of course, $|U_i|$ does not equal $\mathbf{E}|U_i|$ for all i. However, $|U_i|$ does track its expectation closely enough that a consideration of the expectation yields an accurate prediction of the first-order behaviour of $\chi(G(n,p))$. We next explain this in more detail, then proceed to the proof of Theorem 4.4. Write $t(n,p)$ for the largest real solution of $n(1-p)^x = n - x$. We will use the next exercise, the first part of which gives an idea of how $t(n,p)$ behaves when p is moderately small.

Exercise 23 ⊛

(a) Show that $t(n,p) = n \cdot \alpha(n\log(1/(1-p)))$. Conclude that if $p \le n^{-3/4}$ then with $\mu = np$, we have

$$\alpha(\mu)n \le t(n,p) \le \alpha(\mu)n + \frac{2n^{1/2}}{1-p}.$$

(**Hint.** Use Exercise 21(d).)

(b) Show that

$$n(1-p)^s \ge (n-s) + (s-t)(1 + (n-t)\log(1-p)) \text{ for } s > t$$

Write L and S for the sizes of the largest and second largest components of $G(n,p)$, respectively. From time 0 to time $t = t(n,p)$, the search process essentially explores a single component. We thus expect that $L \ge t$. Next, since $n(1-p)^{t+1} > n - (t+1)$ and $n(1-p)^t = n - t$, by the convexity of $(1-p)^s$ we have $n(1-p)^{s+1} \ge n(1-p)^s - 1$ for all $s \ge t$. Exercise 22 then implies that $\mathbf{E}|U_{s+1}| \ge \mathbf{E}|U_s| - 1$ for all integer $s \ge t$. In other words, when exploring a component after time t, the search process on average discovers less than one new vertex in each step. Such an exploration should quickly die out and, indeed, after time t the components uncovered by the search process typically all have size $o(n)$. Together with the first point, this suggests that $L \le t + o(n)$ and $S = o(n)$. Using the bounds on t from Exercise 23(a) and the bounds on $\chi(G)$ from Exercise 19, we are led to predict that

$$\alpha(np)^2 n + o(n) = \frac{L^2}{n} \le \chi(G(n,p)) \le \frac{L^2}{n} + S = \alpha(np)^2 n + o(n).$$

Theorem 4.4 formalizes and sharpens this prediction, and we now proceed to its proof.

Proof of Theorem 4.4 Throughout the proof we assume n is large (which is required for some of the inequalities), and write $t = t(n,p)$, $\alpha = \alpha(np)$.

Case 1: $p \le 1/n + 6/n^{6/5}$ **("subcritical** p**").**

Recall that exploration of $C(1)$ concludes the first time i that $|U_i| \ge n - i$. Letting $t^+ = 21n^{4/5}$, we have $(1-p)^{t^+} \ge 1 - t^+p + (t^+p)^2/2 - (t^+p)^3/6 > 1 - t^+p + (t^+p)^2/3$, and it follows straightforwardly that

$$\mathbf{E}|U_{t^+}| = n(1-p)^{t^+} \ge n(1 - t^+p + (t^+p)^2/3) \ge n - t^+ + 3n^{3/5}.$$

Applying the bound for the lower tail from Theorem 4.6 to $|U_t|$, it follows that

$$\mathbf{P}\left\{|U_{t^+}| \le n - t^+\right\} \le \mathbf{P}\left\{|U_{t^+}| \le \mathbf{E}|U_{t^+}| - 3n^{3/5}\right\} \le e^{-(9/2)n^{1/5}}.$$

At all times i before exploration of the first component concludes we have $|U_i| < n-i$, so the preceding bound yields

$$\mathbf{P}\left\{|C(1)| \geq 21n^{4/5}\right\} \leq e^{-(9/2)n^{1/5}},$$

We always have $\chi(G(n,p)) \leq \max_{i\in[n]} |C(i)|$ so, by a union bound,

$$\mathbf{P}\left\{\chi(G(n,p)) \geq 21n^{4/5}\right\} \leq \mathbf{P}\left\{\max_{i\in[n]} |C(i)| \geq 21n^{4/5}\right\} \leq ne^{-(9/2)n^{1/5}}.$$

For this range of p, by Exercise 21(d) we also have $n\alpha(np)^2 \leq 17n^{3/5}$, and so the bound in Theorem 4.4 follows.

Case 2: $1/n + 6/n^{6/5} < p \leq 1/n^{19/20}$ **("supercritical** p**").**

We begin by explaining the steps of the proof. (I) First, logic similar to that in case 1 shows that the largest component of $G(n,p)$ is unlikely to have size much larger than t. (II) Next, we need a corresponding *lower* tail bound on the size of the largest component; the proof of this relies on Fact 4.5. (III) Finally, we need to know that with high probability there is only one component of large size; after ruling out one or two potential pathologies, this follows from the subcritical case. We treat the three steps in this order. Write $\Delta = n^{3/4}$ and $t^{\pm} = t(n,p) \pm \Delta$.

(I) We claim that

$$n(1-p)^{t^+} \geq n - t^+ + 5n^{11/20}. \tag{4.6}$$

To see this, first use Exercise 23(b) to obtain

$$n(1-p)^{t^+} \geq n - t^+ + \Delta(1 + (n-t)\log(1-p)).$$

Let $c = n\log(1/(1-p))$. By Exercise 23(a),

$$1 + (n-t)\log(1-p) = 1 - c(1-\alpha(c)).$$

Next, as $p \geq 1/n + 6/n^{6/5}$ we have $n\log(1/(1-p)) \geq np \geq 1 + 6/n^{1/5} =: c^*$. By Exercise 21(c) and (d), it follows that

$$c(1-\alpha(c)) \leq c^*(1-\alpha(c^*)) = \left(1 + \frac{6}{n^{1/5}}\right)\left(1 - \frac{(2+o(1))6}{n^{1/5}}\right) \leq 1 - \frac{5}{n^{1/5}},$$

so $1 + (n-t)\log(1-p) \geq 5/n^{1/5}$. (Similar bounds using Exercise 21(c) and (d) crop up again later in the proof). Since $\Delta/n^{1/5} = n^{11/20}$, (4.6) follows. Having established (4.6), essentially the same logic as in Case 1 yields

$$\mathbf{P}\left\{\max_{i\in[n]} |C(i)| \geq t^+\right\} \leq n\mathbf{P}\left\{|C(1)| \geq t^+\right\} \leq ne^{-(25/2)n^{1/10}}. \tag{4.7}$$

(II) We now turn to the lower tail of $\max_{i\in[n]} |C(i)|$. The calculations are similar but slightly more involved. Since $p = o(1)$ and $p\Delta \leq n^{-1/5} = o(1)$, for n large $(1-p)^{-\Delta} \leq 1 + p\Delta + (p\Delta)^2$, so

$$n(1-p)^{t^-} = (n-t)(1-p)^{-\Delta} \leq (n-t)(1+p\Delta+(p\Delta)^2). \tag{4.8}$$

Since $t \geq n\alpha(n,p)$, it follows easily from Exercise 21(c) and (d) that $p(n-t) \leq 1 - \frac{5}{n^{1/5}}$. Using (4.8) and the bound $p\Delta \leq n^{-1/5}$, we thus have

$$\begin{aligned} n(1-p)^{t^-} &\leq (n-t) + \Delta\left(1 - \frac{5}{n^{1/5}}\right)(1+p\Delta) \\ &\leq n - t^- - \frac{4\Delta}{n^{1/5}} \\ &= n - t^- - 4n^{3/5}. \end{aligned} \tag{4.9}$$

Next, basic arithmetic shows that if $m \geq (n+p^{-1})/2$ then $m(1-p) \leq m-1-(np-1)/2 \leq m-(1+6n^{-1/5})$. Furthermore, for p in the range under consideration, $(n+p^{-1})/2 \leq n - 2n^{4/5}$, so

$$n(1-p)^{\Delta} \leq \max(n-2n^{4/5}, n-\Delta-6\Delta n^{-1/5}) = n - \Delta - 6n^{11/20}.$$

Since $n(1-p)^t$ is concave as a function of t, this bound and (4.9) together imply that $n(1-p)^x \leq n-x-6n^{11/20}$ for all $x \in [\Delta, t^-]$. Applying Theorem 4.6 for $t \in [\Delta, t^-]$, and a union bound, yields

$$\mathbf{P}\left\{\exists t \in [\Delta, t^-] : |U_t| \geq n - t - n^{11/20}\right\} \leq (t^- - \Delta)e^{-(25/2)n^{1/10}}.$$

Now suppose that $|U_t| < n - t - n^{11/20}$ for all $t \in [\Delta, t^-]$. In this case, if a component exploration concludes at some time $t \in [\Delta, t^-]$ then by Fact 4.5 there is $i < t$ such that $D_i = \emptyset$ and $|D_{i+1}| > n - t - |U_t| > n^{11/20}$. On the other hand, for all $i \geq 0$, $|D_{i+1} \setminus D_i|$ is stochastically dominated by $\mathrm{Bin}(n,p)$, so by a union bound followed by a Chernoff bound (or an application of Theorem 4.6),

$$\mathbf{P}\left\{\exists i < t^- : D_i = \emptyset, |D_{i+1}| > n^{11/20}\right\} \leq t^-\mathbf{P}\left\{\mathrm{Bin}(n,p) > n^{11/20}\right\} \leq t^- e^{-n^{11/20}}.$$

It follows that

$$\begin{aligned}
&\mathbf{P}\{\text{A component exploration concludes between times } \Delta \text{ and } t^-\} \\
&\leq (t^- - \Delta)e^{-(25/2)n^{1/10}} + t^- e^{-n^{11/20}} \\
&\leq 2ne^{-(25/2)n^{1/10}}. \qquad (4.10)
\end{aligned}$$

(III) Let N be the number of vertices which remain undiscovered the first time after time t^- that the search process finishes exploring a component, and write B for the event that some component whose exploration starts *after* time t^- has size greater than $21n^{4/5}$. Then

$$\begin{aligned}
\mathbf{P}\{B\} &= \mathbf{P}\{B, N > n-(t^- - \Delta)\} + \sum_{m \leq n-(t^- -\Delta)} \mathbf{P}\{B, N = m\} \\
&\leq \mathbf{P}\{N > n-(t^- - \Delta)\} + \sup_{m \leq n-(t^- -\Delta)} \mathbf{P}\{B \mid N = m\}.
\end{aligned}$$

The first probability is at most $2ne^{-(25/2)n^{1/10}}$ by (4.10). To bound the second, note that

$$n-(t^- - \Delta) \leq n + 2n^{3/4} - t \leq n + 3n^{3/4} - n\alpha(np) = n(1-\alpha(np) + 3n^{-1/4}).$$

By Exercise 21(c) and (d), and since $p \leq n^{-19/20}$, for $m \leq n-(t^- - \Delta)$ we therefore have

$$mp \leq np(1-\alpha(np)) + 3n^{3/4}p < 1.$$

For such m, the bound for "subcritical p" from Case 1 thus yields

$$\mathbf{P}\{B \mid N = m\} \leq me^{-(9/2)m^{1/5}}.$$

This is less than $ne^{-(9/2)n^{4/25}}$ for $m \geq n^{4/5}$. If $m \leq n^{4/5}$ then the largest component explored after time m also has size $\leq n^{4/5}$, so $\mathbf{P}\{B \mid N = m\} = 0$. We conclude that

$$\mathbf{P}\{B\} \leq 2ne^{-(25/2)n^{1/10}} + ne^{-(9/2)n^{4/25}} \leq 3ne^{-(25/2)n^{1/10}}. \qquad (4.11)$$

(IV) Now to put the pieces together. The lower bound is easier: by the first inequality from Exercise 19, we have $\chi(G(n,p)) \geq n^{-1}\max_{i\in[n]}|C(i)|^2$. From (4.10) it then follows that

$$\mathbf{P}\left\{\chi(G(n,p)) < \frac{(t^- - \Delta)^2}{n}\right\} \leq \mathbf{P}\left\{\max_{i\in[n]}|C(i)| < \Delta - t^-\right\} \leq 2ne^{-(25/2)n^{1/10}}. \qquad (4.12)$$

By Exercise 23(a),

$$\frac{(t^- - \Delta)^2}{n} = \frac{(t - 2n^{3/4})^2}{n} \geq \frac{(n\alpha - 3n^{3/4})^2}{n} \geq n\alpha^2 - 9n^{1/2},$$

and the lower bound then follows from (4.12).

For the upper bound, any component of $G(n,p)$ whose exploration concludes before step $n^{3/4}$ of the search process has size at most $n^{3/4}$. Write S for the number of vertices of the second-largest component of $G(n,p)$. By (4.11), we then have

$$\mathbf{P}\left\{S \geq 21n^{4/5}\right\} \leq 3ne^{-(25/2)n^{1/10}}.$$

Combined with the second inequality from Exercise 19 and with (4.7), we obtain

$$\mathbf{P}\left\{\chi(G(n,p)) \geq \frac{(t + n^{3/4})^2}{n} + 21n^{4/5}\right\} \leq 4ne^{-(25/2)n^{1/10}}. \tag{4.13}$$

An easy calculation using Exercise 23(a) shows that $(t+n^{3/4})^2/n + 21n^{4/5} \leq n\alpha^2 + 22n^{4/5}$, and the theorem then follows from (4.12) and (4.13). □

To conclude the section, we use Theorem 4.4 to show that $\mathbf{E}\left[\chi(G_m^{(n)})\right]$ is well-approximated by $\alpha(2m/n)$ in a range which covers the most important values of m. (Exercise 24, below, extends this to all $0 \leq m \leq \binom{n}{2}$.)

Lemma 4.7 *For n large, for all $m \leq n^{22/21}$,*

$$\left|\mathbf{E}\left[\chi(G_m^{(n)})\right] - \alpha^2(2m/n)n\right| \leq 23n^{4/5}.$$

Proof Write $x_m = \inf\{p : |E(G(n,p))| = m\}$. Since $\mathbf{E}|E(G(n,p))| = p\binom{n}{2}$, we expect x_m to be near $p_m := m/\binom{n}{2}$. Write $\hat{\alpha} = \alpha(2m/(n-1)) = \alpha(np_m)$, let $\delta = n^{-4/3}$, and let $p_m^{\pm} = p_m \pm \delta$.

In the coupling of $(G(n,p), 0 \leq p \leq 1)$ and $(G_m^{(n)}, 0 \leq m \leq \binom{n}{2})$, if $x_m > p_m^-$ then $G(n, p_m^-)$ is a subgraph of $G_m^{(n)}$ and so $\chi(G_m^{(n)}) \geq \chi(G(n, p_m^-))$. Likewise, if $x_m < p_m^+$ then $\chi(G_m^{(n)}) \leq \chi(G(n, p_m^+))$ We thus have

$$\begin{aligned}
\chi(G_m^{(n)}) &\geq \chi(G(n,p_m^-))\mathbf{1}_{[x_m > p_m^-]} \\
&\geq \chi(G(n,p_m^-)) - n\mathbf{1}_{[x_m \leq p_m^-]}, \qquad \text{and} \\
\chi(G_m^{(n)}) &\leq \chi(G(n,p_m^+))\mathbf{1}_{[x_m < p_m^+]} + n\mathbf{1}_{[x_m \geq p_m^+]} \\
&\leq \chi(G(n,p_m^+)) + n\mathbf{1}_{[x_m \geq p_m^+]}.
\end{aligned}$$

Since α is 2-Lipschitz, $\hat{\alpha} - 2/n^{1/3} \leq \alpha(np_m^-) \leq \alpha(np_m^+) \leq \hat{\alpha} + 2/n^{1/3}$, from which it follows that both $\alpha(np_m^-)^2 n$ and $\alpha(np_m^+)^2 n$ are within $5n^{2/3}$ of $\hat{\alpha}^2 n$. By the preceding lower bound on $\chi(G_m^{(n)})$ and Theorem 4.4 we thus have

$$\begin{aligned}\mathbf{E}\left[\chi(G_m^{(n)})\right] &\geq \mathbf{E}\left[\chi(G(n,p_m^-))\right] - n\mathbf{P}\left\{x_m \leq p_m^-\right\}\\ &\geq \hat{\alpha}^2 n - 5n^{2/3} - 22n^{4/5} - n\mathbf{P}\left\{\text{Bin}\left(\binom{n}{2}, p_m^-\right) \geq m\right\}\\ &\geq \hat{\alpha}^2 n - 5n^{2/3} - 22n^{4/5} - 1\end{aligned}$$

the last inequality holding straightforwardly by a Chernoff bound (note that $\binom{n}{2}p_m^- = m - (n-1)/(2n^{1/3}) \leq m - m^{3/5}/3$). We likewise have

$$\mathbf{E}\left[\chi(G_m^{(n)})\right] \leq \hat{\alpha}^2 n + 5n^{2/3} + 22n^{4/5} + 1.$$

Finally, $2m/(n-1) - 2m/n = 2m/(n(n-1)) = O(n^{-8/9})$, so since α is 2-Lipschitz we have $\alpha(2m/n)^2 = \hat{\alpha}^2 + O(n^{-8/9})$, and the result follows. □

5 Frieze's $\zeta(3)$ Limit for the MST Weight

Before proving Theorem 4.2, we warm up by using the same approach to study the total weight of random MSTs. Throughout the section, $\mathbf{W} = (W_e, e \in E(K_n))$ are exchangeable, distinct, non-negative edge weights. Recall from Sect. 2.3.2 that "Version 2" of the multiplicative coalescent (aka Kruskal's algorithm) considers edges one-by-one in increasing order of weight, adding only edges which connect distinct trees in the forest, and that the result is the minimum spanning tree $T = \text{MST}(K_n, \mathbf{W})$.

Write $w(T) = \sum_{e \in E(T)} W_e$ for the total weight of T. We use susceptibility bounds to approximate $w(T)$ and derive a version of Frieze's famous $\zeta(3)$ limit.

Theorem 5.1 (Frieze [9]) *Write $X_1, \ldots, X_{\binom{n}{2}}$ for the increasing ordering of $\mathbf{W}$. If $\mathbf{E}X_m = (1 + o(1))m$, then $2\mathbf{E}w(MST(K_n, \mathbf{W}))/n^2 \to \zeta(3)$ as $n \to \infty$.*

By $\mathbf{E}X_m = (1 + o(1))m$ we mean that $\lim_{n\to\infty} \sup_{m\in[\binom{n}{2}]} |1 - \mathbf{E}X_m/m| = 0$. This condition can be relaxed, and the proof can be modified to obtain convergence in probability under suitable hypotheses, but for expository reasons we have opted for simplicity over full generality. Before beginning the proof, we first note a special case. Suppose that the weights W_e are independent Uniform[0, 1] random variables. Then $\mathbf{E}[X_k] = k/(\binom{n}{2} + 1)$, $\mathbf{E}\left[X_k \cdot n^2/2\right] = (1 + o(1))k$. The theorem thus implies that for such uniform edge weights, the total weight of the random MST of K_n converges to $\zeta(3)$ without renormalization. This is the most often quoted special case of Frieze's result.

Our proof is based on the following identity for $\mathbf{E}[w(T)]$.

Proposition 5.2 *Write $X_1, \dots, X_{\binom{n}{2}}$ for the increasing ordering of* $\mathbf{W}$. *Then*

$$\mathbf{E}[w(T)] = \sum_{m=0}^{\binom{n}{2}-1} \mathbf{E}[X_{m+1}] \cdot \mathbf{P}\left\{\chi(G_{m+1}^{(n)}) > \chi(G_m^{(n)})\right\}. \tag{5.1}$$

Proof Let $e_1, \dots, e_{\binom{n}{2}}$ be the ordering of $E(K_n)$ by increasing weight, so e_m has weight X_m. In the coupling with the Erdős-Rényi coalescent, Kruskal's algorithm adds edge e_k precisely if e_k joins distinct components of $G_{k-1}^{(n)}$, which occurs if and only if $\chi(G_k^{(n)}) > \chi(G_{k-1}^{(n)})$. For this coupling we thus have

$$w(T) = \sum_{m=0}^{\binom{n}{2}-1} X_{m+1} \cdot \mathbf{1}_{[\chi(G_{m+1}^{(n)}) > \chi(G_m^{(n)})]}.$$

By the exchangeability of $\mathbf{W}$, the vector $(X_1, \dots, X_{\binom{n}{2}})$ is independent of the ordering of $E(K_n)$. The event that $\chi(G_{m+1}^{(n)}) > \chi(G_m^{(n)})$ is measurable with respect to the ordering of $E(K_n)$, so is independent of $(X_1, \dots, X_{\binom{n}{2}})$. The proposition follows on taking expectations. □

We use the result of the following exercise to deduce that terms with $m \geq 5n \log n$ play an unimportant role in the summation (5.1). Fix $1 \leq k \leq \lfloor n/2 \rfloor$ and let N_k be the number of sets $A \subset [n]$ of size k such that, in $G_m^{(n)}$, there are no edges from A to $[n] \setminus A$. Note that $G_m^{(n)}$ is connected precisely if $N_k = 0$ for all $1 \leq k \leq \lfloor n/2 \rfloor$.

Exercise 24 ⊛

(a) Let E_k be the event that there are no edges from $[k]$ to $[n] \setminus [k]$ in $G_m^{(n)}$. With $p = m/\binom{n}{2}$, show that $\mathbf{P}\{E_k\} \leq (1-p)^{k(n-k)}$. Deduce that

$$\mathbf{P}\{N_k > 0\} \leq n^k (1-p)^{k(n-k)} \leq (ne^{-p(n-k)})^k.$$

(b) Show that $\mathbf{P}\left\{G_{\lceil 5n \log n \rceil}^{(n)} \text{ is not connected}\right\} \leq n^{-4}$.

(c) Show that the bound in Lemma 4.7 in fact holds for all $m \in [\binom{n}{2}]$.

Corollary 5.3 *With the notation of Proposition 5.2, we have*

$$\mathbf{E}[w(T)] \geq \sum_{m=0}^{5n \log n} \mathbf{E}[X_{m+1}] \left(1 - \frac{\mathbf{E}\chi(G_m^{(n)})}{n}\right) \quad \textit{and}$$

$$\mathbf{E}[w(T)] \leq \left(1 + \frac{12 \log n}{n}\right) \sum_{m=0}^{5n \log n} \mathbf{E}X_{m+1} \left(1 - \frac{\mathbf{E}\chi(G_m^{(n)})}{n}\right) + \frac{1}{2n^2} \mathbf{E}\left[X_{\binom{n}{2}}\right].$$

Proof Write

$$\mathbf{P}\left\{\chi(G_{m+1}^{(n)}) > \chi(G_m^{(n)})\right\} = \mathbf{E}\left[\mathbf{P}\left\{\chi(G_{m+1}^{(n)}) > \chi(G_m^{(n)}) \mid G_m^{(n)}\right\}\right].$$

We derived an identity for the inner conditional probability in (4.3); using that identity and linearity of expectation, we obtain

$$\mathbf{P}\left\{\chi(G_{m+1}^{(n)}) > \chi(G_m^{(n)})\right\} = \left(1 - \frac{\mathbf{E}\chi(G_m^{(n)})}{n}\right)\left(1 - \frac{n+2m}{n^2}\right)^{-1}. \tag{5.2}$$

The latter is always at least $1 - \mathbf{E}\chi(G_m^{(n)})/n$, and the lower bound then follows from Proposition 5.2 by truncating the sum at $m = 5n\log n$.

For the upper bound, note that if $G_m^{(n)}$ is connected then $\chi(G_m^{(n)}) = n$, so

$$\mathbf{P}\left\{\chi(G_{m+1}^{(n)}) > \chi(G_m^{(n)})\right\} \leq \mathbf{P}\left\{G_m^{(n)} \text{ is not connected}\right\}.$$

Using Exercise 24(b) and the fact that the X_i are increasing, it follows that

$$\begin{aligned}
\mathbf{E}\left[w(T)\right] &\leq \sum_{m=0}^{5n\log n} \mathbf{E}X_{m+1} \cdot \mathbf{P}\left\{\chi(G_{m+1}^{(n)}) > \chi(G_m^{(n)})\right\} \\
&\quad + \sum_{m=5n\log n+1}^{\binom{n}{2}-1} \mathbf{E}X_{m+1} \cdot \mathbf{P}\left\{G_m^{(n)} \text{ is not connected}\right\} \\
&\leq \sum_{m=0}^{5n\log n} \mathbf{E}X_{m+1} \cdot \mathbf{P}\left\{\chi(G_{m+1}^{(n)}) > \chi(G_m^{(n)})\right\} + \binom{n}{2} \cdot \mathbf{E}\left[X_{\binom{n}{2}}\right] \cdot \frac{1}{n^4}.
\end{aligned}$$

For $m \leq 5n\log n$, $(1 - \frac{n+2m}{n^2})^{-1} \leq 1 + 12\log n/n$, and the result follows from (5.2). □

To prove Theorem 5.1, we use Lemma 4.7 and Corollary 5.3 to show that after appropriate rescaling, the sum in Proposition 5.2 is essentially a Riemann sum approximating an appropriate integral. The value of that integral is derived in the following lemma.

Lemma 5.4

$$\int_0^\infty \lambda \cdot (1 - \alpha^2(\lambda))\mathrm{d}\lambda = 2\zeta(3).$$

Proof Aldous and Steele [5] write that "calculation of this integral is quite a pleasing experience"; though the calculation appears in that work, why should we deprive ourselves of the pleasure? Anyway, the proof is short. First, use integration by parts to write

$$\int_0^\infty \lambda \cdot (1-\alpha^2(\lambda))\mathrm{d}\lambda = \int_0^\infty \alpha(\lambda)\alpha'(\lambda)\lambda^2 \mathrm{d}\lambda = \int_1^\infty \alpha(\lambda)\alpha'(\lambda)\lambda^2 \mathrm{d}\lambda\,,$$

the second equality since $\alpha(\lambda) = 0$ for $\lambda < 1$. The identity $1-\alpha(c) = e^{-c\alpha(c)}$ (this is how we *defined* α) implies that $\lambda^2 = (\alpha(\lambda)^{-1}\log(1-\alpha(\lambda)))^2$, so we may rewrite the latter integral as

$$\int_1^\infty \frac{\log^2(1-\alpha(\lambda))}{\alpha(\lambda)} \cdot \alpha'(\lambda)\mathrm{d}\lambda = \int_0^1 \frac{\log^2(1-\alpha)}{\alpha}\mathrm{d}\alpha\,,$$

where we used the obvious change of variables $\alpha = \alpha(\lambda)$. Now a final change of variables: $u = -\log(1-\alpha)$ transforms this into

$$\int_0^\infty u^2 \frac{e^{-u}}{1-e^{-u}}\mathrm{d}u = \int_0^\infty u^2 \sum_{k=1}^\infty e^{-ku}\mathrm{d}u\,.$$

Since $\int_0^\infty u^2 e^{-ku} = 2/k^3$, the final expression equals $\sum_{k=1}^\infty 2/k^3 = 2\zeta(3)$. □

Our final step before the proof is to show that the integrand is well-behaved on the region of integration; the straightforward bound we require is stated in the following exercise. Recall that α is continuous on $[0,\infty)$ and is differentiable except at $x = 1$.

Exercise 25 ⊛ Let $f(x) = x(1-\alpha^2(x))$, where α is as above. Show there exists $C < \infty$ such that $|f'(x)| \le C$ for all $x \ne 1$. (In fact we can take $C = 2$.)

Proof of Theorem 5.1 By the preceding exercise, for all $0 < \epsilon \le x$,

$$\left| \int_{x-\epsilon}^{x} \lambda(1-\alpha^2(\lambda))\mathrm{d}\lambda - \epsilon x(1-\alpha^2(x)) \right| \le C\epsilon^2\,.$$

Taking $\epsilon = 2/n$, $x = 2m/n$ and summing over $m \in \{1,\ldots,5n\log n\}$ we obtain in particular that

$$\begin{aligned}\sum_{m=1}^{5n\log n} \frac{4m}{n^2}(1-\alpha^2(2m/n)) &= \int_0^{10\log n} \lambda(1-\alpha^2(\lambda))\mathrm{d}\lambda + O\left(\frac{\log n}{n^2}\right) \\ &= 2\zeta(3) - o(1)\,,\end{aligned}$$

the second equality by Lemma 5.4. If $\mathbf{E}[X_m] = (1+o(1))m$ then by the preceding equation, Lemma 4.7, and the lower bound in Corollary 5.3, we have

$$\mathbf{E}[w(T)] \ge (1+o(1))\frac{n^2}{2} \cdot \zeta(3)\,,$$

and likewise (this time using the upper bound in Corollary 5.3)

$$\begin{aligned}\mathbf{E}\left[w(T)\right] &\leq (1+o(1))\frac{n^2}{2}\cdot\zeta(3)+\frac{1}{2n^2}\mathbf{E}\left[X_{\binom{n}{2}}\right]\\ &= (1+o(1))\frac{n^2}{2}\cdot\zeta(3)+O(1)\,,\end{aligned}$$

which completes the proof. □

6 Estimating the Empirical Entropy

We already know the broad strokes of the argument, since they are the same as for our proof of Theorem 5.1. Recall that we are trying to approximate $\mathbf{E}\left[\log\hat{Z}_{\mathrm{MC}}(n)\right]=\mathbf{E}\left[\log\hat{Z}^{\rightarrow}_{\mathrm{MC}}(n)\right]-(n-1)\log 2$. Proposition 4.1 reduces this to the study of the sum

$$\Xi=\sum_{m=0}^{\binom{n}{2}-1}\left(1-\frac{n+2m}{n^2}\right)^{-1}\mathbf{E}\left[\left(1-\frac{\chi(G_m^{(n)})}{n}\right)\log\left(1-\frac{\chi(G_m^{(n)})}{n}\right)\right]. \tag{6.1}$$

We use Theorem 4.4 and Exercise 24 to approximate this sum by an integral. Before proceeding to the ϵ's and δ's, we evaluate the integral.

Proposition 6.1 *We have*

$$\int_0^\infty(1-\alpha^2(\lambda))\cdot\log(1-\alpha^2(\lambda))\mathrm{d}\lambda=2(\zeta_{\mathrm{MC}}+\log 2)\,. \tag{6.2}$$

Proof A similar calculation to that of Lemma 5.4, though decidedly less pleasing. Since $\alpha(\lambda)=0$ for $\lambda\leq 1$, we may change the domain of integration to $[1,\infty)$. Then use the identity

$$\alpha'(\lambda)=\frac{\alpha(\lambda)^2(1-\alpha(\lambda))}{\alpha(\lambda)+(1-\alpha(\lambda))\log(1-\alpha(\lambda))}\,,$$

which follows from the fact that $1-\alpha(\lambda)=e^{-\lambda\alpha(\lambda)}$ by differentiation. The integral under consideration thus equals

$$\int_1^\infty(1-\alpha^2(\lambda))\log(1-\alpha^2(\lambda))\cdot\frac{\alpha(\lambda)+(1-\alpha(\lambda))\log(1-\alpha(\lambda))}{\alpha(\lambda)^2(1-\alpha(\lambda))}\cdot\alpha'(\lambda)\mathrm{d}\lambda\,,$$

from which the substitution $\alpha = \alpha(\lambda)$ gives

$$\int_0^1 \frac{(1+\alpha)\log(1-\alpha^2)(\alpha+(1-\alpha)\log(1-\alpha))}{\alpha^2} d\alpha\,.$$

Substituting $u = -\log(1-\alpha)$, we have $1-\alpha = e^{-u}$, $1+\alpha = 2-e^{-u}$ and $\log(1-\alpha^2) = \log(2-e^{-u}) - u$, and the above integral becomes

$$\int_0^1 \frac{(2-e^{-u})(\log(2-e^{-u})-u)\cdot(1-e^{-u}-ue^{-u})}{(1-e^{-u})^2} \cdot e^{-u} du\,.$$

This integral can be calculated with a little effort (or easily, for those who accept computer assisted proofs), and equals

$$\frac{\pi^2}{3} - 6 + 4\log 2 - 2\log^2(2)$$

Comparing with (4.5) completes the proof (recall that $\zeta(2) = \pi^2/6$). □

The next lemma generalizes Lemma 4.7, at the cost of obtaining a non-explicit error bound. We use a slightly different proof technique than for Lemma 4.7; this time we exploit that a binomial random variable is reasonably likely to take values close to its mean (see the following exercise).

Exercise 26 ⊛ Show that

$$\mathbf{P}\left\{\mathrm{Bin}\left(\binom{n}{2}, \frac{2m}{n^2}\right) = m\right\} = \Omega\left(\frac{1}{n}\right)$$

uniformly in $0 \le m \le n^2/4$, in that

$$\liminf_{n\to\infty} \inf_{m\in\{0,1,\ldots,\lfloor n^2/4\rfloor\}} n\cdot\mathbf{P}\left\{\mathrm{Bin}\left(\binom{n}{2}, \frac{2m}{n^2}\right) = m\right\} > 0.$$

For a continuous function $f : [0,1] \to \mathbb{R}$, and $\epsilon \in [0,1]$, write $\gamma_f(\epsilon) = \sup_{|x-y|\le\epsilon} |f(x)-f(y)|$, and $\|f\| = \sup_{x\in[0,1]} |f(x)|$.

Lemma 6.2 *Let $f : [0,1] \to \mathbb{R}$ be continuous. Then*

$$\sup_{m\in[\binom{n}{2}]} \left|\mathbf{E}\left[f(\chi(G_m^{(n)})/n)\right] - f(\alpha(2m/n)^2)\right| = O(\gamma_f(n^{-1/5}))) + O(n^{-4}).$$

Proof First suppose $m \ge n^2/4$. Then $\alpha(2m/n) \ge \alpha(n/2) \ge 1 - 2e^{-n/2}$ by Exercise 21(b), so we have $|f(\alpha^2(2m/n)) - f(1)| \le \gamma_f(2e^{-n/2}) \le \gamma_f(n^{-1/5})$, the latter for n large. Next, since $\chi(G) = |G|$ whenever G is connected, by

Exercise 24(b) we have

$$|\mathbf{E}\left[f(\chi(G_m^{(n)})/n)\right] - f(1)| \le \|f\|\mathbf{P}\left\{\chi(G_m^{(n)}) \neq n\right\} = O(n^{-4}).$$

This handles the case $m \ge n^2/4$, so we now assume $0 \le m \le n^2/4$.

Let $p = 2m/n^2$, so $np = 2m/n$. By Exercise 20, we have

$$\mathbf{E}\left[f(\chi(G_m^{(n)})/n)\right] = \mathbf{E}\left[f(\chi(G(n,p))/n) \mid |E(G(n,p))| = m\right].$$

By Exercise 26, there is $C > 0$ such that for all $m \le n^2/4$,

$$\begin{aligned}
&\mathbf{P}\left\{|\chi(G(n,p)) - n\alpha(2m/n)^2| > 22n^{4/5} \mid |E(G(n,p))| = m\right\} \\
\le& Cn\mathbf{P}\left\{|\chi(G(n,p)) - n\alpha(2m/n)^2| > 22n^{4/5}\right\} \\
=& O(n^2 e^{-12n^{1/10}}),
\end{aligned}$$

the last line by Theorem 4.4. It follows that

$$\begin{aligned}
&\mathbf{E}\left[f(\chi(G_m^{(n)})/n)\right] \\
\ge& \inf_{|a-n\alpha(2m/n)^2|\le 22n^{4/5}} f(a/n)\cdot \mathbf{P}\left\{|\chi(G(n,p)) - n\alpha(2m/n)^2| \le 22n^{4/5} \mid |E(G(n,p))| = m\right\} \\
&- \|f\|\cdot \mathbf{P}\left\{|\chi(G(n,p)) - n\alpha(2m/n)^2| > 22n^{4/5} \mid |E(G(n,p))| = m\right\} \\
=& \inf_{|a-n\alpha(2m/n)^2|\le 22n^{4/5}} f(a/n) - O(n^2 e^{-12n^{1/10}}) \\
\ge& f(\alpha(2m/n)^2) - \beta(f, 22n^{-1/5}) - O(n^2 e^{-12n^{1/10}}).
\end{aligned}$$

We likewise have $\mathbf{E}\left[f(\chi(G_m^{(n)})/n)\right] \le f(\alpha(2m/n)^2) + \beta(f, 22n^{-1/5}) + O(n^2 e^{-12n^{1/10}})$. Since $\gamma_f(22n^{-1/5}) \le 22\gamma_f(n^{-1/5})$ and $n^2 e^{-12n^{1/10}} = o(n^{-4})$, the result follows. □

In what follows we only apply the preceding lemma with $m = o(\binom{n}{2})$, but it seems more satisfying to prove the estimate over the full range of possibilities; handling larger m only added a few lines to the proof. We are now ready to wrap things up. Define a function $f : [0, 1] \to \mathbb{R}$ by $f(x) = (1-x)\log(1-x)$ for $x \in [0, 1)$; recalling the convention that $0 \log 0 = 0$, we see that f is a continuous function. The next exercise, the last of the notes, asks the reader to establish two basic properties of f.

Exercise 27

(a) Show that $\|f\| = 1/e$.
(b) Show that for all $x, y \in [0, 1]$, $|f(x) - f(y)| \le |x-y| \log(1/|x-y|)$.

Proof of Theorem 4.2 By Exercises 24(b) and 27(a), for $m \geq 5n\log n$ we have

$$\mathbf{E}\left[\left(1-\frac{\chi(G_m^{(n)})}{n}\right)\log\left(1-\frac{\chi(G_m^{(n)})}{n}\right)\right] \leq \|f\| \cdot \mathbf{P}\left\{G_m^{(n)} \text{ is not connected}\right\} \leq \frac{1}{en^4}.$$

Summing in (6.1) over indices $m \geq 5n\log n$ and using that $1 \leq (1-(n+2m)/n^2)^{-1} = 1 + O(\log n/n)$ for $m \leq 5n\log n$, this implies that

$$\left|\Xi - \sum_{m=0}^{5n\log n}\mathbf{E}\left[\left(1-\frac{\chi(G_m^{(n)})}{n}\right)\log\left(1-\frac{\chi(G_m^{(n)})}{n}\right)\right]\right| = O\left(\log^2 n\right).$$

By Lemma 6.2 we have

$$\begin{aligned}&\sum_{m=0}^{5n\log n}\mathbf{E}\left[\left(1-\frac{\chi(G_m^{(n)})}{n}\right)\log\left(1-\frac{\chi(G_m^{(n)})}{n}\right)\right]\\ &= \sum_{m=0}^{5n\log n}\left(1-\alpha^2(2m/n)\right)\log\left(1-\alpha^2(2m/n)\right) + O\left(n\log n\cdot\left(\gamma_f(n^{-1/5})+n^{-4}\right)\right).\end{aligned}$$

Exercise 27(b) implies that $\gamma_f(n^{-1/5}) \leq (\log n)/(5n^{1/5})$, so the preceding two bounds yield

$$\left|\Xi - \sum_{m=0}^{5n\log n}\left(1-\alpha^2(2m/n)\right)\log\left(1-\alpha^2(2m/n)\right)\right| = O(n^{4/5}\log^2 n). \tag{6.3}$$

Next, by Exercise 21(d) we know that $\alpha'(x) < 2$ for $x > 1$. Also, $\alpha(x) = 0$ for $x \leq 1$. It follows that for $x, y \geq 0$ and $\epsilon \in (0,1)$, if $|x-y| < \epsilon$ then $\alpha^2(x) - \alpha^2(y) \leq 4\epsilon(\alpha(x)+\epsilon) < 8\epsilon$. By Exercise 27(b), this implies that for $x \in [0,\infty)$ and $\epsilon \in (0,1)$,

$$\left|\epsilon\cdot f(\alpha^2(x)) - \int_{x-\epsilon}^{x} f(\alpha^2(\lambda))\mathrm{d}\lambda\right| \leq 8\epsilon^2\log(1/8\epsilon).$$

We then have, as in Theorem 5.1, that

$$\begin{aligned}&\sum_{m=1}^{5n\log n}\left(1-\alpha^2(2m/n)\right)\log\left(1-\alpha^2(2m/n)\right)\\ &= \frac{n}{2}\int_0^{10\log n}(1-\alpha^2(\lambda))\log(1-\alpha^2(\lambda))\mathrm{d}\lambda + O\left(\frac{\log^2 n}{n}\right)\\ &= n(\zeta_{\mathrm{MC}} + \log 2) + o(n),\end{aligned}$$

The second equality holds since $\int_0^\infty (1-\alpha^2(\lambda))\log(1-\alpha^2(\lambda))\mathrm{d}\lambda = 2(\zeta_{\mathrm{MC}}+\log 2)$ and the integrand is negative, so $\lim_{x\to\infty}\int_x^\infty (1-\alpha^2(\lambda))\log(1-\alpha^2(\lambda))\mathrm{d}\lambda = 0$.

Combining this with (6.3) and Proposition 4.1, we obtain

$$\mathbf{E}\left[\log \hat{Z}_{\mathrm{MC}}(n)\right] = \mathbf{E}\left[\log \hat{Z}_{\overrightarrow{\mathrm{MC}}}(n)\right] - (n-1)\log 2 = n\cdot(2\log n + \zeta_{\overrightarrow{\mathrm{MC}}} + o(1)),$$

which is the assertion of the theorem. □

7 Unanswered Questions

The partition function of the multiplicative coalescent provides an interesting avenue by which to approach the probabilistic study of the process. It connects up nicely with other perspectives, and offers its own insights and challenges. We saw above that the empirical partition function of the multiplicative coalescent is a subtle and interesting random variable. Here are a few questions related to $\hat{Z}_{\mathrm{MC}}(n)$, and more generally to the multiplicative coalescent, that occurred to me in the course of writing these notes and which I believe deserve investigation.

- The large deviations of $\log \hat{Z}_{\mathrm{MC}}(n)$ should be interestingly non-trivial. Can a large deviations rate function be derived? This should be related to large deviations for component sizes in the random graph process. Such results exist for fixed p [7, 12], but not (so far as I am aware) for the process as p varies. (Considering the following sort of problem would be a step in the right direction. Let E_n be the event that the largest component of $G(n,p)$ has at least $0.1n$ fewer vertices than average, for all $p \in [2/n, 3/n]$. Find a law of large numbers for $\log \mathbf{P}\{E_n\}$.)
- Relatedly, what partition chains are responsible for the large value of $\mathbf{E}\left[\hat{Z}_{\mathrm{MC}}(n)\right]$? It is not too hard to show the following: to maximize $\prod_{i=1}^{n-1} n^2(1-\chi(F_i)^2/n)$ one should keep the component sizes as small as possible. In particular, if $n=2^p$ then one maximizes this product by first pairing all singletons to form trees of size two, then pairing these trees to form trees of size 4, etcetera. This shows that for $n=2^p$,

$$\operatorname{ess\,sup} \hat{Z}_{\mathrm{MC}}(n) = 2^{-(n-1)} \prod_{k=1}^{p=1} \prod_{j=0}^{n/2^k-1} \left(n^2 - 2^{k-1}(n+j\cdot 2^k)\right).$$

which is within a factor 4 of $2^{-(n-1)}n^{2(n-1)}e^{-\log_2 n}$. On the other hand, a straightforward calculation shows the probability of choosing two minimal trees to pair at every step is around $e^{-(1+o(1))2n}$, so the contribution to $\mathbf{E}\hat{Z}_{\mathrm{MC}}(n)$ from such paths is $n^{2n}e^{-(1+o(1))(2+\log 2)n}$. This is exponentially small compared to $n^{n-2}(n-1)!$, so the lion's share of the expected value comes from elsewhere.

- Suppose we condition $\hat{Z}_{\mathrm{MC}}(n)$ to be close to $n^{n-2}(n-1)! = \mathbf{E}\left[\hat{Z}_{\mathrm{MC}}(n)\right]$; we know by Corollary 4.3 that this event has exponentially small probability. Perhaps, under this conditioning, the tree $T_1^{(n)}$ built by the multiplicative coalescent might be similar to that built by the additive coalescent? At any rate, it would certainly be interesting to study, e.g., $\mathbf{E}\left[\mathrm{height}(T_1^{(n)}) \mid \hat{Z}_{\mathrm{MC}}(n) \geq n^{n-2}(n-1)!\right]$, or more generally to study observables of $T_1^{(n)}$ under unlikely conditionings of $\hat{Z}_{\mathrm{MC}}(n)$.
- Condition $T_1^{(n)}$ to have exactly k leaves. After rescaling distances appropriately, $T_1^{(n)}$ should converge in the Gromov-Hausdorff sense. What is the limit? Write $\mathbf{E}_k$ for the coresponding conditional expectation; then we should have, for example, $\mathbf{E}_k[\mathrm{diam}(T_1^{(n)})/n] \to f(k)$ for some function $f(k)$. How does f behave as $k \to \infty$? It is known [1] that without conditioning, $\mathbf{E}\left[\mathrm{diam}(T_1^{(n)})\right] = \Theta(n^{1/3})$.
- Pitman's coalescent, Kingman's coalescent, and the multiplicative coalescent correspond to rate kernels $\kappa(x,y) = x + y$, $\kappa(x,y) = 1$, and $\kappa(x,y) = xy$, respectively. Baur [6] has shown that for any integer $k \geq 3$, the coalescent with kernel $\kappa(x_1,\ldots,x_k) = x_1 + \ldots + x_k + k/(k-2)$ admits a representation in terms of $(k-1)$-ary forests. Are there further rate kernels that may be naturally enriched to form interesting forest-valued coalescent processes?

Acknowledgements My thanks go to a very careful and astute referee. I would also like to thank several people who saw me present parts of this material in the form of mini-courses, for thought-provoking questions and feedback. During the preparation of this work I was supported by funding from both NSERC and the FRQNT.

References

1. L. Addario-Berry, N. Broutin, B. Reed, Critical random graphs and the structure of a minimum spanning tree. Random Struct. Algorithms **35**, 323–347 (2009). http://www.cs.mcgill.ca/~nbrout/pub/knmst_rsa.pdf
2. L. Addario-Berry, N. Broutin, C. Goldschmidt, G. Miermont, The scaling limit of the minimum spanning tree of the complete graph (2013). arXiv:1301.1664 [math.PR]
3. D.J. Aldous, A random tree model associated with random graphs. Random Struct. Algorithms **1**(4), 383–402 (1990). http://stat-www.berkeley.edu/~aldous/Papers/me49.pdf
4. D.J. Aldous, The continuum random tree. II. An overview, in *Stochastic Analysis (Durham, 1990)*. Volume 167 of London Mathematical Society Lecture Note Series (Cambridge University Press, Cambridge, 1991), pp. 23–70. http://www.stat.berkeley.edu/~aldous/Papers/me55.pdf
5. D.J. Aldous, J.M. Steele, The objective method: probabilistic combinatorial optimization and local weak convergence. Probab. Discret. Struct. **110**, 1–72 (2004). http://www.stat.berkeley.edu/~aldous/Papers/me101.pdf
6. E. Baur, On a ternary coalescent process. ALEA Lat. Am. J. Probab. Math. Stat. **10**(2), 561–589 (2013). http://arxiv.org/pdf/1301.0409.pdf
7. M. Biskup, L. Chayes, S.A. Smith, Large-deviations/thermodynamic approach to percolation on the complete graph. Random Struct. Algorithms **31**(3), 354–370 (2007). http://arxiv.org/abs/math/0506255

8. L. Devroye, A note on the height of binary search trees. J. Assoc. Comput. Mach. **33**(3), 489–498 (1986). http://luc.devroye.org/devroye_1986_univ_a_note_on_the_height_of_binary_search_trees.pdf
9. A.M. Frieze, On the value of a random minimum spanning tree problem. Discret. Appl. Math. **10**, 47–56 (1985). http://www.math.cmu.edu/~af1p/Texfiles/MST.pdf
10. J.-F. Le Gall, Random trees and applications. Probab. Surv. **2**, 245–311 (electronic) (2005). http://www.i-journals.org/ps/viewarticle.php?id=53
11. C. McDiarmid, On the method of bounded differences, in *Surveys in Combinatorics (Norwich, 1989)*. Volume 141 of London Mathematical Society Lecture Note Series (Cambridge University Press, Cambridge, 1989), pp. 148–188. http://www.stats.ox.ac.uk/people/academic_staff/colin_mcdiarmid/?a=4113
12. N. O'Connell, Some large deviation results for sparse random graphs. Probab. Theory Relat. Fields **110**(3), 277–285 (1998). http://homepages.warwick.ac.uk/~masgas/pubs/noc98b.pdf
13. J. Pitman, Coalescent random forests. J. Combin. Theory Ser. A **85**, 165–193 (1999). http://www.stat.berkeley.edu/~pitman/457.pdf
14. A. Rényi, Some remarks on the theory of trees. Magyar Tud. Akad. Mat. Kutató Int. Közl. **4**, 73–85 (1959)
15. A. Rényi, G. Szekeres, On the height of trees. J. Austral. Math. Soc. **7**, 497–507 (1967). http://dx.doi.org/10.1017/S1446788700004432

Asymptotic Spectral Distributions of Distance-*k* Graphs of Star Product Graphs

Octavio Arizmendi and Tulio Gaxiola

Abstract Let G be a finite connected graph and let $G^{[\star N,k]}$ be the distance-k graph of the N-fold star power of G. For a fixed $k \geq 1$, we show that the large N limit of the spectral distribution of $G^{[\star N,k]}$ converges to a centered Bernoulli distribution, $1/2\delta_{-1} + 1/2\delta_1$. The proof is based in a fourth moment lemma for convergence to a centered Bernoulli distribution.

Keywords Star product • distance-k graph • Fourth moment • Limit theorem • Spectral distribution

Mathematics Subject Classification (2000). Primary 05C50; Secondary 05C12,47A10.

1 Introduction and Statement of Results

The interest in asymptotic aspects of growing combinatorial objects has increased in recent years. In particular, the asymptotic spectral distribution of graphs has been studied from the quantum probabilistic point of view, see Hora [9] and Hora and Obata [10]. Moreover, as observed in Accardi, A. Ben Ghorbal and Obata [2], Obata [16] and Accardi, Lenczewski and Salapata [3], the cartesian, star, rooted and free products of graphs correspond to natural independences in non-commutative probability, see [5, 14, 18]. This has lead to state central limit theorems for these product of graphs by reinterpreting the classical, free [6, 20], Boolean [19] and monotone [15] central limit theorems.

T. Gaxiola was supported by Conacyt Master's Scholarship No. 45710.

O. Arizmendi (✉) • T. Gaxiola
Research Center for Mathematics, CIMAT, Aparatado Postal 402, Guanajuato, GTO 36240, Mexico
e-mail: octavius@cimat.mx; marco.gaxiola@cimat.mx

R.H. Mena et al. (eds.), *XI Symposium on Probability and Stochastic Processes*, Progress in Probability 69, DOI 10.1007/978-3-319-13984-5_2

More recently, in a series of papers [7, 8, 11–13, 17] the asymptotic spectral distribution of the distance-k graph of the N-fold power of the cartesian product was studied. These investigations finally lead to the following theorem which generalizes the central limit theorem for cartesian products of graphs.

Theorem 1.1 (Hibino, Lee and Obata [8]) *Let $G = (V, E)$ be a finite connected graph with $|V| \geq 2$. For $N \geq 1$ and $k \geq 1$ let $G^{[N,k]}$ be the distance-k graph of $G^N = G \times \cdots \times G$ (N-fold Cartesian power) and $A^{[N,k]}$ its adjacency matrix. Then, for a fixed $k \geq 1$, the eigenvalue distribution of $N^{-k/2}A^{[N,k]}$ converges in moments as $N \to \infty$ to the probability distribution of*

$$\left(\frac{2|E|}{|V|}\right)^{k/2} \frac{1}{k!}\tilde{H}_k(g), \tag{1.1}$$

where $\tilde{H}_k$ is the monic Hermite polynomial of degree k and g is a random variable obeying the standard normal distribution $N(0, 1)$.

In this note we study the distribution (with respect to the vacuum state) of the distance-k graph of the star product of graphs. It is worth noting that the distance-k graph of the star product of a graph G is *not* equivalent to the star product of the distance-k graph of the graph G.

More precisely, we prove the analog of Theorem 1.1 by changing the cartesian product by the star product.

Theorem 1.2 *Let $G = (V, E, e)$ be a locally finite connected graph and let $k \in \mathbb{N}$ be such that $G^{[k]}$ is not trivial. For $N \geq 1$ and $k \geq 1$ let $G^{[\star N,k]}$ be the distance-k graph of $G^{\star N} = G \star \cdots \star G$ (N-fold star power) and $A^{[\star N,k]}$ its adjacency matrix. Furthermore, let $\sigma = V_e^{[k]}$ be the number of neighbors of e in the distance-k graph of G, then the distribution with respect to the vacuum state of $(N\sigma)^{-1/2}A^{[\star N,k]}$ converges in distribution as $N \to \infty$ to a centered Bernoulli distribution. That is,*

$$\frac{A^{[\star N,k]}}{\sqrt{N\sigma}} \longrightarrow \frac{1}{2}\delta_{-1} + \frac{1}{2}\delta_1,$$

weakly.

The limit distribution above is universal in the sense that it is independent of the details of a factor G, but also in this case the limit **does not** depend on k. The proof of Theorem 1.2 is based in a fourth moment lemma for convergence to a centered Bernoulli distribution.

Apart from the introduction, this note is organized as follows. Section 2 is devoted to preliminaries. In Sect. 3 we prove a fourth moment lemma for convergence to a centered Bernoulli distribution. Finally in Sect. 4 we prove Theorem 1.2.

2 Preliminaries

In this section we give very basic preliminaries on graphs, the Cauchy transform, Jacobi parameters and non-commutative probability. The reader familiar with these objects may skip this section.

2.1 Graphs

A *directed graph* or *digraph* is a pair $G = (V, E)$ such that V is a non-empty set and $E \subseteq V \times V$. The elements of V and E are called the *vertices* and the *edges* of the digraph G, respectively. Two vertices $x, y \in V$ are *adjacent,* or *neighbors* if $(x, y) \in E$.

We call *loop* an edge of the form (v, v) and we say that a graph is *simple* if has no loops. A digraph is called *undirected* if $(v, w) \in E$ implies $(w, v) \in E$.

We will work with simple undirected digraphs and use the word *graph* for a simple undirected digraph without any further reference.

The *adjacency matrix* of G is the matrix indexed by the vertex set V, where $A_{xy} = 1$ when $(x, y) \in E$ and $A_{xy} = 0$ otherwise.

A *path* is a graph $P = (V, E)$ with vertex set $V = \{v_1, \ldots, v_k\}$ and edges $E = \{(v_1, v_2), \ldots, (v_{k-1}, v_k)\}$. A *walk* is a path that can repeat edges. We say that a graph is *connected* if every pair of distinct vertices $x, y \in V$ are connected by a walk (or equivalently by a path).

In this note we focus on specific type of graphs coming from the distance-k graphs of the star product of finite rooted graph.

For a given graph $G = (V, E)$ and a positive integer k the *distance k-graph* is defined to be a graph $G^{[k]} = \left(V, E^{[k]}\right)$ with

$$E^{[k]} = \{(x, y) : x,\ y \in V,\ \partial_G (x, y) = k\},$$

where $\partial_G (x, y)$ is the graph distance. Figure 1 shows the distance-2 graph induced by the 3 dimensional cube.

A *rooted* graph is a graph with a labeled vertex $o \in V$. Finally we define the star product of rooted graphs. For $G_1 = (V_1, E_1)$ and $G_2 = (V_2, E_2)$ be two graph with

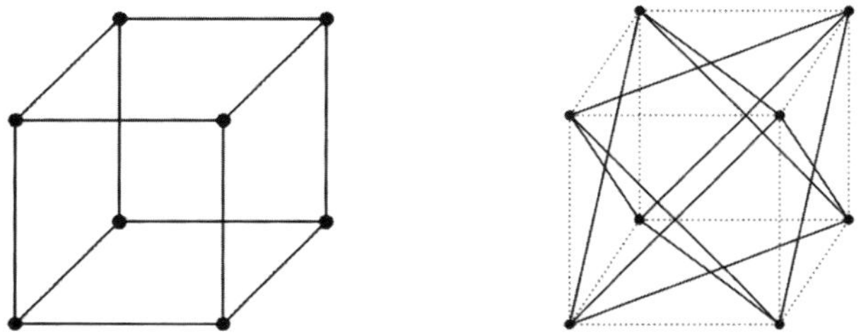

Fig. 1 3-Cube and its distance-2 graph

distinguished vertices $o_1 \in V_1$ and $o_2 \in V_2$, the *star product graph* of G_1 with G_2 is the graph $G_1 \star G_2 = (V_1 \times V_2, E)$ such that for (v_1, w_1), $(v_2, w_2) \in V_1 \times V_2$ the edge $e = (v_1, w_1) \sim (v_2, w_2) \in E$ if and only if one of the following holds:

1. $v_1 = v_2 = o_1$ and $w_1 \sim w_2$
2. $v_1 \sim v_2$ and $w_1 = w_2 = o_2$.

As we can see, the star product is a graph obtained by gluing two graphs at their distinguished vertices o_1 and o_2 (Fig. 2).

2.2 The Cauchy Transfom

We denote by $\mathcal{M}$ the set of Borel probability measures on $\mathbb{R}$. The upper half-plane and the lower half-plane are respectively denoted as $\mathbb{C}^+$ and $\mathbb{C}^-$.

For a measure $\mu \in \mathcal{M}$, the Cauchy transform $G_\mu : \mathbb{C}^+ \to \mathbb{C}^-$ is defined by the integral

$$G_\mu(z) = \int_{\mathbb{R}} \frac{\mu(dx)}{z - x}, \quad z \in \mathbb{C}^+$$

The Cauchy transform is an important tool in non-commutative probability. For us, the following relation between weak convergence and the Cauchy Transform will be important.

Proposition 2.1 *Let μ_1 and μ_2 be two probability measures on $\mathbb{R}$ and*

$$d(\mu_1, \mu_2) = \sup\left\{\left|G_{\mu_1}(z) - G_{\mu_2}(z)\right| ; \Im(z) \geq 1\right\}. \tag{2.1}$$

Then d is a distance which defines a metric for the weak topology of probability measures. Moreover, $|G_\mu(z)|$ is bounded in $\{z : \Im(z) \geq 1\}$ by 1.

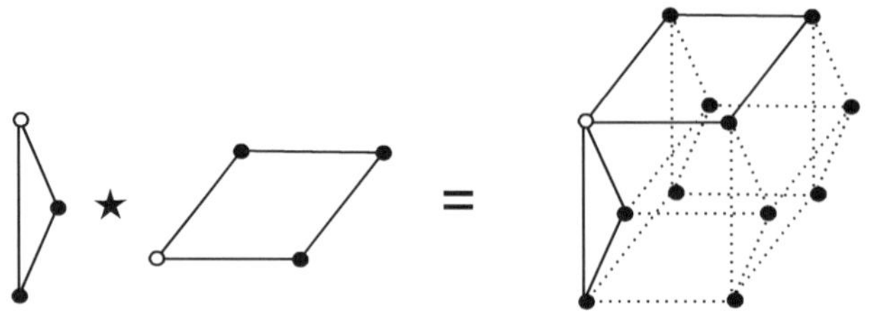

Fig. 2 Star product of two cycles

In other words, a sequence of probability measures $\{\mu_n\}_{n\geq 1}$ on $\mathbb{R}$ converges weakly to a probability measure μ on $\mathbb{R}$ if and only if for all z with $\Im(z) \geq 1$ we have

$$\lim_{n\to\infty} G_{\mu_n}(z) = G_\mu(z).$$

2.3 The Jacobi Parameters

Let μ be a probability measure with all moments, that is $m_n(\mu) := \int_{\mathbb{R}} x^n \mu(dx) < \infty$. The Jacobi parameters $\gamma_m = \gamma_m(\mu) \geq 0, \beta_m = \beta_m(\mu) \in \mathbb{R}$, are defined by the recursion

$$xP_m(x) = P_{m+1}(x) + \beta_m P_m(x) + \gamma_{m-1} P_{m-1}(x),$$

where the polynomials $P_{-1}(x) = 0$, $P_0(x) = 1$ and $(P_m)_{m\geq 0}$ is a sequence of orthogonal monic polynomials with respect to μ, that is,

$$\int_{\mathbb{R}} P_m(x) P_n(x) \mu(dx) = 0 \quad \text{if } m \neq n.$$

A measure μ is supported on m points iff $\gamma_{m-1} = 0$ and $\gamma_n > 0$ for $n = 0, \ldots, m-2$.

The Cauchy transform may be expressed as a continued fraction in terms of the Jacobi parameters, as follows.

$$G_\mu(z) = \int_{-\infty}^{\infty} \frac{1}{z-t} \mu(dt) = \cfrac{1}{z - \beta_0 - \cfrac{\gamma_0}{z - \beta_1 - \cfrac{\gamma_1}{z - \beta_2 - \cdots}}}$$

An important example for this paper is the Bernoulli distribution $\frac{1}{2}\delta_{-1} + \frac{1}{2}\delta_1$ for which $\beta_0 = 0, \gamma_0 = 1$, and $\beta_n = \gamma_n = 0$ for $n \geq 1$. Thus, the Cauchy transform is given by

$$G_{\mathbf{b}}(z) = \frac{1}{z - 1/z}.$$

In the case when μ has $2n+2$-moments we can still make an orthogonalization procedure until the level n. In this case the Cauchy transform has the form

$$G_\mu(z) = \cfrac{1}{z - \beta_0 - \cfrac{\gamma_0}{z - \beta_1 - \cfrac{\gamma_1}{\ddots \atop z - \beta_n - \gamma_n G_\nu(z)}}} \tag{2.2}$$

where ν is a probability measure.

2.4 Non-commutative Probability Spaces

A C^*-*probability space* is a pair $(\mathcal{A}, \varphi)$, where $\mathcal{A}$ is a unital C^*-algebra and $\varphi : \mathcal{A} \to \mathbb{C}$ is a positive unital linear functional. The elements of $\mathcal{A}$ are called (non-commutative) random variables. An element $a \in \mathcal{A}$ such that $a = a^*$ is called self-adjoint.

The functional φ should be understood as the expectation in classical probability.

For $a_1, \ldots, a_k \in \mathcal{A}$, we will refer to the values of $\varphi(a_{i_1} \cdots a_{i_n})$, $1 \leq i_1, \ldots i_n \leq k$, $n \geq 1$, as the *mixed moments* of $a_1, \ldots, a_k$.

For any self-adjoint element $a \in \mathcal{A}$, by the Riesz-Markov-Kakutani representation theorem, there exists a unique probability measure μ_a (its spectral distribution) with the same moments as a, that is,

$$\int_{\mathbb{R}} x^k \mu_a(dx) = \varphi(a^k), \quad \forall k \in \mathbb{N}.$$

We say that a sequence $a_n \in \mathcal{A}_n$ *converges in distribution* to $a \in \mathcal{A}$ if μ_{a_n} converges in distribution to μ_a.

In this note we will only consider the C^*-probability spaces $(\mathcal{M}_n, \phi_1)$, where $\mathcal{M}_n$ is the set of matrices of size $n \times n$ and for a matrix $M \in \mathcal{M}_n$ the functional ϕ_1 evaluated in M is given by

$$\phi_1(M) = M_{11}.$$

Let $G = (V, E, 1)$ be a finite rooted graph with vertex set $\{1, \ldots, n\}$ and let A_G be the adjacency matrix. We denote by $A(G) \subset \mathcal{M}_n$ be the adjacency algebra, i.e., the $*$-algebra generated by A_G.

It is easy to see that the k-th moment of A with respect to the ϕ_1 is given the number of walks in G of size k starting and ending at the vertex 1. That is,

$$\phi_1(A^k) = |\{(v_1, \ldots, v_k) : v_1 = v_k = 1 \text{ and } (v_i, v_{i+1}) \in E\}|.$$

Thus one can get combinatorial information of G from the values of ϕ_1 in elements of $A(G)$ and vice versa.

3 The Fourth Moment Lemma

The following lemma which shows that the first, second and fourth moments are enough to ensure convergence to a Bernoulli distribution was observed in [4]. We give a new proof in terms of Jacobi parameters for the convenience of the reader.

Lemma 3.1 *Let $\{X_n\}_{n\geq 1} \subset (\mathcal{A}, \varphi)$, be a sequence of self-adjoint random variables in some non-commutative probability space, such that $\varphi(X_n) = 0$ and $\varphi(X_n^2) = 1$. If*

$\varphi\left(X_n^4\right) \to 1$, *as* $n \to \infty$, *then* μ_{X_n} *converges in distribution to a symmetric Bernoulli random variable* **b**.

Proof Let $(\{\gamma_i(\mu_{X_n})\}, \{\beta_i(\mu_{X_n})\})$ be the Jacobi parameters of the measures μ_{X_n}. The first moments $\{m_n\}_{n\geq 1}$ are given in terms of the Jacobi Parameters as follows, see [1].

$$\begin{aligned}
m_1 &= \beta_0 \\
m_2 &= \beta_0^2 + \gamma_0 \\
m_3 &= \beta_0^3 + 2\beta_0\gamma_0 + \beta_1\gamma_0 \\
m_4 &= \beta_0^4 + 3\beta_0^2\gamma_0 + 2\beta_1\beta_0\gamma_0 + \beta_1^2\gamma_0 + \gamma_0^2 + \gamma_0\gamma_1.
\end{aligned}$$

Since $m_1(\mu_{X_n}) = 0$ and $m_2(\mu_{X_n}) = 1$ we have

$$\beta_0(\mu_{X_n}) = 0 \quad \text{and} \quad \gamma_0(\mu_{X_n}) = 1 \qquad \forall n \geq 1.$$

Hence,

$$m_4(\mu_{X_n}) = \beta_1^2(\mu_{X_n}) + 1 + \gamma_1(\mu_{X_n}). \tag{3.1}$$

Now, since $m_4(\mu_{X_n}) \to 1$ and $\gamma_1 \geq 0$ we have the convergence

$$\beta_1(\mu_{X_n}) \underset{n\to\infty}{\to} 0 \quad \text{and} \quad \gamma_1(\mu_{X_n}) \underset{n\to\infty}{\to} 0.$$

Let G_{μ_n} be the Cauchy transform of μ_n. By (2.2) we can expand G_μ as a continued fraction as follows

$$G_{\mu_n}(z) = \cfrac{1}{z - \cfrac{1}{z - \beta_1 - \gamma_1 G_{\nu_n}(z)}},$$

where ν_n is some probability measure. Now, recall that $|G_{\nu_n}(z)|$ is bounded by 1 in the set $\{z|; \Im(z) \geq 1\}$ and thus, since $\gamma_1 \to 0$ and $\beta_1 \to 0$ we see that $\gamma_n G_{\nu_n}(z) \to 0$. This implies the point-wise convergence

$$G_{\mu_n}(z) \to \cfrac{1}{z - \cfrac{1}{z}},$$

in the set $\{z|; \Im(z) \geq 1\}$, which then implies the weakly convergence $\mu_n \to \mathbf{b}$. □

From the proof of the previous lemma one can give a quantitative version in terms of the distance given in (2.1).

Proposition 3.2 *Let μ be a probability measure such that $m_4 := m_4(\mu)$ is finite. Then*

$$d(\mu, \frac{1}{2}\delta_1 + \frac{1}{2}\delta_{-1}) \leq 4\sqrt{m_4 - 1}.$$

Proof If $m_4 - 1 > 1/16$ then the statement is trivial since $d(\mu, 1/2\delta_1 + 1/2\delta_{-1}) \leq 1$ for any measure μ. Thus we may assume that $(m_4 - 1) \leq 1/16$.

Denoting by $f(z) = \beta_1 - \gamma_1 G_{\nu_n}(z)$ we have

$$|G_\mu(z) - G_b(z)| = \left| \frac{1}{z - \frac{1}{z}} - \frac{1}{z - \frac{1}{z - f(z)}} \right| = \left| \frac{f(z)}{(z^2 - 1)(z^2 - 1 - f(z)z)} \right|.$$

From (3.1) we get the inequalities $\sqrt{m_4 - 1} \geq |\beta_1|$ and $\sqrt{m_4 - 1} \geq m_4 - 1 \geq \gamma_1$. Since, for $Im(z) > 1$, we have that, $|G_\nu(z)| < 1$ we see that $|f(z)| = |\beta_1 - \gamma_1 G_\nu(z)| \leq 2\sqrt{m_4 - 1} \leq 1/2$, from where we can easily obtain the bound $|\frac{1}{z^2 - 1 - f(z)z}| \leq 2$. Also, for $\Im(z) > 0$ we have the bound $|\frac{1}{(z^2 - 1)}| < 1$. Thus we have

$$|G_\mu(z) - G_b(z)| = \left| \frac{f(z)}{(z^2 - 1)(z^2 - 1 - f(z)z)} \right| \tag{3.2}$$

$$= |f(z)| \left| \frac{1}{(z^2 - 1)} \right| \left| \frac{1}{z^2 - 1 - f(z)z} \right| \tag{3.3}$$

$$\leq 2|f(z)| \leq 4\sqrt{m_4 - 1}, \tag{3.4}$$

as desired. □

4 Proof of Theorem 1.2

Before proving Theorem 1.2 we will prove a lemma about the structure of distance-k graph of the iterated star product of a graph.

Lemma 4.1 *Let $G = (V, E, e)$ be a connected finite graph with root e and k such that $G^{[k]}$ is a non-trivial graph. Let $G^{\star N^{[k]}}$ be the distance-k graph of the N-th star product of G, then $G^{\star N^{[k]}}$ admits a decomposition of the form*

$$G^{\star N^{[k]}} = (G^{[k]})^{\star N} \cup \hat{G},$$

where $\hat{G}$ is a graph with same vertex set as G and $\partial G(z, e) < k$ for all $z \in \hat{G}$.

Proof Let $G_1, G_2, \ldots, G_N$ be the N copies of G, that form the star product graph $G^{\star N}$ by gluing them at e. For $x, y \in G_i$, the distance between x and y is given by

$$\partial_{G^{\star N}}(x,y) = \partial_{G_i}(x,y) = \partial_G(x,y),$$

hence

$$(x,y) \in E\left(G_i^{[k]}\right) \text{ if and only if } (x,y) \in E\left(\left(G^{\star N}\right)^{[k]}\right),$$

therefore we have $\left(G^{[k]}\right)^{\star N} \subseteq \left(G^{\star N}\right)^{[k]}$.

Now, if $x \in G_i$ and $y \in G_j$ with $j \neq i$, by definition all the paths in $G^{\star N}$ from x to y must pass throw e, then we have

$$\partial_{G^{\star N}}(x,y) = \partial_{G_i}(x,e) + \partial_{G_j}(y,e),$$

thus

$$(x,y) \in E\left(\left(G^{\star N}\right)^{[k]}\right) \text{ if and only if } \partial_{G_i}(x,e) + \partial_{G_j}(y,e) = k.$$

Since $\partial_{G_i}(x,e)$, $\partial_{G_j}(y,e) > 0$, we obtain the desired result. □

Now, we are in position to prove the main theorem of the paper.

Proof of Theorem 1.2 Consider the non-commutative probability space $(\mathcal{A}, \phi_1)$ with $\phi_1(M) = M_{11}$, for $M \in \mathcal{A}$ (see Sect. 2). Then, recall that, if A is an adjacency matrix, $\phi_1\left(A^k\right)$ equals the number of walks of size k starting and ending at the vertex 1.

Since G is a simple graph, it has no loops and then $G^{\star N}$ is also a simple graph. Thus,

$$\phi_1\left(\frac{A^{[\star N,k]}}{\sqrt{N|V_e^{[k]}|}}\right) = 0.$$

Now, observe that since the graph $G^{\star N}$ has no loops, the only walks in G of size 2 which start in e and end in e are of the form (exe), where x is a neighbor of e in $\left(G^{\star N}\right)^{[k]}$. The number of neighbors of e is exactly $N|V_e^{[k]}|$, thus

$$\begin{aligned}\phi_1\left(\left(\frac{A^{[\star N,k]}}{\sqrt{N|V_e^{[k]}|}}\right)^2\right) &= \frac{1}{N|V_e^{[k]}|}\phi_1\left(\left(A^{[\star N,k]}\right)^2\right) \\ &= \frac{1}{N|V_e^{[k]}|}N|V_e^{[k]}| = 1.\end{aligned}$$

Thus, we have seen that $\phi(A_N) = 0$ and $\phi(A_N^2) = 1$. Hence, it remains to show that $\phi(A_N^4) \to 1$ as $N \to \infty$.

We are interested in counting the number of walk of size 4 that start and finish at e in $\left(G^{\star N}\right)^{[k]}$. We will divide this walks in two types (Fig. 3).

Type 1. The first type of walk is of the form *exeye*. That is, the walk starts at e, then visits a neighbor x of e to then come back to e, this can be done in $N|V_e^{[k]}|$ ways. After this, he again visits a neighbour y (which could be again x) of e to finally come back to e. Again, this second step can be done in $N|V_e^{[k]}|$ different ways, so there is $\left(N|V_e^{[k]}|\right)^2$ walks of this type.

Type 2. Let $G_x^{[k]}$ be the copy of $G^{[k]}$ in the distance-k graph of the star product $\left(G^{[k]}\right)^{\star N}$ which contains x. The second type of walks is as follows. From e it goes to some $x \in V_e^{[k]}$ (which can be chosen in $N|V_e^{[k]}|$ different ways), and then from x then he goes to some $y \neq e$. This y should belong to $G_x^{[k]}$. Indeed, since $\partial_{(G^{\star N})}(e,x) = k$, if y would be in another copy of $G^{[k]}$ the distance $\partial_{(G^{\star N})}(y,x)$, between y and x would be bigger than k. The number of ways of choosing y is bounded by the number of neighbors of x in $G^{[k]}$.

For the next step of the walk, from y we can only go to a neighbor of e, say $z \in V_e^{[k]}$ (since in the last step it must come back to e). This z indeed must also belong to $G_x^{[k]}$. If this wouldn't be the case and $z \notin G_x^{[k]}$, then we would have that $\partial_{(G^{\star N})}(e,z) \neq k$, which is a contradiction because of Lemma 4.1 (Fig. 4).

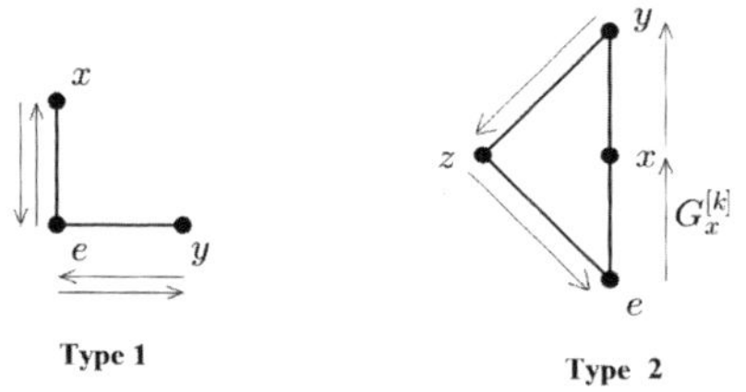

Fig. 3 Types of walks of size 4

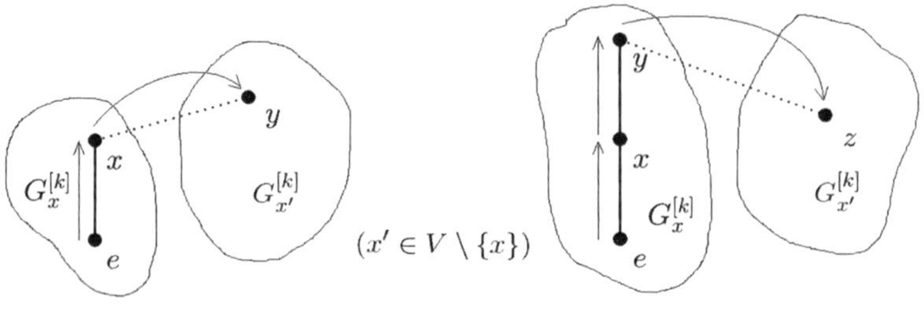

Fig. 4 Obstructions

Finally, let $M = \max_{x \in V} |V_x^{[k]}|$. Then, from the above considerations we see that the number of walks of Type 2 is bounded by $M \left(N|V_e^{[k]}|\right)\left(|V_e^{[k]}|\right)$, from where

$$\phi_1\left(\left(\frac{A^{[\star N,k]}}{\sqrt{N|V_e^{[k]}|}}\right)^4\right) \leq \frac{\left(N|V_e^{[k]}|\right)^2}{\left(N|V_e^{[k]}|\right)^2} + \frac{N|V_e^{[k]}|M|V_e^{[k]}|}{\left(N|V_e^{[k]}|\right)^2}$$
$$= 1 + \frac{M}{N} \underset{N\to\infty}{\longrightarrow} 1,$$

since M does not depend on N. Thanks to Lemma 3.1 we obtain the desired result.
□

References

1. L. Accardi, M. Bozejko, Interacting Fock spaces and Gaussianization of probability measures. Infin. Dimen. Anal. Quantum Probab. Relat. Top. **1**, 663–670 (1998)
2. L. Accardi, A.B. Ghorbal, N. Obata, Monotone independence, comb graphs and Bose-Einstein condensation. Infin. Dimen. Anal. Quantum Probab. Relat. Top. **7**, 419–435 (2004)
3. L. Accardi, R. Lenczewski, R. Salapata, Decompositions of the free product of graphs. Infin. Dimen. Anal. Quantum Probab. Relat. Top. **10**, 303–334 (2007)
4. O. Arizmendi, Convergence of the fourth moment and infinitely divisibility. Probab. Math. Stat. **33**(2), 201–212 (2013)
5. A. Ben Ghorbal, M. Schürman, Non-commutative notions of stochastic independence. Math. Proc. Camb. Philos. Soc. **133**, 531–561 (2002)
6. M. Bożejko, On $\Lambda(p)$ sets with minimal constant in discrete noncommutative groups. Proc. Am. Math. Soc. **51**, 407–412 (1975)
7. Y. Hibino, Asymptotic spectral distributions of distance-k graphs of Hamming graphs. Commun. Stoch. Anal. **7**(2), 275–282 (2013)
8. Y. Hibino, H.H. Lee, N. Obata, Asymptotic spectral distributions of distance-k graphs of Cartesian product graphs. Colloq. Math. **132**(1), 35–51 (2013)
9. A. Hora, Central limit theorem for the adjacency operators on the infinite symmetric group. Commun. Math. Phys. **195**, 405–416 (1998)
10. A. Hora, N. Obata, *Quantum Probability and Spectral Analysis of Graphs* (Springer, Berlin, 2007)
11. J. Kurihara, Asymptotic spectral analysis of a distance k-graph of N-fold direct product of a graph. Commun. Stoch. Anal. **6**, 213–221 (2012)
12. J. Kurihara, Y. Hibino, Asymptotic spectral analysis of a generalized N-cube by a quantum probabilistic method. Infin. Dimen. Anal. Quantum Probab. Relat. Top. **14**, 409–417 (2011)
13. H.H. Lee, N. Obata, Distance-k graphs of hypercube and q-Hermite polynomials. Infin. Dimens. Anal. Quantum Probab. Relat. Top. **16**(2) 1350011, 12pp. (2013)
14. N. Muraki, The five independences as natural products. Infin. Dimen. Anal. Quantum Probab. Relat. Top. **6**, 337–371 (2003)
15. N. Muraki, Monotonic independence, monotonic central limit theorem and monotonic law of small numbers. Infin. Dimen. Anal. Quantum Probab. Relat. Top. **4**, 39–58 (2001)
16. N. Obata, Quantum probabilistic approach to spectral analysis of star graphs. Interdisciplin. Inf. Sci. **10**, 41–52 (2004)

17. N. Obata, Asymptotic spectral distributions of distance k-graphs of large-dimensional hypercubes. Banach Center Publ. **96**, 287–292 (2012)
18. R. Speicher, On universal products, in *Free Probability Theory*, ed. by D. Voiculescu. Fields Institute Communications, vol. 12 (American Mathematical Society, Providence, 1997), pp. 257–266
19. R. Speicher, R. Woroudi, Boolean convolution, in *Free Probability Theory*, ed. by D. Voiculescu. Fields Institute Communications, vol. 12 (American Mathematical Society, Providence, 1997), pp. 267–280
20. D. Voiculescu, Symmetries of some reduced free product C*-algebras, in *Operator Algebras and Their Connections with Topology and Ergodic Theory*. Lecture Notes in Mathematics, vol. 1132 (Springer, Berlin, 1985), pp. 556–588

Stochastic Differential Equations Driven by Loops

Fabrice Baudoin

Abstract We study stochastic differential equations of the type

$$X_t = x + \sum_{i=1}^{d} \int_0^t V_i(X_s^x) \circ dM_s^i,\ 0 \le t \le T,$$

where $(M_s)_{0 \le s \le T}$ is a semimartingale generating a loop in the free Carnot group of step N and show how the properties of the random variable X_T^x are closely related to the Lie subalgebra generated by the commutators of the V_i's with length greater than $N + 1$. It is furthermore shown that if f is a smooth function, then

$$\lim_{T \to 0} \frac{\mathbb{E}(f(X_T^x)) - f(x)}{T^{N+1}} = (\Delta_N f)(x),$$

where Δ_N is a second order operator related to the $V_i' s$.

Keywords Brownian loops • Carnot groups • Chen development • Holonomy operator • Hörmander's type theorems

1 Introduction

Let us consider a stochastic differential equations on $\mathbb{R}^n$ on the type

$$X_t^{x_0} = x_0 + \sum_{i=1}^{d} \int_0^t V_i(X_s^{x_0}) \circ dM_s^i,\ 0 \le t \le T, \tag{1.1}$$

F. Baudoin (✉)
Department of Mathematics, Purdue University, West Lafayette, IN, USA
e-mail: fbaudoin@purdue.edu

R.H. Mena et al. (eds.), *XI Symposium on Probability and Stochastic Processes*, Progress in Probability 69, DOI 10.1007/978-3-319-13984-5_3

where:

1. $x_0 \in \mathbb{R}^n$;
2. $V_1, \ldots, V_d$ are C^∞ bounded vector fields on $\mathbb{R}^n$;
3. $\circ$ denotes Stratonovitch integration;
4. $(M_t)_{0 \le t \le T} = (M_t^1, \ldots, M_t^d)_{0 \le t \le T}$ is a d-dimensional continuous semimartingale.

It is well-known that if $(M_t)_{0 \le t \le T}$ is a standard Brownian motion then for any smooth function $f : \mathbb{R}^n \to \mathbb{R}$ we have in $\mathbf{L}^2$,

$$\lim_{t \to 0} \frac{\mathbf{P}_t f - f}{t} = \frac{1}{2} \left(\sum_{i=1}^{d} V_i^2 \right) f,$$

where $\mathbf{P}_t$ is the semigroup associated with (1.1) which is defined by

$$\mathbf{P}_t f(x) = \mathbb{E} \left(f(X_t^x) \right).$$

In that case, it is furthermore known from Hörmander's theorem that $\mathbf{P}_t$ has a smooth transition kernel with respect to the Lebesgue measure as soon as for all $x_0 \in \mathbb{R}^n$, $\mathfrak{L}(x_0) = \mathbb{R}^n$ where $\mathfrak{L}$ is the Lie algebra generated by the vector fields V_i's. So, when $(M_t)_{0 \le t \le T}$ is a Brownian motion, the properties of $(X_t^{x_0})_{0 \le t \le T}$ are closely related to the diffusion operator $\sum_{i=1}^d V_i^2$ and the Lie algebra $\mathfrak{L}$.

In this paper, we show that there are other choices of the driving semimartingale $(M_t)_{0 \le t \le T}$ for which the solution $(X_t^{x_0})_{t \ge 0}$ is naturally associated to other diffusion operators.

Let us roughly describe our approach. The Chen-Strichartz expansion theorem (see [5, 12, 23]) states that, formally, the stochastic flow $(\Phi_t)_{0 \le t \le T}$ associated with the stochastic differential equations (1.1) can be written as

$$\Phi_t = \exp \left(\sum_{k \ge 1} \sum_{i_1, \ldots, i_k \in \{1, \ldots, d\}} F_{i_1, \ldots, i_k} \left(\int_{0 \le t_1 \le \ldots \le t_k \le t} \circ dM_{t_1}^{i_1} \ldots \circ dM_{t_k}^{i_k} \right)_{anti} \right), \quad t \le T,$$

where $F_{i_1, \ldots, i_k}$ are universal Lie polynomials in $V_1, \ldots, V_d$, which depend on the choice of a Hall basis in the free Lie algebra with d generators, and $\left(\int_{0 \le t_1 \le \ldots \le t_k \le t} \circ dM_{t_1}^{i_1} \ldots \circ dM_{t_k}^{i_k} \right)_{anti}$ are universal antisymmetrizations of the iterated integrals of the semimartingale $(M_t)_{t \ge 0}$.

Consider now $N \ge 0$, and take for $(M_t^1, \ldots, M_t^d)_{0 \le t \le T}$ a d-dimensional standard Brownian motion conditioned by

$$\left(\int_{0 \le t_1 \le \ldots \le t_k \le T} \circ dM_{t_1}^{i_1} \ldots \circ dM_{t_k}^{i_k} \right)_{anti} = 0, \; i_1, \ldots, i_k \in \{1, \ldots, d\}, \; 1 \le k \le N.$$

It is shown that such a process, that we call a N-step Brownian loop, is a semimartingale up to time T and can be constructed from a diffusion in the loop space over the free Carnot group of step N.

For this choice of $(M_t)_{t\geq 0}$, the Chen development for Φ_T writes

$$\Phi_T = \exp\left(\sum_{k\geq N+1} \sum_{i_1,\ldots,i_k\in\{1,\ldots,d\}} F_{i_1,\ldots,i_k} \left(\int_{0\leq t_1\leq\ldots\leq t_k\leq t} \circ dM_{t_1}^{i_1} \ldots \circ dM_{t_k}^{i_k} \right)_{anti} \right),$$

and thus only involves the Lie subalgebra $\mathfrak{L}^{N+1}$, where $\mathfrak{L}$ is the Lie algebra generated by the vector fields V_i's and for $p \geq 2$, $\mathfrak{L}^p$ is inductively defined by

$$\mathfrak{L}^p = \{[X,Y],\ X \in \mathfrak{L}^{p-1}, Y \in \mathfrak{L}\},\ \mathfrak{L}^1 = \mathfrak{L}.$$

Hence, we can expect that the properties of the random variable X_T^x, where $(X_t^x)_{0\leq t\leq T}$ is the solution of (1.1) with initial condition x, are closely related to this Lie subalgebra $\mathfrak{L}^{N+1}$.

Precisely, we show that:

- If $f : \mathbb{R}^n \to \mathbb{R}$ is a smooth function which is compactly supported, then in L^2,

$$\lim_{T\to 0} \frac{\mathcal{H}_T^N f - f}{T^{N+1}} = \Delta_N f,$$

  where Δ_N is an homogeneous second order differential operator that belongs to the universal enveloping algebra of $\mathfrak{L}^{N+1}$ and $\mathcal{H}_T^N$ the N-step holonomy operator that we define by

$$\mathcal{H}_T^{N+1} f(x) = \mathbb{E}\left(f(X_T^x)\right).$$

- If $\mathfrak{L}^{N+1} = 0$, then for all $x \in \mathbb{R}^n$, almost surely $X_T^x = x$;
- If for all $x \in \mathbb{R}^n$, $\mathfrak{L}^{N+1}(x) = \mathbb{R}^n$, then for all $x \in \mathbb{R}^n$, X_T^x has a smooth density $p_T(x)$ with respect to the Lebesgue measure. Moreover, in that case $p_T(x) \sim_{T\to 0} m(x)T^{-\frac{D(x)}{2}}$, where m is a smooth non negative function and $D(x)$ an integer (the graded dimension of $\mathfrak{L}^{N+1}(x)$);

We stress the fact that from a geometrical point of view the family of operators $(\Delta_N)_{N\geq 0}$ is quite interesting and is an important invariant of the intrinsic geometry of the differential system generated by the V_i's. For instance we shall see that, for some positive constant C,

$$\Delta_1 = C \sum_{i,j=1}^{d} [V_i, V_j]^2$$

and, in a way, this operator sharply measures the curvature of the differential system generated by the vector fields V_i.

Let us mention that, since the seminal work of Rotschild and Stein [22], the study Carnot groups arise naturally in PDE's theory and is now an active research field. Our interest in SDE's driven by loops in Carnot groups originally comes from the study of the Brownian holonomy on sub-Riemannian manifolds. The understanding of this holonomy is closely related to the construction of parametrices for hypoelliptic Schrödinger equations on vector bundles (see [4]).

The paper is organized as follows. The second section is here for the sake of clarity of the paper since the framework is quite simple but the results already interesting: We study stochastic differential equations driven by Brownian loops. All the results presented in this section will be later generalized. In the third section, we introduce the notion of free Carnot group of step N and define a fundamental diffusion on it. We then study the coupling of this diffusion with the solution of a generic stochastic differential equation driven by Brownian motions. In particular we establish an Hörmander type theorem for the existence of a smooth density for the joint law of this coupling. The fourth section constitutes the heart of this paper and gives the proofs of the results presented above.

Some results of the paper were already announced in the note [2] and the book [5].

2 Stochastic Differential Equations Driven by Brownian Loops and Bridges

We consider first on $\mathbb{R}^n$ stochastic differential equations of the type

$$X_t = x_0 + \sum_{i=1}^{d} \int_0^t V_i(X_s) \circ dP^i_{s,T},\ t \leq T \tag{2.2}$$

where:

1. $x_0 \in \mathbb{R}^n$;
2. $V_1, \ldots, V_d$ are C^∞ bounded vector fields on $\mathbb{R}^n$;
3. $(P^1_{t,T}, \ldots, P^d_{t,T})_{0 \leq t \leq T}$ is a given d -dimensional Brownian bridge from 0 to 0 with length $T > 0$.

Notice that since $(P_{t,T})_{0 \leq t \leq T}$ is known to be a semimartingale up to time T, the notion of solution for (2.2) is well-defined up to time T.

Proposition 2.1 *For every $x_0 \in \mathbb{R}^n$, there is a unique solution $(X_t^{x_0})_{0 \leq t \leq T}$ to (2.2). Moreover there exists a stochastic flow $(\Phi_t, 0 \leq t \leq T)$ of smooth diffeomorphisms $\mathbb{R}^n \to \mathbb{R}^n$ associated to the equations (2.2).*

Proof We refer to the book of Kunita [18], where the questions of existence and uniqueness of a smooth flow for stochastic differential equations driven by general continuous semimartingales are treated (cf. Theorem 3.4.1. p. 101 and Theorem 4.6.5. p. 173). □

The random variable X_T where $(X_t)_{0\le t\le T}$ is a solution of (2.2), is of particular interest: It is closely related to the commutations properties of the vector fields $V_1, \dots, V_d$. Indeed, let us consider the following family of operators $(\mathcal{H}_T)_{T\ge 0}$ defined on the space of compactly supported smooth functions $f : \mathbb{R}^n \to \mathbb{R}$ by

$$(\mathcal{H}_T f)(x) = \mathbb{E}\left(f(X_T^x)\right), \ x \in \mathbb{R}^n.$$

Obviously, the family of operators $(\mathcal{H}_T)_{T\ge 0}$ does not satisfy the semigroup property. It is interesting that in some cases, we can explicitly compute $(\mathcal{H}_T)_{T\ge 0}$.

Theorem 2.2 *Assume that the Lie algebra generated by the vector fields V_i is two-step nilpotent (that is, any commutator with length greater than 3 is 0) then*

$$\mathcal{H}_T = \det\left(\frac{T\Omega}{2\sinh(\frac{1}{2}T\Omega)}\right)^{\frac{1}{2}},$$

where Ω is the $d \times d$ matrix such that $\Omega_{i,j} = [V_i, V_j]$.

Before turning to the proof, we mention that the above expression for $\mathcal{H}_T$ is understood in the sense of pseudo-differential operators. Namely, the expression

$$\det\left(\frac{\left(x_{i,j}\right)_{1\le i,j\le d}}{2\sinh(\frac{1}{2}\left(x_{i,j}\right)_{1\le i,j\le d})}\right)^{\frac{1}{2}}$$

defines an analytic function

$$\Phi(\left(x_{i,j}\right)_{1\le i,j\le d})$$

and the above theorem says that

$$\mathcal{H}_T = \int_{\mathbb{R}^{\frac{d(d-1)}{2}}} \hat{\Phi}(\xi) e^{iT\sum_{i<j}\xi_{i,j}[V_i,V_j]} d\xi$$

where $\hat{\Phi}$ denotes the Fourier transform of Φ. For further details on pseudo-differential operators we refer to the Chapter 7 of [24].

Proof Itô's formula shows that in that two-nilpotent case,

$$f(X_T^x) = \left(\exp\left(\frac{1}{2} \sum_{1 \le i < j \le d} [V_i, V_j] \int_0^t P^i_{s,T} dP^j_{s,T} - P^j_{s,T} dP^i_{s,T} \right) f \right)(x).$$

But, from Gaveau-Lévy's area formula see [14], if A is a $d \times d$ skew-symmetric matrix valued in a commutative ring, then,

$$\mathbb{E}\left(e^{i \int_0^T (AP_{s,T}, dP_{s,T})} \right) = \det\left(\frac{tA}{\sin tA} \right)^{\frac{1}{2}}.$$

This completes the proof. □

It seems difficult to find a closed expression for $\mathcal{H}_T$ in the general case, we can nevertheless compute a small-time asymptotics:

Theorem 2.3 *Let $f : \mathbb{R}^n \to \mathbb{R}$ be a smooth function which is compactly supported. In $\mathbf{L}^2$,*

$$\lim_{T \to 0} \frac{\mathcal{H}_T f - f}{T^2} = \frac{1}{24} \left(\sum_{1 \le i < j \le d} [V_i, V_j]^2 \right) f.$$

Proof We refer to the proof of Theorem 4.8 which is more general. We however show how the constant $\frac{1}{24}$ is obtained. The proof of Theorem 4.8 shows that there is a universal constant C such that

$$\lim_{T \to 0} \frac{\mathcal{H}_T f - f}{T^2} = C \left(\sum_{1 \le i < j \le d} [V_i, V_j]^2 \right).$$

Since this constant is universal, in order to compute it, it suffices to look at the two-step nilpotent case. In that case, from the previous theorem

$$\mathcal{H}_T = \det\left(\frac{T\Omega}{2\sinh(\frac{1}{2}T\Omega)} \right)^{\frac{1}{2}}.$$

Therefore

$$\mathcal{H}_T \sim_{T \to 0} \det\left(\mathbf{1} - \frac{1}{24} T^2 \Omega^2 \right)^{\frac{1}{2}},$$

and the computation is easily done. □

We study now sufficient conditions which ensure that the operator $\mathcal{H}_T$ has a smooth kernel (in the two variables) with respect to the Lebesgue measure of $\mathbb{R}^n$. To answer this question, it is enough to decide under which conditions the random variable X_T^x has a smooth density.

As noted before, we denote by $\mathfrak{L} = Lie(V_1, \cdots, V_d)$ the Lie algebra generated by the vector fields V_i's. On one hand, we have the following result:

Theorem 2.4 *Assume that* $[\mathfrak{L}, \mathfrak{L}] = 0$, *then for any solution* $(X_t^{x_0})_{0 \le t \le T}$ *of* (2.2) *we have almost surely* $X_T^{x_0} = x_0$.

Proof For $i = 1, \dots, d$, let us denote $(e^{tV_i})_{t \in \mathbb{R}}$ the one-parameter flow associated with the complete vector field V_i. Since the V_i's are assumed to commute, an iterative application of Itô's formula shows that the process

$$\left(\left(e^{P^1_{s,T} V_1} \circ \dots \circ e^{P^d_{s,T} V_d} \right) (x_0) \right)_{0 \le s \le T}$$

solves the equation (2.2) with initial condition x_0. By uniqueness, we have hence

$$\Phi_s(x_0) = \left(e^{P^1_{s,T} V_1} \circ \dots \circ e^{P^d_{s,T} V_d} \right) (x_0), \; 0 \le s \le T.$$

In particular,

$$\Phi_T(x_0) = x_0,$$

which is the expected result. □

In general, the weaker condition $[\mathfrak{L}, \mathfrak{L}](x_0) = 0$ is not enough to conclude that for the solution $(X_t^{x_0})_{0 \le t \le T}$ of (2.2) we have almost surely $X_T^{x_0} = x_0$. For instance, consider in dimension 2,

$$V_1 = \begin{pmatrix} 1 \\ 0 \end{pmatrix}, \text{ and } V_2 = \begin{pmatrix} 0 \\ f(x), \end{pmatrix}$$

where f is a smooth function whose Taylor development at 0 is 0 (by e.g. $f(x) = e^{-\frac{1}{x^2}} \mathbf{1}_{x>0}$). Nevertheless, if the vector fields V_i's are assumed to be analytic on whole $\mathbb{R}^n$, $[\mathfrak{L}, \mathfrak{L}](x_0) = 0$ implies that $[\mathfrak{L}, \mathfrak{L}] = 0$ and therefore that almost surely $X_T^{x_0} = x_0$.

Theorem 2.5 *Assume that* $[\mathfrak{L}, \mathfrak{L}](x_0) = \mathbb{R}^n$, *then for the solution* $(X_t^{x_0})_{0 \le t \le T}$ *of* (2.2) *the random variable* $X_T^{x_0}$ *has a smooth density with respect to the Lebesgue measure of* $\mathbb{R}^n$.

Proof Let us consider the solution $(Y_t)_{t\geq 0}$ of the following stochastic differential equation:

$$Y_t = x_0 + \sum_{i=1}^{d} \int_0^t V_i(Y_s) \circ dB_s^i, \ t \geq 0,$$

where $(B_t^1, \ldots, B_t^d)_{t\geq 0}$ is a d-dimensional standard Brownian motion. Since $[\mathfrak{L}, \mathfrak{L}](x_0) = \mathbb{R}^n$, it easily seen that $(Y_t, B_t)_{t\geq 0}$ is a diffusion process whose infinitesimal generator satisfies the (strong) Hörmander's condition at $(x_0, 0)$. Therefore, the random variable

$$(Y_T, B_T)$$

has a smooth density with respect to the Lebesgue measure on $\mathbb{R}^n \times \mathbb{R}^n$. This implies the existence of a smooth function $p : \mathbb{R}^n \to \mathbb{R}$ such that for all bounded measurable function $f : \mathbb{R}^n \to \mathbb{R}$

$$\mathbb{E}(f(Y_T) \mid B_T = 0) = \int_{\mathbb{R}^n} f(y)p(y)dy.$$

Now, since in law the process $(P_{t,T})_{0\leq t\leq T}$, is identical to the Brownian motion $(B_t)_{0\leq t\leq T}$ conditioned by $B_T = 0$, the function p is actually exactly the density of the random variable $X_T^{x_0}$ where $(X_t^{x_0})_{0\leq t\leq T}$ is the solution of (2.2) with initial condition x_0. □

Another proof of this result may be given by using standard Malliavin calculus tools (see Chapter 2 of Nualart's book [20], whose notations below are taken).

We work in the d-dimensional Wiener space and define the Brownian loop $(P_{t,T}^1, \ldots, P_{t,T}^d)_{0\leq t\leq T}$ as the Wiener integral

$$P_{t,T} = (T-t)\int_0^t \frac{dW_s}{T-s}, \ t < T, \text{ and } P_{T,T} = 0,$$

where W is the d-dimensional Wiener process. In this setting, it is not difficult to prove that if $(X_t)_{0\leq t\leq T}$ is a solution of (2.2), then $X_T \in \mathbb{D}^{\infty}$. Moreover, a direct computation shows that for any $0 \leq s \leq T$, the Malliavin derivative is given

$$\mathbf{D}_s X_T = \mathbf{J}_{0\to T}\left(\mathbf{J}_{0\to s}^{-1}\sigma(X_s) - \frac{1}{T-s}\int_s^T \mathbf{J}_{0\to u}^{-1}\sigma(X_u)du\right),$$

where $(\mathbf{J}_{0\to t})_{0\leq t\leq T}$ is the first variation process defined by

$$\mathbf{J}_{0\to t} = \frac{\partial \Phi_t}{\partial x},$$

and σ the $n \times d$ matrix field $\sigma = (V_1, \dots, V_d)$. From this, we can deduce that the Malliavin matrix of X_T must be invertible. Indeed, if not, we could find a non zero vector $h \in \mathbb{R}^d$ and a finite stopping time $\theta > 0$ such that $\mathbf{D}_s X_T \cdot h = 0$ for $0 \leq s \leq \theta$. This would lead to the conclusion that $(\mathbf{J}_{0\to s}^{-1}\sigma(X_s) \cdot h)_{0\leq s\leq T}$ must be constant.

Observe now that

$$\mathbf{J}_{0\to s}^{-1} V_i(X_s) = \Phi_s^* V_i,$$

where Φ denotes the stochastic flow associated with equation (2.2), and where $\Phi_s^* V_i$ denotes the pull-back action of Φ on V_i. Therefore, according to the Itô's formula, we obtain that for $t < \theta$,

$$\sum_{j=1}^{d} \int_0^t {}^T h \left(\Phi_s^*[V_j, V_i] \right)(x_0) \circ dP_{s,T}^j, \quad i = 1, \dots, d.$$

is constant. Therefore, for $0 \leq s < \theta$,

$${}^T h \left(\Phi_s^*[V_j, V_i] \right)(x_0) = 0, \quad i, j = 1, \dots, d.$$

By applying this at $s = 0$, we obtain then

$${}^T h \, [V_j, V_i](x_0) = 0, \quad i, j = 1, \dots, d.$$

New iterations of the Itô's formula show then that, we actually have

$${}^T h \, U(x_0) = 0, \quad U \in \mathfrak{L}^2(x_0),$$

so that $h = 0$.

We mention that the problem of the existence of densities for stochastic differential equations driven by Brownian bridges was also discussed in [10]. But unlike our case, the existence of the density is discussed for times $t < T$ and under the standard Hörmander's condition.

3 Free Carnot Groups and Hörmander's Type Theorems

In this section we now state some basic facts about Carnot groups.

3.1 Free Carnot Groups

We introduce the notion of Carnot groups. Such Lie groups appear as tangent spaces to hypoelliptic diffusions (see [3]). For more details on the material presented in this section, we refer to the Chapter 2 of [5]. Let $N \geq 1$. A Carnot group of depth (or step) N is a simply connected Lie group $\mathbb{G}$ whose Lie algebra can be written

$$\mathcal{V}_1 \oplus \ldots \oplus \mathcal{V}_N,$$

where

$$[\mathcal{V}_i, \mathcal{V}_j] = \mathcal{V}_{i+j}$$

and

$$\mathcal{V}_s = 0, \text{ for } s > N.$$

Example 3.1 (Heisenberg Group) The Heisenberg group $\mathbb{H}$ can be represented as the set of 3×3 matrices:

$$\begin{pmatrix} 1 & x & z \\ 0 & 1 & y \\ 0 & 0 & 1 \end{pmatrix}, \quad x, y, z \in \mathbb{R}.$$

The Lie algebra of $\mathbb{H}$ is spanned by the matrices

$$D_1 = \begin{pmatrix} 0 & 1 & 0 \\ 0 & 0 & 0 \\ 0 & 0 & 0 \end{pmatrix}, \; D_2 = \begin{pmatrix} 0 & 0 & 0 \\ 0 & 0 & 1 \\ 0 & 0 & 0 \end{pmatrix} \text{ and } D_3 = \begin{pmatrix} 0 & 0 & 1 \\ 0 & 0 & 0 \\ 0 & 0 & 0 \end{pmatrix},$$

for which the following equalities hold

$$[D_1, D_2] = D_3, \; [D_1, D_3] = [D_2, D_3] = 0.$$

Thus

$$\mathfrak{h} \sim \mathbb{R} \oplus [\mathbb{R}, \mathbb{R}],$$

and, therefore, $\mathbb{H}$ is a (free) two-step Carnot group.

Let us now take a basis $U_1, \dots, U_d$ of the vector space $\mathcal{V}_1$. The vectors U_i's can be seen as left invariant vector fields on $\mathbb{G}$ so that we can consider the following stochastic differential equation on $\mathbb{G}$:

$$d\tilde{B}_t = \sum_{i=1}^{d} \int_0^t U_i(\tilde{B}_s) \circ dB_s^i, \ t \geq 0, \tag{3.3}$$

where $(B_t)_{t\geq 0}$ is a standard Brownian motion. This equation is easily seen to have a unique (strong) solution $(\tilde{B}_t)_{t\geq 0}$ associated with the initial condition $\tilde{B}_0 = 0_{\mathbb{G}}$.

Definition 3.2 The process $(\tilde{B}_t)_{t\geq 0}$ is called the lift of the standard Brownian motion $(B_t)_{t\geq 0}$ in the group $\mathbb{G}$ with respect to the basis $(U_1, \dots, U_d)$.

Notice that $(\tilde{B}_t)_{t\geq 0}$ is a Markov process with generator $\frac{1}{2}\sum_{i=1}^{d} U_i^2$. This second-order differential operator is, by construction, left-invariant and hypoelliptic. For $(\tilde{B}_t)_{t\geq 0}$, we actually have an explicit expression. To give this expression, we first have to introduce some notations. If $I = (i_1, \dots, i_k) \in \{1, \dots, d\}^k$ is a word, we denote $\mid I \mid = k$ its length and by U_I the commutator defined by

$$U_I = [U_{i_1}, [U_{i_2}, \dots, [U_{i_{k-1}}, U_{i_k}] \dots].$$

The group of permutations of the index set $\{1, \dots, k\}$ is denoted $\mathfrak{S}_k$. If $\sigma \in \mathfrak{S}_k$, we denote $e(\sigma)$ the cardinality of the set

$$\{j \in \{1, \dots, k-1\}, \sigma(j) > \sigma(j+1)\}.$$

As a direct consequence of the Chen-Strichartz development theorem (see Proposition 2.3 of [5], or [12, 23]), we have

Proposition 3.3 *We have*

$$\tilde{B}_t = \exp\left(\sum_{k=1}^{N} \sum_{I=(i_1,\dots,i_k)} \Lambda_I(B)_t U_I \right), \ t \geq 0,$$

where:

$$\Lambda_I(B)_t = \sum_{\sigma\in\mathfrak{S}_k} \frac{(-1)^{e(\sigma)}}{k^2 \binom{k-1}{e(\sigma)}} \int_{0\leq t_1 \leq \dots \leq t_k \leq t} \circ dB_{t_1}^{\sigma^{-1}i_1} \circ \dots \circ dB_{t_k}^{\sigma^{-1}i_k}.$$

For instance:

1. The component of the process $(\ln(\tilde{B}_t))_{t\geq 0}$ in $\mathcal{V}_1$ is simply

$$\sum_{i=1}^{d} U_i B^i_t,\ t \geq 0.$$

2. The component in $\mathcal{V}_2$ is

$$\frac{1}{2}\sum_{1\leq i<j\leq d} [U_i, U_j]\left(\int_0^t B^i_s dB^j_s - B^j_s dB^i_s\right),\ t \geq 0.$$

The Carnot group $\mathbb{G}$ is said to be free if $\mathfrak{g}$ is isomorphic to the free Lie algebra with d generators with the relations that all brackets of length more than N vanish. In that case, $\dim \mathcal{V}_j$ is the number of Hall words of length j in the free algebra with d generators. We thus have, according to Bourbaki [9] (see also Reutenauer [21] pp. 96):

$$\dim \mathcal{V}_j = \frac{1}{j}\sum_{i|j} \mu(i) d^{\frac{j}{i}},\ j \leq N,$$

where μ is the Möbius function. We easily deduce from this that when $N \to +\infty$,

$$\dim \mathfrak{g} \sim \frac{d^N}{N}.$$

An important algebraic point is that, up to an isomorphism there is one and only one free Carnot with a given depth and a given dimension for the basis. Let us denote $m = \dim \mathbb{G}$. Choose now a Hall family and consider the $\mathbb{R}^m$-valued process $(B^*_t)_{t\geq 0}$ obtained by writing the components of $(\ln(\tilde{B}_t))_{t\geq 0}$ in the corresponding Hall basis of $\mathfrak{g}$. It is easily seen that $(B^*_t)_{t\geq 0}$ solves a stochastic differential equation that can be written

$$B^*_t = \sum_{i=1}^{d} \int_0^t D_i(B^*_s) \circ dB^i_s,$$

where the D_i's are polynomial vector fields on $\mathbb{R}^m$ (for an explicit form of the D_i's, which depend of the choice of the Hall basis, we refer to Vershik-Gershkovich [15] pp. 27). With these notations, we have the following proposition (see also [15]).

Proposition 3.4 *On $\mathbb{R}^m$, there exists a unique group law $\star$ which makes the vector fields $D_1, \dots, D_d$ left invariant. This group law is polynomial of degree N and we have moreover*

$$(\mathbb{R}^m, \star) \sim \mathbb{G}.$$

The group $(\mathbb{R}^m, \star)$ is called the free Carnot group of step N over $\mathbb{R}^d$. It shall be denoted $\mathbb{G}_N(\mathbb{R}^d)$. The process B^* shall be called the lift of B in $\mathbb{G}_N(\mathbb{R}^d)$.

Remark 3.5 Notice that $\mathbb{G}_N(\mathbb{R}^d)$ is, by construction, endowed with the basis of vector fields $(D_1, \dots, D_d)$. These vector fields agree at the origin with $\left(\frac{\partial}{\partial x_1}, \cdots, \frac{\partial}{\partial x_d}\right)$.

3.2 Hörmander's Type Theorems

Consider now on $\mathbb{R}^n$ stochastic differential equations of the type

$$X_t = x_0 + \sum_{i=1}^{d} \int_0^t V_i(X_s) \circ dB_s^i, \ t \geq 0, \tag{3.4}$$

where:

1. $x_0 \in \mathbb{R}^n$;
2. $V_1, \dots, V_d$ are C^∞ bounded vector fields on $\mathbb{R}^n$;
3. $\circ$ denotes Stratonovitch integration;
4. $(B_t^1, \dots, B_t^d)_{t \geq 0}$ is a d-dimensional standard Brownian motion.

It is well-known that for every $x_0 \in \mathbb{R}^n$, there is a unique solution $(X_t^{x_0})_{t \geq 0}$ to (3.4) and moreover that there exists a stochastic flow $(\Phi_t, t \geq 0)$ of smooth diffeomorphisms $\mathbb{R}^n \to \mathbb{R}^n$ associated to the equations (3.4). As before, let us denote by $\mathfrak{L}$ the Lie algebra generated by the vector fields V_i and for $p \geq 2$, by $\mathfrak{L}^p$ the Lie subalgebra defined by

$$\mathfrak{L}^p = \{[X, Y], \ X \in \mathfrak{L}^{p-1}, Y \in \mathfrak{L}\}.$$

Moreover if $\mathfrak{a}$ is a subset of $\mathfrak{L}$, we denote

$$\mathfrak{a}(x) = \{V(x), V \in \mathfrak{a}\}, \ x \in \mathbb{R}^n.$$

In this framework, we have the following:

Theorem 3.6 *Let $x_0 \in \mathbb{R}^n$. If $\mathfrak{L}^{N+1}(x_0) = \mathbb{R}^n$, then for any $t > 0$, the random variable*

$$(X_t^{x_0}, B_t^*)$$

has a smooth density with respect to any Lebesgue measure on $\mathbb{R}^n \times \mathbb{G}_N(\mathbb{R}^d)$*, where* $(X_t^{x_0})_{t\geq 0}$ *is the solution of* (3.4) *with initial condition* x_0 *and* $(B_t^*)_{t\geq 0}$ *the lift of* $(B_t)_{t\geq 0}$ *in* $\mathbb{G}_N(\mathbb{R}^d)$.

Proof With a slight abuse of notation, we still denote V_i (resp. D_i) the extension of V_i (resp. D_i) to the space $\mathbb{R}^n \times \mathbb{G}_N(\mathbb{R}^d)$. The process $(X_t^{x_0}, B_t^*)_{t\geq 0}$ is easily seen to be a diffusion process in $\mathbb{R}^n \times \mathbb{G}_N(\mathbb{R}^d)$ with infinitesimal generator

$$\frac{1}{2}\sum_{i=1}^{d}(V_i + D_i)^2.$$

Thus, to prove the theorem, it is enough to check the Hörmander's condition for this operator at the point $(x_0, 0)$. Now, notice that $[\mathfrak{L}, \mathfrak{g}_N(\mathbb{R}^d)] = 0$, so that

$$\mathbf{Lie}(V_1 + D_1, \dots, V_n + D_n)(x_0, 0) \simeq \mathbb{R}^n \oplus \mathfrak{g}_N(\mathbb{R}^d),$$

because $\mathfrak{g}_N(\mathbb{R}^d)$ is step N nilpotent. We denoted $\mathbf{Lie}(V_1 + D_1, \dots, V_n + D_n)$ the Lie algebra generated by $(V_1 + D_1, \dots, V_n + D_n)$. The conclusion follows readily. □

Example 3.7 For $N = 0$, we have $\mathbb{G}_0(\mathbb{R}^d) = \{0\}$ and Theorem 3.6 is the classical Hörmander's theorem.

Example 3.8 For $N = 1$, we have $\mathbb{G}_1(\mathbb{R}^d) \simeq \mathbb{R}^d$ and Theorem 3.6 gives a sufficient condition for the existence of a smooth density for the variable

$$(X_t^{x_0}, B_t).$$

Example 3.9 For $N = 2$, we have $\mathbb{G}_2(\mathbb{R}^d) \simeq \mathbb{R}^d \times \mathbb{R}^{\frac{d(d-1)}{2}}$ and Theorem 3.6 gives a sufficient condition for the existence of a smooth density for the variable

$$(X_t^{x_0}, B_t, \wedge B_t).$$

where

$$\wedge B_t = \left(\frac{1}{2}\int_0^t B_s^i dB_s^j - B_s^j dB_s^i\right)_{1\leq i<j\leq d}.$$

4 Stochastic Differential Equations Driven by N-Step Brownian Loops

In this section, we now enter into the heart of our study.

4.1 N-Step Brownian Loops

On the free Carnot group $\mathbb{G}_N(\mathbb{R}^d)$, consider the fundamental process $(B_t^*)_{t\geq 0}$ defined as the solution of the stochastic differential equation

$$B_t^* = \sum_{i=1}^{d} \int_0^t D_i(B_s^*) \circ dB_s^i, \ t \geq 0.$$

As a consequence of Hörmander's theorem, the diffusion with generator

$$\frac{1}{2} \sum_{i=1}^{d} D_i^2$$

has a smooth transition kernel $p_t(x, y)$, $t > 0$ with respect to the Lebesgue measure.

Proposition 4.1 *Let $T > 0$. There exists a unique $\mathbb{R}^d$-valued continuous process $(P_{t,T}^N)_{0\leq t\leq T}$ such that*

$$P_{t,T}^{N,i} = B_t^i + \int_0^t D_i \ln p_{T-s}\left(P_{s,T}^{N,*}, 0_{\mathbb{G}_N(\mathbb{R}^d)}\right) ds, \quad t < T, \quad i = 1, \ldots, d, \tag{4.5}$$

where $(P_{t,T}^{N,*})_{0\leq t\leq T}$ denotes the lift of $(P_{t,T}^N)_{0\leq t\leq T}$ in $\mathbb{G}_N(\mathbb{R}^d)$. It enjoys the following properties:

1. *$P_{T,T}^{N,*} = 0_{\mathbb{G}_N(\mathbb{R}^d)}$, almost surely;*
2. *For any predictable and bounded functional F,*

$$\mathbb{E}\left(F\left((B_t)_{0\leq t\leq T}\right) \mid B_T^* = 0_{\mathbb{G}_N(\mathbb{R}^d)}\right) = \mathbb{E}\left(F\left((P_{t,T}^N)_{0\leq t\leq T}\right)\right);$$

3. *$(P_{t,T}^N)_{0\leq t\leq T}$ is a semimartingale up to time T.*

Proof The construction of the bridge over a given diffusion is very classical and very general (see for example [1, 7, 13] and [16] p. 142), so that we do not present the details. The only delicate point in the previous statement is the semimartingale property up to time T. In the elliptic case Bismut [7] deals with the end point singularity by proving an estimate of the logarithmic derivatives of the heat kernel. For the heat kernel on Carnot groups, such an estimate can directly be obtained from [17] and [8]. Namely, for $g \in \mathbb{G}_N(\mathbb{R}^d)$, $t > 0$,

$$\mid D_i \ln p_t(g, 0) \mid \leq \frac{C}{\sqrt{t}}$$

where $C > 0$. Now, to prove that $(P^{N,*}_{t,T})_{0 \le t \le T}$ is a semimartingale up to time T, we need to show that for any $1 \le i \le d$,

$$\int_0^T \mid D_i \ln p_{T-s}(P^{N,*}_{s,T}, 0) \mid ds < +\infty$$

with probability 1, which follows therefore from the above estimate. □

The semimartingale $(P^N_{t,T})_{0 \le t \le T}$ shall be called a Brownian loop of step N.

Example 4.2 The process $(P^1_{t,T})_{0 \le t \le T}$ is simply the d-dimensional Brownian bridge from 0 to 0 with length T.

Example 4.3 The process $(P^2_{t,T})_{0 \le t \le T}$ is the d-dimensional standard Brownian motion $(B_t)_{0 \le t \le T}$ conditioned by $(B_T, \wedge B_T) = 0$.

Remark 4.4 Notice that in law,

$$(P^N_{t,T})_{0 \le t \le T} = (\sqrt{T} P^N_{\frac{t}{T},1})_{0 \le t \le T}. \tag{4.6}$$

4.2 SDEs Driven by N-Step Brownian Loops

Consider now on $\mathbb{R}^n$ stochastic differential equations of the type

$$X_t = x_0 + \sum_{i=1}^{d} \int_0^t V_i(X_s) \circ dP^{i,N}_{s,T}, \; t \le T \tag{4.7}$$

where:

1. $x_0 \in \mathbb{R}^n$;
2. $V_1, \ldots, V_d$ are C^∞ bounded vector fields on $\mathbb{R}^n$;
3. $(P^{1,N}_{t,T}, \ldots, P^{d,N}_{t,T})_{0 \le t \le T}$ is a d -dimensional N-step Brownian loop from 0 to 0 with length $T > 0$.

Proposition 4.5 *For every $x_0 \in \mathbb{R}^n$, there is a unique solution $(X^{x_0}_t)_{0 \le t \le T}$ to (4.7). Moreover there exists a stochastic flow $(\Phi_t, 0 \le t \le T)$ of smooth diffeomorphisms $\mathbb{R}^n \to \mathbb{R}^n$ associated to the equations (4.7).*

We consider now the following family of operators $(\mathcal{H}^N_T)_{T \ge 0}$ defined on the space of compactly supported smooth functions $f : \mathbb{R}^n \to \mathbb{R}$ by

$$(\mathcal{H}^N_T f)(x) = \mathbb{E}\left(f(X^x_T)\right), \; x \in \mathbb{R}^n.$$

The operator $\mathcal{H}^N_T$ shall be called the depth N holonomy operator.

Remark 4.6 Of course, as for $N = 1$ which is the case treated in Sect. 2, the operator $\mathcal{H}_T^N$ does not satisfy a semi-group property.

A relevant intrinsic property of $\mathcal{H}_T^N$ is the following:

Proposition 4.7 *$\mathcal{H}_T^N$ does not depend on the particular free Carnot group $\mathbb{G}_N(\mathbb{R}^d)$.*

Proof Consider the stochastic differential equation

$$X_t = x_0 + \sum_{i=1}^{d} \int_0^t V_i(X_s) \circ d\tilde{P}_{s,T}^{i,N},\ t \leq T \tag{4.8}$$

where $(\tilde{P}_{t,T}^{1,N}, \dots, \tilde{P}_{t,T}^{d,N})_{0 \leq t \leq T}$ generates a loop in a free Carnot group $\mathbb{G}$ of step N. Let $(B_t)_{t \geq 0}$ denote a d-dimensional standard Brownian motion, and let $(B_t^*)_{t \geq 0}$ (resp. $(\tilde{B}_t)_{t \geq 0}$) denote a lift in $\mathbb{G}_N(\mathbb{R}^d)$ (resp. $\mathbb{G}$). Thanks to the proposition 2.10 of [5], there exists a Carnot group isomorphism

$$\phi : \mathbb{G}_N(\mathbb{R}^d) \to \mathbb{G}$$

such that $\tilde{B} = \phi(B^*)$. Therefore almost surely, $\tilde{B}_T = 0$ is equivalent to $B_T^* = 0$ and thus

$$(\tilde{P}_{t,T}^{1,N}, \dots, \tilde{P}_{t,T}^{d,N})_{0 \leq t \leq T} =^{law} (P_{t,T}^{1,N}, \dots, P_{t,T}^{d,N})_{0 \leq t \leq T}. \square$$

Theorem 4.8 *Let $f : \mathbb{R}^n \to \mathbb{R}$ be a smooth, compactly supported function. In $\mathbf{L}^2$,*

$$\lim_{T \to 0} \frac{\mathcal{H}_T^N f - f}{T^{N+1}} = \Delta_N f,$$

where Δ_N is a second order differential operator.

Proof Before we start the proof, let us precise some notations we already used. If $I = (i_1, \dots, i_k) \in \{1, \dots, d\}^k$ is a word, we denote $\mid I \mid = k$ its length and by U_I the commutator defined by

$$U_I = [U_{i_1}, [U_{i_2}, \dots, [U_{i_{k-1}}, U_{i_k}] \dots].$$

The group of permutations of the index set $\{1, \dots, k\}$ is denoted $\mathfrak{S}_k$. If $\sigma \in \mathfrak{S}_k$, we denote $e(\sigma)$ the cardinality of the set

$$\{j \in \{1, \dots, k-1\}, \sigma(j) > \sigma(j+1)\}.$$

Finally, we denote

$$\Lambda_I(P^N_{\cdot,T})_t = \sum_{\sigma \in \mathfrak{S}_k} \frac{(-1)^{e(\sigma)}}{k^2 \binom{k-1}{e(\sigma)}} \int_{0 \le t_1 \le \ldots \le t_k \le t} \circ dP^{N,\sigma^{-1} i_1}_{t_1,T} \circ \ldots \circ dP^{N,\sigma^{-1} i_k}_{t_k,T}.$$

Due to the scaling property

$$(P^N_{t,T})_{0 \le t \le T} = (\sqrt{T} P^N_{\frac{t}{T},1})_{0 \le t \le T},$$

we can closely follow the article of Strichartz [23] (see also Castell [11]), to obtain the following asymptotic development of $f(X^x_T)$:

$$f(X^x_T) = \left(\exp\left(\sum_{k=1}^{2N+2} \sum_{I=(i_1,\ldots,i_k)} \Lambda_I(P^N_{\cdot,T})_T V_I \right) f \right)(x) + T^{\frac{2N+3}{2}} \mathbf{R}_{2N+3}(T,f,x),$$

where the remainder term satisfies when $T \to 0$,

$$\sup_{x \in \mathbb{R}^n} \sqrt{\mathbb{E}\left(\mathbf{R}_{2N+3}(T,f,x)^2\right)} \le C$$

for some non negative constant C. By definition of $(P^N_{t,T})_{0 \le t \le T}$, we actually have

$$\sum_{k=1}^{2N+2} \sum_{I=(i_1,\ldots,i_k)} \Lambda_I(P^N_{\cdot,T})_T V_I = \sum_{k=N+1}^{2N+2} \sum_{I=(i_1,\ldots,i_k)} \Lambda_I(P^N_{\cdot,T})_T V_I,$$

so that

$$f(X^x_T) = \left(\exp\left(\sum_{k=N+1}^{2N+2} \sum_{I=(i_1,\ldots,i_k)} \Lambda_I(P^N_{\cdot,T})_T V_I \right) f \right)(x) + T^{\frac{2N+3}{2}} \mathbf{R}_{2N+3}(T,f,x).$$

Therefore, since f is assumed to be compactly supported

$$\mathcal{H}^N_T f(x) = \mathbb{E}\left(\left(\exp\left(\sum_{k=N+1}^{2N+2} \sum_{I=(i_1,\ldots,i_k)} \Lambda_I(P^N_{\cdot,T})_T V_I \right) f \right)(x) \right) + T^{\frac{2N+3}{2}} \tilde{\mathbf{R}}_{2N+3}(T,f,x),$$

where

$$\tilde{\mathbf{R}}_{2N+1}(T,f,x) = \mathbb{E}\left(\mathbf{R}_{2N+1}(T,f,x)\right).$$

Since, by symmetry, we always have

$$\mathbb{E}\left(\Lambda_I(P^N_{.,T})_T\right) = 0,$$

we have to go at the order 2 in the asymptotic development of the exponential when $T \to 0$. By neglecting the terms which have order more than $T^{\frac{2N+3}{2}}$, we obtain

$$\mathcal{H}^N_T f(x) = f(x) + \sum_{\substack{I=(i_1,\dots,i_{N+1})\\ J=(j_1,\dots,j_{N+1})}} \frac{1}{2}\mathbb{E}\left(\Lambda_I(P^N_{.,T})_T \Lambda_J(P^N_{.,T})_T\right)(V_I V_J f)(x)$$
$$+T^{\frac{2N+3}{2}} \mathbf{R}^*_{2N+3}(T,f,x),$$

where the remainder term $\mathbf{R}^*_{2N+3}(T,f,x)$ is bounded in $\mathbf{L}^2$ when $T \to 0$. This leads to the expected result. □

Example 4.9 We have

$$\Delta_0 = \frac{1}{2}\sum_{i=1}^{d} V_i^2,$$

and, as already seen in Sect. 2,

$$\Delta_1 = \frac{1}{24}\sum_{1\le i<j\le d} [V_i, V_j]^2.$$

Example 4.10 It is easily seen that the distribution of $(P^N_{t,T})_{0\le t\le T}$ is invariant by the action of the orthogonal group, that is for any $d \times d$ orthogonal matrix M, we have $(MP^N_{t,T})_{0\le t\le T} =^{law} (P^N_{t,T})_{0\le t\le T}$. As a consequence Δ_N is a second order differential operator which is itself invariant by rotations of the vector fields V_i's. From this observation, we can deduce that on the Lie group $\mathbf{SU}(2)$, for which $d = 3$ and $[V_1, V_2] = V_3$, $[V_2, V_3] = V_1$ and $[V_3, V_1] = V_2$ we have $\Delta_N = c_N(V_1^2 + V_2^2 + V_3^2)$ where c_N is a non negative constant. Indeed, since $\mathbf{SU}(2)$ is a rank-one symmetric space, it is a well-know result that the only second order left-invariant operators which are invariant by rotations are the multiple of the Laplace-Beltrami operator.

We now have the following generalization of Theorem 2.4 and Theorem 2.5:

Theorem 4.11 *Assume that $\mathfrak{L}^{N+1} = 0$, then for any solution $(X^{x_0}_t)_{0\le t\le T}$ of (4.7) we have almost surely $X^{x_0}_T = x_0$. On the other hand, assume that $\mathfrak{L}^{N+1}(x_0) = \mathbb{R}^n$, then for the solution $(X^{x_0}_t)_{0\le t\le T}$ of (4.8) the random variable $X^{x_0}_T$ has a smooth density with respect to the Lebesgue measure of $\mathbb{R}^n$.*

Proof Assume that $\mathfrak{L}^{N+1} = 0$, then there exists a smooth map

$$F : \mathbb{R}^n \times \mathbb{G}_N(\mathbb{R}^d) \to \mathbb{R}^n$$

such that, for $x_0 \in \mathbb{R}^n$, the solution $(X_t^{x_0})_{0 \le t \le T}$ of the SDE (4.8) can be written

$$X_t^{x_0} = F(x_0, Q_{t,T}^N),$$

which implies immediately the expected result.

Assume now that $\mathfrak{L}^{N+1}(x_0) = \mathbb{R}^n$. Let us consider the solution $(Z_t)_{t \ge 0}$ of the following stochastic differential equation:

$$Z_t = x_0 + \sum_{i=1}^d \int_0^t V_i(Z_s) \circ dB_s^i, \ t \ge 0,$$

where $(B_t^1, \dots, B_t^d)_{t \ge 0}$ is a d-dimensional standard Brownian motion. From Theorem 3.6, the random variable

$$(Z_T, B_T^*)$$

has a smooth density with respect to any Lebesgue measure on $\mathbb{R}^n \times \mathbb{G}_{d,N}$. It implies in the same way as in the proof of Theorem 2.5 that $X_T^{x_0}$ has a density with respect to the Lebesgue measure because the density of B_T^* does not vanish at 0 (see [6]). □

In the case of the existence of a density for $X_T^{x_0}$, we can moreover give an equivalent of this density when the length of the loop tends to 0. To this end, let us precise some notations.

We set for $x \in \mathbb{R}^n$ and $k \ge N$,

$$\mathcal{U}_k(x) = \mathbf{span}\{V_I, \ N \le\mid I \mid\le k\}.$$

In the case where $\mathfrak{L}^{N+1}(x) = \mathbb{R}^n$, if k is big enough then $\mathcal{U}_k(x) = \mathbb{R}^n$. We denote $d(x)$ the smallest integer $k \ge N+1$ for which this equality holds and define the graded dimension

$$\dim_{\mathcal{H}} \mathfrak{L}^{N+1}(x) := \sum_{k=N+1}^{d(x)} k \left(\dim \mathcal{U}_k(x) - \dim \mathcal{U}_{k-1}(x)\right).$$

Theorem 4.12 *Assume that for any $x \in \mathbb{R}^n$, $\mathfrak{L}^{N+1}(x) = \mathbb{R}^n$. Let us denote $p_T(x)$ the density of X_T^x with respect to the Lebesgue measure. We have*

$$p_T(x) \sim_{T \to 0} \frac{m(x)}{T^{\frac{\dim_{\mathcal{H}} \mathfrak{L}^{N+1}(x)}{2}}},$$

where m is a smooth non negative function.

Proof Let us, once time again, consider the solution $(Z_t^x)_{t\geq 0}$ of the following stochastic differential equation:

$$Z_t^x = x + \sum_{i=1}^{d} \int_0^t V_i(Z_s^x) \circ dB_s^i, \ t \geq 0,$$

where $(B_t^1, \dots, B_t^d)_{t\geq 0}$ is a d-dimensional standard Brownian motion. From [6] (see also [19]), the density at $(x, 0)$ of the random variable (Z_T^x, B_T^*) behaves when T goes to zero like

$$\frac{\tilde{m}(x)}{T^{\frac{\dim_{\mathcal{H}} \mathbb{G}_N(\mathbb{R}^d) + \dim_{\mathcal{H}} \mathfrak{L}^{N+1}(x)}{2}}},$$

where $\dim_{\mathcal{H}} \mathbb{G}_N(\mathbb{R}^d) = \sum_{j=1}^N j \dim \mathcal{V}_j$ is the graded dimension of $\mathbb{G}_N(\mathbb{R}^d)$, and $\tilde{m}$ a smooth non negative function. Always from [6], the density of the random variable Y_T behaves when T goes to zero like

$$\frac{C}{T^{\frac{\dim_{\mathcal{H}} \mathbb{G}_N(\mathbb{R}^d)}{2}}},$$

where C is a non negative constant. The conclusion follows readily. □

References

1. F. Baudoin, Conditioned stochastic differential equations: theory, examples and applications to finance. Stoch. Process. Appl. **100**, 109–145 (2002)
2. F. Baudoin, Equations différentielles stochastiques conduites par des lacets dans les groupes de Carnot. (French. English, French summary) [Stochastic differential equations driven by loops in Carnot groups]. C. R. Math. Acad. Sci. Paris **338**(9), 719–722 (2004)
3. F. Baudoin, The tangent space to a hypoelliptic diffusion and applications, in *Séminaire de Probabilités XXXVIII* (Springer, 2005)
4. F. Baudoin, A Bismut type theorem for subelliptic semigroups. C. R. Math. Acad. Sci. Paris **34**(12), 765 (2007)
5. F. Baudoin, *An Introduction to the Geometry of Stochastic Flows* (Imperial College Press, 2005)
6. G. Ben Arous, Développement asymptotique du noyau de la chaleur hypoelliptique sur la diagonale. Annales de l'institut Fourier, tome **39**(1), 73–99 (1989)
7. J.M. Bismut, *Large Deviations and the Malliavin Calculus* (Birkhauser, 1984)
8. A. Bonfiglioli, E. Lanconelli, F. Uguzzoni, Uniform Gaussian estimates of the fundamental solutions for heat operators on Carnot groups. Adv. Differ. Equ. **7**, 1153–1192 (2002)
9. N. Bourbaki, *Groupes et Algèbres de Lie*, chaps. 1–3 (Hermann, 1972)
10. T. Cass, P. Friz, Densities for rough differential equations under Hörmander's condition. Ann. Math. (2) **171**(3), 2115–2141 (2010)
11. F. Castell, Asymptotic expansion of stochastic flows. Probab. Rel. Fields **96**, 225–239 (1993)

12. K.T. Chen, Integration of paths, geometric invariants and a generalized Baker-Hausdorff formula. Ann. Math. **65**(1), (1957)
13. P. Fitzsimmons, J.W. Pitman, M. Yor, Markov bridges, construction, palm interpretation and splicing, in *Seminar on Stochastic Processes* (Birkhauser, 1993), pp. 101–134
14. B. Gaveau, Principe de moindre action, propagation de la chaleur et estimées sous-elliptiques sur certains groupes nilpotents. Acta Math. **139**(1–2), 95–153 (1977)
15. V.Ya. Gershkovich, A.M. Vershik, Nonholonomic dynamical systems, geometry of distributions and variational problems, in *Dynamical Systems VII*, ed. by V.I. Arnold, S.P. Novikov. Encyclopaedia of Mathematical Sciences, vol. 16 (1994)
16. E.P. Hsu, *Stochastic Analysis on Manifolds*. Graduate Texts in Mathematics, vol. 38 (AMS, 2002)
17. D.S. Jerison, A. Sanchez-Calle, Estimates for the heat kernel for a sum of squares of vector fields. Ind. Math. J. **35**(4), (1986)
18. H. Kunita, *Stochastic Flows and Stochastic Differential Equations*. Cambridge Studies in Advanced Mathematics, vol. 24 (1990)
19. R. Léandre, Développement asymptotique de la densité d'une diffusion dégénérée. Forum Math. **4**(1), 45–75
20. D. Nualart, *The Malliavin Calculus and Related Topics* (Springer, Berlin/Heidelberg/New York, 1995)
21. C. Reutenauer, *Free Lie Algebras*. London Mathematical Society Monographs, New series, vol. 7 (1993)
22. L.P. Rotschild, E.M. Stein, Hypoelliptic differential operators and Nilpotent Groups. Acta Math. **137**, 247–320 (1976)
23. R.S. Strichartz, The Campbell-Baker-Hausdorff-Dynkin formula and solutions of differential equations. J. Funct. Anal. **72**, 320–345 (1987)
24. M.E. Taylor, *Partial Differential Equations: Qualitative Studies of Linear Equations*. Applied Mathematical Sciences, vol. 116 (Springer, 1996)

Genealogy of a Wright-Fisher Model with Strong SeedBank Component

Jochen Blath, Bjarki Eldon, Adrián González Casanova, and Noemi Kurt

Abstract We investigate the behaviour of the genealogy of a Wright-Fisher population model under the influence of a strong seedbank effect. More precisely, we consider a simple seedbank age distribution with two atoms, leading to either classical or long genealogical jumps (the latter modeling the effect of seed-dormancy). We assume that the length of these long jumps scales like a power N^{β} of the population size N, thus giving rise to a 'strong' seedbank effect. For a certain range of β, we prove that the ancestral process of a sample of n individuals converges under a non-classical time-scaling to Kingman's $n-$coalescent. Further, for a wider range of parameters, we analyze the time to the most recent common ancestor of two individuals analytically and by simulation.

Keywords Seedbanks • Wright-Fisher model • Kingman's coalescent

Mathematics Subject Classification (2000). Primary 60K35; Secondary 92D15.

1 Introduction

We consider a generalization of the classical Wright-Fisher model in the following sense: Consider a neutral, haploid population of fixed size N that reproduces asexually in discrete generations indexed by the natural numbers. In contrast to the Wright-Fisher model, at each new generation, while most of the population stems from direct reproduction of individuals of the previous generation, a few remaining individuals obtain their type from a parent having lived in the (possibly far) past. In the retrospective viewpoint, an individual alive in generation 0, instead of selecting its parent uniformly from the previous generation as in the classical Wright-Fisher model, uses some probability measure μ on $\mathbb{N}$ to sample the distance in generations

J. Blath • B. Eldon • A. González Casanova (✉) • N. Kurt
Institut für Mathematik, Sekr. MA 7-5, Straße des 17. Juni 136, 10623 Berlin, Germany
e-mail: blath@math.tu-berlin.de; eldon@math.tu-berlin.de; adriangcs@hotmail.com; kurt@math.tu-berlin.de

R.H. Mena et al. (eds.), *XI Symposium on Probability and Stochastic Processes*, Progress in Probability 69, DOI 10.1007/978-3-319-13984-5_4

to its parent, and then picks its ancestor uniformly among the individuals at the sampled distance. The biological interpretation of this mechanism is that it allows old genes to become re-activated in a population after some time as the result of a *seedbank effect*. Such an effect could be seen as an ‘evolutionary force’ that may have to be taken into account for populations that produce dormant forms, such as plant seeds or bacterial spores. Some dormant forms may remain inactive for a long time, and, after becoming active again, potentially re-introduce old genetic material into the present population, thus increasing genetic variability [4, 6, 12, 13]. In the case where the dormancy is on a timescale which is non-negligible with respect to the population size, this is expected to lead to drastic changes in the genealogy of such a population. In the present paper we analyze a simple mathematical model which illustrates this effect.

Informally, the model can be described as follows. Fix the population size $N \in \mathbb{N}$ and a probability measure μ on the natural numbers. This measure determines the generation of the immediate ancestor of an individual backward in time, meaning that an individual living k generations before the present, for $k \in \mathbb{N}$, has its immediate ancestor $k + l$ generations before the present, for $l \in \mathbb{N}$, with probability $\mu(l)$. Such a genealogical process, in the case where μ has *finite* support independent of N, was introduced and analyzed in [5], where it was shown that the genealogy converges, after classical rescaling by the population size, to a constant time change of Kingman's coalescent. The Kingman coalescent is the continuous-time Markov process taking values in the partitions of $\mathbb{N}$ that is characterized by the property that each pair of blocks coalesces independently at rate 1 (see for example [1] for an introduction). It is well-known that the Kingman Coalescent arises as a universal limit of genealogical processes in many population genetic models, such as the classical Wright-Fisher and the Moran model, and also in many situations with additional structure, cf. [14].

In [2], the case of a stronger seedbank effect with unbounded measure μ was considered. More precisely, power law distributions of the form

$$\mu_\alpha(\{n, n+1, \ldots\}) = L(n)n^{-\alpha} \tag{1.1}$$

for some $\alpha > 0$ and some slowly varying function $L(n)$ were investigated. Three regimes concerning the time to the most recent common ancestor were identified: If $\alpha > 1$, then the expected time to the most recent common ancestor is of order N, and the ancestral process converges to a constant time change of Kingman's coalescent under classical rescaling by the population size. For $1/2 < \alpha < 1$, the time to the most recent common ancestor is finite almost surely, but the expectation does not exist for any N. If $\alpha < 1/2$, then there might be no common ancestor at all. The boundary cases $\alpha = 1$ and $\alpha = 1/2$ depend on the choice of $L(n)$.

In the present paper, we investigate a rather natural framework that was not considered in [2]. We choose μ to be of the following form: For $N \in \mathbb{N}$ fixed, $\beta > 0$ and $\varepsilon \in (0, 1)$, let

$$\mu_N = (1-\varepsilon)\delta_1 + \varepsilon\delta_{N^\beta}. \tag{1.2}$$

This means that in each new generation, a proportion $(1-\varepsilon)$ of the total population obtains its genetic type from the previous generation, whereas a fraction ε of the population gets its type from generation N^β in the past. The most important difference to previously studied models is the fact that the expected length of a genealogical jump is equal to $1+\varepsilon(N^\beta-1)$, and hence is finite for fixed N, but diverges as $N\to\infty$. This puts us in a situation that is outside the scope of [5] or [2].

Our main result shows that for $0<\beta<1/4$, after rescaling time by $\varepsilon^2N^{2\beta+1}$, the ancestral process of a sample from such a population converges to Kingman's n–coalescent, showing that the relevant timescale in N is indeed much larger than in the Wright-Fisher model, namely $N^{2\beta+1}$ as opposed to N. This is particularly relevant if one considers scenarios including mutation. Moreover, we show that for any $\beta>0$ the time to the most recent common ancestor of two individuals is always of an order that is strictly greater than N, and we provide some simulations that support the conjecture that $N^{2\beta+1}$ is the relevant timescale also for (at least some) $\beta>1/4$.

This paper deals for the first time with seedbanks for which the distribution μ_N of the genealogical distance between parent and offspring depends on the population size N. The model gives rise to many questions outside the scope of this paper. For example, it would be interesting to investigate the existence of non-degenerate limits if in (1.2) we let ε depend on N and converge to 0 as $N\to\infty$. The answer to this question is expected to depend on the speed at which $\varepsilon(N)$ converges to zero. Our result suggests that the scaling $\varepsilon=N^{-\beta}$ should lead to the standard Kingman coalescent. However, the methods of proof in this paper do not immediately carry over to a situation where ε depends on N. A more general question concerns the existence of a seedbank, with a sequence μ_N, such that its genealogy converges to a non-trivial scaling limit other than the Kingman coalescent. Such a regime can indeed be found, and is investigated in an forthcoming paper, [3].

2 Model and Main Results

The formal construction of our model follows [2, 5]. Fix $\beta>0, \varepsilon\in(0,1)$. For each $N\in\mathbb{N}$ we choose μ_N as defined in (1.2), that is,

$$\mu_N := (1-\varepsilon)\delta_1+\varepsilon\delta_{N^\beta}. \tag{2.1}$$

In order to simplify notation, we will assume throughout this paper that N^β is a natural number, otherwise imagine it replaced by $\lfloor N^\beta \rfloor$. We call the probability measure μ_N the *seedbank age distribution*. Fix once and for all a reference generation 0, from which time in discrete generations runs backwards. Fix a sample size $m\geq 2$ and a sampling measure γ for the generations of the original sample on the integers $\mathbb{N}$. We will usually assume that γ has finite support (independent of N), an important example being $\gamma=\delta_0$. The ancestral lineages of m sampled individuals indexed by $i \in \{1, \ldots, m\}$ in the seedbank process, who lived in generations sampled according to γ with respect to reference time 0, are constructed as follows. For each $i \in \{1, \ldots, m\}$, let $(S_n^{(i)})_{n \in \mathbb{N}}$ be a Markov chain independent of $\{(S_n^{(j)})_n, j \in \{1, \ldots, m\}, j \neq i\}$, whose state space is the non-negative integers $\mathbb{N}_0$, with $S_0^{(i)} \sim \gamma$, and homogeneous transition probabilities

$$\mathbb{P}\big(S_1^{(i)} = k' \,\big|\, S_0^{(i)} = k\big) = \mu_N(k' - k),\ 0 \leq k < k',\ i = 1, \ldots, m.$$

The interpretation is that $S_0^{(i)}$ represents the generation of individual i, and $S_1^{(i)}$ the generation of its parent (backward in time), and so on. The set $\{S_0^{(i)}, S_1^{(i)}, \ldots\} \subset \mathbb{N}_0$ is thus the set of generations of all ancestors of individual i, including the individual itself.

In order to construct the ancestral process of several individuals, we introduce interaction between ancestral lines as follows. Within the population of size N, in any fixed generation k, the individuals are labeled from 1 to N. Let $(U_n^{(i)})_{n \in \mathbb{N}}, i \in \{1, \ldots, m\}$ denote m independent families of independent random variables distributed uniformly on $\{1, \ldots, N\}$, independent of $\{(S_n^{(i)}), i = 1, \ldots, m\}$. We think of $U^{(i)}_{S_n^{(i)}}$ as the label within the population of size N of the nth ancestor of individual i. This means that the label of each ancestor of each individual is picked uniformly at random in each generation that the ancestral line of this individual visits, exactly as it is done in the Wright-Fisher model. The difference is that an ancestral line in the Wright-Fisher model visits every generation, while in our seedbank model it does not. Note that of course all the random variables introduced up to now depend on the population size N.

The *time to the most recent common ancestor* of two individuals i and j, denoted by $T_{MRCA}(2)$, is defined as

$$T_{MRCA}(2) := \inf\big\{k > 0 : \exists n, m \in \mathbb{N}, k = S_n^{(i)} = S_m^{(j)} \text{ and } U_k^{(i)} = U_k^{(j)}\big\}. \tag{2.2}$$

In other words, $T_{MRCA}(2)$ is the first generation back in time (counted from 0 on) in which two randomly sampled individuals ('initial' generations sampled according to γ) have an ancestor, and both ancestors have the same label U, hence, it is indeed the first generation back in time that i and j have the *same* ancestor.

It should be clear how to generalize this construction to lead to a full ancestral process of $m \geq 2$ individuals: Construct the process $(S_n^{(i)}, U_n^{(i)})_{n \in \mathbb{N}}$ independently for each individual, and couple the lines of individual i and individual j at the time of their most recent common ancestor by letting them evolve together from this time onward, as represented in Fig. 1.

A precise construction is given in the following way: let

$$T_1 := \inf\big\{k > 0 : \exists n, l \in \mathbb{N}, i \neq j \in \{1, \ldots, m\} : k = S_n^{(i)} = S_l^{(j)}, U_k^{(i)} = U_k^{(j)}\big\}, \tag{2.3}$$

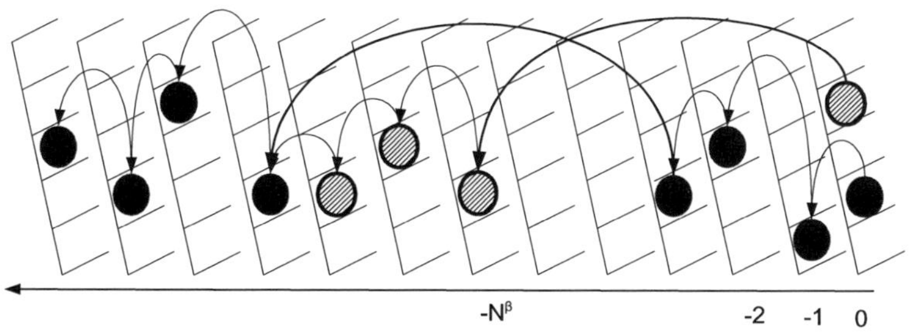

Fig. 1 Ancestral line of one individual. The numbers indicate the generations, counted backward from generation 0

be the time of the first coalescence of two (or more) lines, and let the set of individuals whose lines participate in a coalescence at time T_1 be denoted by

$$I_1 := \{i \in \{1, \ldots, m\} : \exists n, l \in \mathbb{N}, j \neq i : T_1 = S_n^{(i)} = S_l^{(j)}, U_{T_1}^{(i)} = U_{T_1}^{(j)}\}. \tag{2.4}$$

I_1 can be further divided into (possibly empty) pairwise disjoint sets

$$I_1^p := \{i \in I_1 : U_{T_1}^{(i)} = p\}, \quad p = 1, \ldots, N. \tag{2.5}$$

Note that by construction there is at least one p such that I_1^p is non-empty, which actually means that any such I_1^p contains at least two elements. Let

$$i_1^p := \min I_1^p,$$

and let

$$J_1 := \bigcup_{p: I_1^p \neq \emptyset} \{i_1^p\}.$$

After time T_1 we discard all $S^{(j)}$ for $j \in I_1^p, j \neq i_1^p$, and only keep $S^{(i_1^p)}$ for every $p = 1, \ldots, N$. We interpret this as merging the ancestral lineages of all individuals from I_1^p into one lineage at time T_1, separately for every p with $I_1^p \neq \emptyset$. In case there are several non-empty I_1^p, we observe simultaneous mergers. For $r \geq 2$ we define now recursively (we will use $\overline{A}$ to denote the complement of a set A)

$$T_r := \inf\{k > T_{r-1} : \exists n, l \in \mathbb{N}, \exists i, j \in \overline{I_{r-1}} \cup J_{r-1}, i \neq j : k = S_n^{(i)} = S_l^{(j)}, U_k^{(i)} = U_k^{(j)}\}, \tag{2.6}$$

and

$$I_r := \{i \in \overline{I_{r-1}} \cup J_{r-1} : \exists n, l \in \mathbb{N}, \exists j \neq i, j \notin I_{r-1} \cup J_{r-1} : T_r = S_n^{(i)} = S_l^{(j)}, U_{T_r}^{(i)} = U_{T_r}^{(j)}\}, \tag{2.7}$$

and similarly $I_r^p := \{i \in I_r : U_{T_r}^{(i)} = p\}, i_r^p = \min I_r^p, p = 1, \ldots, N$, and $J_r = \cup_{p:I_r^p \neq \emptyset}\{i_r^p\}$. We stop the recursive construction as soon as $\overline{I_r} = \emptyset$, which happens after finitely many r. Now we can finally define the main object of interest of this paper.

Definition 2.1 Fix $N \in \mathbb{N}, \beta > 0$, and $\varepsilon > 0$. Fix $m \ll \mathbb{N}$ and an initial distribution γ on $\mathbb{N}_0$. Define a partition valued process $(A_k^N)_{k\in\mathbb{N}_0}$, starting with $A_0^N = \{\{1\}, \ldots, \{m\}\}$, by setting $A_k^N = A_{k-1}^N$ if $k \notin \{T_1, T_2, ..\}$, and construct the $A_{T_r}^N, r = 1, 2, \ldots$ in the following way: For each $p \in \{1, \ldots, N\}$ such that $I_r^p \neq \emptyset$, the blocks of $A_{T_r-1}^N$ that contain at least one element of I_r^p, are merged. Such merging is done separately for every p with $I_r^p \neq \emptyset$, and the other blocks are left unchanged. The resulting process $(A_k^N)_{k\in\mathbb{N}}$ is called the *ancestral process* of m individuals in the Wright-Fisher model with seedbank age distribution μ_N and initial distribution γ. The time to the most recent common ancestor of the m individuals is defined as

$$T_{MRCA}^N(m) := \inf\{k \in \mathbb{N} : A_k^N = \{1, \ldots, m\}\}. \tag{2.8}$$

It is important to note that A^N is not a Markov process: The probability of a coalescence at time k depends on more than just the configuration A_{k-1}^N. In fact, it depends on the values $\max\{S_n^{(i)} : S_n^{(i)} \leq k-1\}, i = 1, \ldots, m$, that is, on the generation of the last ancestor of each individual before generation k.

An equivalent construction of (A_k^N) in terms of renewal processes was given in [2]. In the present paper, we denote by P_γ the law of $(S_n^{(1)})$, indicating the initial distribution of the generations of the individuals. We write $P_{\otimes\gamma^m}$ for the law of the process (A_k^N) if the generation of each of the m sampled individuals is chosen independently according to γ. We abbreviate by slight abuse of notation both P_{δ_0} and $P_{\otimes\delta_0^m}$ by P_0. In the main result below we will assume γ to have finite support independent of N. In this case, the fact that the individuals may be sampled from different generations will become negligible after rescaling time appropriately. However, for the construction of (A_k^N) this assumption is not necessary.

The aim of this paper is to understand the non-trivial scaling limit of (A_k^N) and the corresponding time-scaling as $N \to \infty$. We write $\mathcal{M}_1(\mathbb{N}_0)$ for the probability measures on $\mathbb{N}_0$ and recall that Kingman's $m-$coalescent is the restriction of Kingman's coalescent to partitions of $\{1, \ldots, m\}$. We are now ready to state our main result.

Theorem 2.2 *Let* $0 < \beta < 1/4$. *For all* $m > 0$ *and* $\gamma \in \mathcal{M}_1(\mathbb{N}_0)$ *with finite support, the process* $(A_{\lfloor \varepsilon^2 N^{1+2\beta} t\rfloor}^N)_{t\geq 0}$ *converges weakly as* $N \to \infty$ *on the Skorohod space of càdlàg paths to Kingman's m−coalescent.*

This result should be compared to the classical result for the ancestral process of the Wright-Fisher model: in that case, convergence to Kingman's coalescent occurs on the timescale N corresponding to the population size. Our result shows that the seedbank effect drastically changes the timescale on which coalescences occur. The intuitive reason for this is that each generation is visited with probability $\approx \varepsilon^{-1}N^{-\beta}$ by the ancestral line of any given individual, and whenever two ancestral lines visit the same generation, they merge with probability $1/N$. Making this intuition precise requires some work, which is carried out in the next two sections of this paper. As a direct consequence of this theorem, the time to the most recent common ancestor of m individuals is of strictly larger order than in the Wright-Fisher model.

Corollary 2.3 *Let $0 < \beta < 1/4$. For all $m > 0$ and $\gamma \in \mathcal{M}_1(\mathbb{N}_0)$ with finite support,*

$$\lim_{N\to\infty} \frac{E_{\otimes\gamma^m}[T^N_{MRCA}(m)]}{\varepsilon^2 N^{1+2\beta}} = 2\Big(1 - \frac{1}{m}\Big). \tag{2.9}$$

Proof Recalling that for Kingman's $m-$coalescent the expected time to the most recent common ancestor is $2(1 - 1/m)$, this result follows from Theorem 2.2. □

As β increases above $1/4$, one expects the seedbank effect to become even more pronounced. However, our methods of proof, relying on mixing time arguments, don't yet allow us to extend the result to arbitrary $\beta > 0$. It is nevertheless not difficult to see that in any case, the expected time to the most recent common ancestor is of course of order greater than N for any choice of $\beta > 0$, and that T^N_{MRCA} is of order at least $N^{2\beta+1}$ with probability tending to 1 if $0 < \beta < 1/3$.

Proposition 2.4 *Fix $m \in \mathbb{N}$, and $\gamma \in \mathcal{M}_1(\mathbb{N}_0)$ with finite support.*

(i) *Let $\beta > 0$. For all $N \in \mathbb{N}$ large enough,*

$$E_{\otimes\gamma^m}\big[T^N_{MRCA}(m)\big] \geq \varepsilon N^{1+\beta} \vee N. \tag{2.10}$$

(ii) *Let $0 < \beta < 1/3$. For all $\delta > 0$,*

$$\lim_{N\to\infty} P_{\otimes\gamma^m}\big(T^N_{MRCA}(m) < \varepsilon^2 N^{1+2\beta-\delta}\big) = 0. \tag{2.11}$$

We will prove these results in Sect. 4. In the next section, we give an alternative construction of the model in terms of an auxiliary Markov process that will be useful in the proof.

In Sect. 5 we present some simulations of $T^N_{MRCA}(2)$ for certain choices of ε and β, where both lines are sampled from the same generation. The simulations show that also for $\beta = 1/3$ and $\beta = 1/2$, the empirical distribution of $T^N_{MRCA}(2)$, scaled by a factor $\varepsilon^2 N^{2\beta+1}$, exhibits a very good fit to an exponential random variable with parameter 1.

3 Construction of Auxiliary Processes

In [5] and [2], an auxiliary urn Markov process plays a crucial role. We present this process now in a set-up that is useful for this paper, and derive some properties that will serve us in the proof of our main results.

Fix N, ε, β and μ_N as in Eq. (1.2). Fix a probability measure γ on $\mathbb{N}$ with finite support, and assume that N is large enough such that $\text{supp}(\gamma) \subseteq \{0, \ldots, N^\beta\}$. Let $\mathbb{P}_\gamma$ be the law of a Markov chain $(X_k)_{k\in\mathbb{N}_0}$ on $\{0, \ldots, N^\beta - 1\}$ with initial distribution γ that moves according to the following rules: for $k \geq 0$, depending on the current state X_k, we have transitions

$$X_k \mapsto \begin{cases} 0 & \text{with probability } (1-\varepsilon)1_{\{X_k=0\}}, \\ N^\beta - 1 & \text{with probability } \varepsilon 1_{\{X_k=0\}}, \\ X_{k-1} - 1 & \text{with probability } 1_{\{X_k\in\{1,\ldots,N^\beta-1\}\}}. \end{cases} \tag{3.1}$$

As in [5], we call this process an *urn process*, because we think of X_k as the position (urn) of a ball that is moved among N^β urns. Figure 2 shows the possible jumps of (X_k).

How does this new process connect to the original processes $(S_k^{(i)})$ resp. our ancestral process (A_k^N)? We can couple X and $S^{(i)}$ such that the successive times that X visits urn 0 are exactly the successive values visited by the process $S^{(i)}$, that is, the generations in which individual i has an ancestor. This coupling is achieved as follows: Define

$$M_0 := \inf\{k \geq 0 : X_k = 0\}, \quad M_n = \inf\{k > M_{n-1} : X_k = 0\}, \; n \geq 1. \tag{3.2}$$

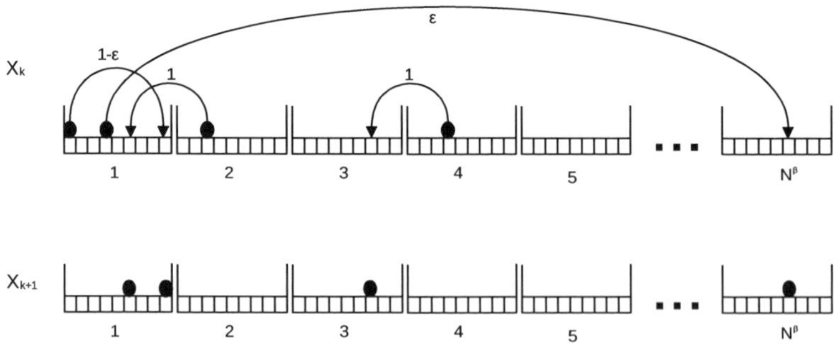

Fig. 2 The possible jumps of X_k : From urn 0 it stays in 0 with probability $1-\varepsilon$, jumps from urn 0 to urn $N^\beta - 1$ with probability ε, and deterministically downwards to the previous urn from any other urn

Then we have

Lemma 3.1 *Let (X_k) be the above urn process with initial distribution γ on $\{0, \ldots, N^\beta - 1\}$, and let $(M_n)_{n\in\mathbb{N}_0}$ be defined as in* (3.2)*. Then the process $(\tilde{S}_n)_{n\in\mathbb{N}_0}$ defined by*

$$\tilde{S}_0 := M_0, \quad \tilde{S}_n := \tilde{S}_{n-1} + M_n \tag{3.3}$$

has the same distribution as $(S_n^{(i)})$ started in γ, and for all $k \in \mathbb{N}$,

$$P_\gamma(\exists n : S_n^{(i)} = k) = \mathbb{P}_\gamma(X_k = 0). \tag{3.4}$$

Proof Immediate by construction. □

Note that, conversely, given $S_n = \tilde{S}_n, n \in \mathbb{N}_0$, Eq. (3.3) uniquely determines $M_n, n \in \mathbb{N}_0$. If N is large enough such that the distribution γ of S_0 satisfies $\mathrm{supp}(\gamma) \subseteq \{0, \ldots, N^\beta - 1\}$, then it is a possible initial distribution for (X_k). Further, since the urn process (X_k) is determined by the successive times it visits urn 0, Lemma 3.1 yields a one-to-one correspondence between ancestral lines in the seedbank process, and the above urn process. As in the construction of (A_k^N), for the process (X_k) we do not need to assume that γ has finite support independent of N, indeed it will actually be useful to allow for initial distributions $\gamma = \gamma_N$ with $\mathrm{supp}(\gamma_N) = \{0, \ldots, N^\beta - 1\}$ depending on N, for each $N \in \mathbb{N}$.

Let E_{μ_N} denote the expectation of μ_N, i.e. $E_{\mu_N} = 1 + \varepsilon(N^\beta - 1)$.

Lemma 3.2 *The probability measure ν_N on $\{0, \ldots, N^\beta - 1\}$ defined by*

$$\nu_N(0) = \frac{1}{1 + \varepsilon(N^\beta - 1)}, \quad \text{and} \quad \nu_N(k) = \frac{\varepsilon}{1 + \varepsilon(N^\beta - 1)}, \quad k \in \{1, \ldots, N^\beta - 1\} \tag{3.5}$$

is the unique stationary distribution of the urn process (X_k).

Proof The proof of this Lemma follows a straight forward computation, observing that

$$\nu_N(k) = \frac{\mu_N\big(\{k, k+1, \ldots\}\big)}{E_{\mu_N}}, \quad k = 0, \ldots, N^\beta - 1.$$

In particular, this is a special case of Lemma 1 in [5] (where the urn process is shifted by one, i.e. takes values in $\{1, \ldots, m\}$ for some m corresponding to our N^β, instead of $\{0, \ldots, m - 1\}$). □

In view of Lemma 3.1, an important quantity in this paper will be the probability that $X_k = 0$, which under stationarity is equal to

$$\nu_N(0) = \frac{1}{1 + \varepsilon(N^\beta - 1)} \sim \frac{1}{\varepsilon N^\beta} \quad \text{as } N \to \infty. \tag{3.6}$$

The following lemma will be crucial in the proof of our main results. By $\|\cdot\|_{TV}$ we denote the total variation distance, which for two probability measures μ, ν on a measureable space $(\Omega, \mathcal{F})$ is given by

$$\|\mu - \nu\|_{TV} = \sup_{A\in\mathcal{F}} |\mu(A) - \nu(A)|.$$

Lemma 3.3 *Let (X_k) be the urn process and ν_N its stationary distribution. For $\beta > 0$ let $\mathcal{P}_{N^\beta}$ denote the set of probability measures on $\{0, \dots, N^\beta - 1\}$.*

(i) *For all $\lambda > 3\beta > 0$, there exist $\delta > 0$ and $N_0 \in \mathbb{N}$, such that for all $N \geq N_0$*

$$\sup_{\mu\in\mathcal{P}_{N^\beta}} \|\mathbb{P}_\mu(X_{N^\lambda} \in \cdot\,) - \nu_N\|_{TV} \leq e^{-N^\delta}.$$

(ii) *Let $\tau = \tau(N)$ be a geometric random variable with parameter $1/N$ independent of (X_k). If $0 < \beta < 1/4$, then there exist $\delta > 0$ and $N_0 \in \mathbb{N}$ such that for all $N \geq N_0$*

$$\sup_{\mu\in\mathcal{P}_{N^\beta}} \|\mathbb{P}_\mu(X_\tau \in \cdot\,) - \nu_N\|_{TV} \leq N^{-(\beta+\delta)}.$$

Proof Fix $\mu \in \mathcal{P}_{N^\beta}$. Let $(Z_n)_{n\in\mathbb{N}_0}$ be a realization of the urn process started in the invariant distribution ν_N independent of $(X_n)_{n\in\mathbb{N}_0}$. We couple (X_n) and (Z_n) by a Doeblin coupling in the following way: Let $\sigma_0 := \inf\{n \in \mathbb{N}_0 : X_n = Z_n\}$. Define

$$\tilde{X}_n := \begin{cases} X_n & \text{if } n \leq \sigma_0, \\ Z_n & \text{if } n > \sigma_0. \end{cases}$$

Write $\mathbb{P} := \mathbb{P}_{\mu\otimes\nu_N}$. Then $\mathbb{P}(\tilde{X}_n = k) = \mathbb{P}_\mu(X_n = k)$ for all $n \in \mathbb{N}_0, k \in \{0, \dots, N^\beta - 1\}$. By Proposition 4.7 of [7], we have

$$\|\mathbb{P}_\mu(X_n \in \cdot\,) - \nu_N\|_{TV} \leq \mathbb{P}(\tilde{X}_n \neq Z_n) = \mathbb{P}(\sigma_0 > n). \tag{3.7}$$

Our aim is therefore to bound $\mathbb{P}(\sigma_0 > n)$. To this end we consider the difference of the two processes at particular times. Define $m_0 := \inf\{n \geq 0 : X_n = 0\}$, $l_0 := \inf\{n \geq 0 : Z_n = 0\}$, and let recursively, for $i \geq 1$,

$$m_i := \inf\{n > m_{i-1} : X_n = 0, X_{n-1} = 1\}$$

and

$$l_i := \inf\{n > l_{i-1} : Z_n = 0, Z_{n-1} = 1\}.$$

Note that for all $i \geq 0$ we have $Z_{m_i} - X_{m_i} \geq 0$ and $X_{l_i} - Z_{l_i} \geq 0$. Without loss of generality we can assume that $Z_0 - X_0 > 0$, which implies $m_0 < l_0$. Since the difference of the two processes remains constant as long as none of the two processes is in urn 0, we see that

$$\sigma_0 \in \{m_i : i \geq 2\} \cup \{l_i : i \geq 1\}, \tag{3.8}$$

i.e. the coupling always happens in urn 0, and it happens if either process (Z_n) jumps from 1 to 0 while (X_n) remains in 0 or vice versa.

Define for $i \geq 0$

$$V_i := \left|\{n \in \{m_i, \ldots, m_{i+1} - 1\} : X_n = 0\}\right|,$$

and

$$W_i := \left|\{n \in \{l_i, \ldots, l_{i+1} - 1\} : Z_n = 0\}\right|,$$

the number of visits in urn 0 of either of the process during one 'cycle' (note that between m_i and m_{i+1} the process (X_k) has exactly one jump of lenght N^β). By construction, $(V_i)_{i \geq 0}$ and $(W_i)_{i \geq 0}$ are independent sequences of iid geometric random variables with parameter ε, and

$$(Z_{m_i} - X_{m_i}) - (Z_{m_{i-1}} - X_{m_{i-1}}) = W_{i-1} - V_{i-1}, \tag{3.9}$$

$$(X_{l_i} - Z_{l_i}) - (X_{l_{i-1}} - Z_{l_{i-1}}) = V_{i-1} - W_{i-1}, \tag{3.10}$$

$i \geq 1$. Moreover we note that

$$m_{i+1} - m_i = V_i + N^\beta, \quad l_{i+1} - l_i = W_i + N^\beta. \tag{3.11}$$

The random sequence $(\sum_{i=0}^k (V_i - W_i))_{k \geq 0}$ is a random walk with centered increments whose variance (depending on ε but not on N) is finite. Moreover, σ_0 can be controlled by the first time this random walk exits the set $\{-N^\beta + 1, \ldots, N^\beta - 1\}$, since this event corresponds to either (Z_n) 'catching up' with (X_n), or vice versa. More precisely, defining

$$R := \inf\{k \geq 0 : |\sum_{i=0}^{k} (V_i - W_i)| \geq N^\beta\},$$

we see from (3.9) and (3.10) that

$$\sigma_0 \leq m_R. \tag{3.12}$$

Equation (3.12) implies that for any $\lambda > 0$,

$$\begin{aligned}\mathbb{P}\big(\sigma_0 > N^\lambda\big) &\le \mathbb{P}\big(\sum_{i=1}^{R}(m_i - m_{i-1}) > N^\lambda\big)\\ &= 1 - \mathbb{P}\big(\sum_{i=1}^{R}(m_i - m_{i-1}) \le N^\lambda\big)\\ &\le 1 - \mathbb{P}\big(\{R < \frac{1}{2}N^{\lambda-\beta}\} \cap \{m_i - m_{i-1} \le 2N^\beta \ \forall i = 1, \dots, \frac{1}{2}N^{\lambda-\beta}\}\big)\\ &\le \mathbb{P}\big(\{R > \frac{1}{2}N^{\lambda-\beta}\} \cup \{\exists 1 \le i \le \frac{1}{2}N^{\lambda-\beta} : m_i - m_{i-1} > 2N^\beta\}\big)\\ &\le \mathbb{P}\big(R > \frac{1}{2}N^{\lambda-\beta}\big) + \mathbb{P}\big(\exists 1 \le i \le N^{\lambda-\beta} : m_i - m_{i-1} > 2N^\beta\big).\end{aligned} \tag{3.13}$$

To control the first term on the right hand side, we use classical bounds on the exit time from an interval of symmetric random walks with finite variance, see e.g. Theorem 23.2 of [11]. This provides that for every $\delta' > 0$ there exists $\delta > 0$ such that

$$\mathbb{P}\big(R > N^{2\beta+\delta'}\big) \le e^{-N^\delta}. \tag{3.14}$$

For $\lambda > 3\beta$, we can choose $\delta' > 0$ such that $2\beta + \delta' < \lambda - \beta$, hence we find the bound

$$\mathbb{P}\big(R > \frac{1}{2}N^{\lambda-\beta}\big) \le e^{-N^\delta}. \tag{3.15}$$

To bound the second term in (3.13), by (3.11) and a union bound we find, for N large enough,

$$\begin{aligned}\mathbb{P}\big(\exists 1 \le i \le N^{\lambda-\beta} : m_i - m_{i-1} > 2N^\beta\big) &\le N^{\lambda-\beta}\mathbb{P}\big(V_1 > N^\beta\big)\\ &= N^{\lambda-\beta}(1-\varepsilon)^{N^\beta} \le N^{\lambda-\beta}e^{-\varepsilon N^\beta} \le e^{-N^{\beta/2}}.\end{aligned} \tag{3.16}$$

In view of (3.7), together the bounds (3.13), (3.15) and (3.16) prove (i). For (ii), recall (see e.g. [7], Chapter 4), that for any $k \ge l$ and $\mu \in \mathcal{P}_{N^\beta}$,

$$\|\mathbb{P}_\mu(X_k \in \cdot) - \nu_N\|_{TV} \le \|\mathbb{P}_\mu(X_l \in \cdot) - \nu_N\|_{TV}.$$

Using this, we see that for any $0 < \lambda < 1$,

$$\begin{aligned} &\|P_\mu(X_\tau \in \cdot) - \nu_N\|_{TV} \\ &\quad \le \|\mathbb{P}_\mu(X_{N^\lambda} \in \cdot) - \nu_N\|_{TV} \cdot \mathbb{P}(\tau \ge N^\lambda) + \|\mathbb{P}_\mu(X_0 \in \cdot) - \nu_N\|_{TV} \cdot \mathbb{P}(\tau < N^\lambda) \\ &\quad \le \|\mathbb{P}_\mu(X_{N^\lambda} \in \cdot) - \nu_N\|_{TV} + \mathbb{P}(\tau < N^\lambda). \end{aligned} \tag{3.17}$$

If $3\beta < \lambda < 1$, we can bound the first term using (i), and second term by choosing N large enough such that the Bernoulli inequality yields

$$\mathbb{P}(\tau < N^\lambda) = 1 - \big(1 - \frac{1}{N}\big)^{N^\lambda} \le N^{\lambda-1}. \tag{3.18}$$

Since we assumed $\beta < 1/4$, there exists $\delta > 0$ such that we can chose λ in a way that $0 < 3\beta < \lambda < 1 - \beta - \delta < 1$, so that

$$\mathbb{P}(\tau < N^\lambda) \le N^{\lambda-1} \le N^{-(\beta+\delta)}.$$

Plugging this together with (3.17) into (i) completes the proof. □

To put this result into a context, recall that the *mixing time* of a Markov chain (Y_n) with invariant distribution ν is often defined as

$$\tau_{mix} := \inf\big\{n > 0 : \sup_\mu \|P_\mu(Y_n = \cdot\,) - \nu\|_{TV} \le \frac{1}{4}\big\}$$

(cf. for example [7]). It thus follows immediately from the previous lemma that for the urn process of the seedbank model,

$$\tau_{mix} \le N^{3\beta+\delta} \tag{3.19}$$

for all $\delta > 0$. This observation allows us to justify that, if $\beta < 1/4$, each generation is visited with probability $\approx \varepsilon^{-1}N^{-\beta} \sim \nu_N(0)$. Indeed, if β is small enough, then for N large the urn process is close to stationarity before the first coalescence, or actually, as we will see in the next section, before the first *attempt* to coalesce. Note that Lemma 3.3 is crucial to achieve this, and to prove this lemma we needed $\beta < 1/4$.

4 Proof of the Main Results

We are now going to use the urn process from the previous section in order to prove our main results. The crucial step is to calculate the time until the coalescence of two lines in terms of this urn process.

Let $(X_k)_{k\geq 0}$ and $(Y_k)_{k\geq 0}$ be two independent copies of the urn process, corresponding in the sense of Lemma 3.1 to the ancestral lines of two individuals i and j. Let $\tau_0 = 0$ and let $\tau_k, k \geq 1$, be such that $(\tau_k - \tau_{k-1})_{k\geq 1}$ is a sequence of independent geometric random variables with parameter $1/N$, independent of X and Y. Let

$$\kappa := \inf\left\{k \geq 1 : X_{\tau_k} = Y_{\tau_k} = 0\right\}. \tag{4.1}$$

Lemma 4.1 *Assume that X, Y independently follow the urn dynamics started from initial distribution γ with support in $\{0, \ldots, N^\beta - 1\}$. Then for $S^{(1)}, S^{(2)}$ started from γ,*

$$T_{MRCA}(2) \stackrel{d}{=} \tau_\kappa. \tag{4.2}$$

Proof Let two sequences $(U_k^{(1)})_{k\geq 1}$ and $(U_k^{(2)})_{k\geq 1}$ of independent uniform random variables on $\{1, \ldots, N\}$, independent of S, X and Y be defined as in Sect. 2. Without loss of generality, due to independence, we can assume that $\tau_1 = \inf\left\{k \geq 0 : U_k^{(1)} = U_k^{(2)}\right\}$ and $\tau_k = \inf\left\{k > \tau_{k-1} : U_k^{(1)} = U_k^{(2)}\right\}, k \geq 2$. Hence

$$T_{MRCA}(2) = \inf\left\{\tau_k > 0 : \exists n, m \in \mathbb{N} : \tau_k = S_n^{(1)} = S_m^{(2)}\right\}.$$

Now the claim follows from Eq. (3.4) and the independence of the two lines. □

Before we turn to the proof of the main result, let us recall some well-known facts about Kingman's m–coalescent. Since it is *exchangeable* (cf. [10]), its law is determined by the *block-counting* process, which is the pure death process on $\{1, \ldots, m\}$ with death rate $\binom{m}{2}$. This will be used in our proof.

Proof of Theorem 2.2 Fix $0 < \beta < 1/4$ and $m \geq 2$. Clearly for all $N \in \mathbb{N}$ the process $(A_k^N)_{k\in\mathbb{N}}$ is exchangeable. In particular, its dynamics do not depend on the respective sizes or order of blocks, or their elements, but only on the number of blocks. Therefore it is sufficient to consider the block-counting process $(|A_k^N|)_{k\in\mathbb{N}_0}$, which is a pure death process started from a fixed $m \in \mathbb{N}$ whose distribution is uniquely determined by the sequence of times at which a coalescence happens, and the specification of the type of coalescence at each of these times. Hence, to prove our main result, it is sufficient to show that the inter-coalescence times of (A_k^N) are asymptotically (as $N \to \infty$) independent and exponentially distributed with the right parameters, and that multiple and simultaneous mergers are negligible.

In order to determine the distribution of the inter-coalescence times, we use Lemma 4.1, which tells us that the time until two given lines coalesce – ignoring the influence of the other lines – is distributed as τ_κ. Since the differences $(\tau_k - \tau_{k-1})$ are geometric, we only need to determine the distribution of κ. By Lemma 3.3 (ii) there exists $\delta > 0$ and r_N such that $|r_N| \leq N^{-(\beta+\delta)}$, and

$$\mathbb{P}_\gamma(X_{\tau_1} = 0) = \nu_N(0) + r_N.$$

Since $\nu_N(0) \sim \varepsilon^{-1}N^{-\beta}$ as $N \to \infty$, we have $\lim_{N\to\infty} r_N/\nu_N(0) = 0$ (note that ε is independent of N), and hence

$$\begin{aligned}\mathbb{P}_{\gamma\otimes\gamma}(\kappa = 1) &= \mathbb{P}_{\gamma\otimes\gamma}(X_{\tau_1} = Y_{\tau_1} = 0) = \mathbb{P}_\gamma(X_{\tau_1} = 0)\mathbb{P}_\gamma(Y_{\tau_1} = 0)\\ &= \nu_N(0)^2(1 + o(1)).\end{aligned} \tag{4.3}$$

Since the result of Lemma 3.3 is uniform in the initial distribution μ, we see that for all $k \in \mathbb{N}$, by the Markov property of (X_k) and (Y_k),

$$\mathbb{P}_{\gamma\otimes\gamma}(\kappa = k \,|\, \kappa > k-1) = \nu_N(0)^2(1 + o(1)), \tag{4.4}$$

with the $o(1)$–notation independent of k.

This shows that κ is asymptotically geometric distributed with parameter $\nu_N(0)^2$. More precisely, for any $c > 0$ and geometric random variables G_1 with parameter $\nu_N(0)^2(1+c)$ and G_2 with parameter $\nu_N(0)^2(1-c)$ there exists $N_0 \in \mathbb{N}$ such that for all $N \geq N_0$ we have that $G_2 \leq \kappa \leq G_1$ stochastically. Since

$$T^N_{MRCA}(2) \stackrel{d}{=} \tau_\kappa = \sum_{k=1}^{\kappa}(\tau_k - \tau_{k-1}), \tag{4.5}$$

we get

$$\sum_{i=1}^{G_2}(\tau_k - \tau_{k-1}) \stackrel{d}{\leq} T^N_{MRCA}(2) \stackrel{d}{\leq} \sum_{k=1}^{G_1}(\tau_k - \tau_{k-1}), \tag{4.6}$$

$\stackrel{d}{\leq}$ indicating stochastic dominance. An easy calculation using moment generating functions shows that a sum of a geometric number of geometric random variables is again geometrically distributed (with a parameter which is the product of the two parameters). Hence

$$\sum_{k=1}^{G_1}(\tau_k - \tau_{k-1}) \sim \mathrm{Geo}(N^{-1}\nu_N(0)^2(1+c)),$$

and similarly for the sum up to G_2. From this we conclude that for any $t > 0$,

$$\begin{aligned}&\lim_{N\to\infty} P_{\gamma\otimes\gamma}\left(T^N_{MRCA}(2) > \lfloor \varepsilon^2 N^{2\beta+1}\rfloor t\right) \geq \lim_{N\to\infty}\left(1 - \frac{1}{N}\nu_N(0)^2(1+c)\right)^{\lfloor \varepsilon^2 N^{2\beta+1}\rfloor t}\\ &= e^{-t(1+c)},\end{aligned} \tag{4.7}$$

and similarly

$$\lim_{N\to\infty} P_{\gamma\otimes\gamma}\big(T^N_{MRCA}(2) > \lfloor \varepsilon^2 N^{2\beta+1}\rfloor t\big) \le e^{-t(1-c)}. \tag{4.8}$$

Since $c > 0$ was arbitrary, we see that asymptotically as $N \to \infty$, the time until two lines coalesce converges in distribution to the time of coalescence in Kingman's 2-coalescent. Further, if we consider $m \geq 2$ lineages, the first coalescence is dominated by (resp. dominates) the minimum of $\binom{m}{2}$ independent geometric variables with parameters $N^{-1}\nu_N(0)(1 \pm c)$ each. Therefore, denoting by $T_{coal}(m)$ the time of the first coalescence event among m lines, we have in a similar fashion

$$\begin{aligned}\lim_{N\to\infty} P_{\otimes\gamma^m}\big(T_{coal}(m) > \lfloor \varepsilon^2 N^{2\beta+1}\rfloor t\big) &= \lim_{N\to\infty}\Big(1 - \frac{1}{N}\nu_N(0)^2\Big)^{\binom{m}{2}\cdot\lfloor \varepsilon^2 N^{2\beta+1}\rfloor t} \\ &= e^{-t\cdot\binom{m}{2}}.\end{aligned} \tag{4.9}$$

This shows that the time of the *first* coalescence is exponentially distributed with the right parameter, independent of the choice of initial distribution γ supported on $\{0,\ldots,N^\beta\}$. In order to see that the subsequent coalescence times are independent of $T_{coal}(m)$ and likewise exponentially distributed, we note that $(A^N_{T_{coal}(m)+k})_{k\in\mathbb{N}_0}$ has the law of $(A^N_k)_{k\in\mathbb{N}_0}$ started in $m' < m$ distinct blocks, corresponding to independent ancestral lineages $(S^{(i)}_n)$, $1 \le i \le m'$, started from some unknown initial distribution γ' supported on $\{0,\ldots,N^\beta\}$. Hence, by the same argument as before, *all* inter-coalescence times are asymptotically independent and converge to those of Kingman's coalescent. In order to see that there are no multiple or simultaneous (multiple) mergers, we note that due to Lemma 3.3 for any τ_i, the probability of a triple merger

$$P_\gamma(X_{\tau_i} = 1, Y_{\tau_i} = 1, Z_{\tau_i} = 1) = \nu_N(0)^3(1 + o(1)),$$

which is negligible compared to $\nu_N(0)^2$. Similarly we see that two simultaneous double mergers are negligible. Standard arguments using exchangeability [8, 9] imply that any multiple or simultaneous (multiple) mergers are negligible. This proves our main result. □

Proof of Proposition 2.4 First of all, observe that at each time step a coalescence happens with probability at most $1/N$, hence the time to the most recent common ancestor of two lines is a priori bounded by the time to the most recent common ancestor in the Wright-Fisher model, hence $E[T^N_{MRCA}(2)] \geq N$. Since clearly $T_{MRCA}(m) \geq T_{MRCA}(2)$ almost surely, we restrict ourselves to the case $m = 2$. Assume first that $\gamma = \delta_0$, which means that $X_0 = Y_0 = 0$. Let $J_0 := \inf\{k \geq 1 : X_k = N^\beta - 1\}$. By construction, $J_0 \sim \text{Geo}(\varepsilon)$. Therefore

$$P_0\big(J_0 < T^N_{MRCA}(2)\big) \geq \frac{\varepsilon}{\varepsilon + 1/N} \geq 1 - \frac{1}{\varepsilon N}, \tag{4.10}$$

noting that the time until the first coalescence while both processes are in 0 is geometric with parameter $1/N$. By construction, the dynamics of X consists of excursions away from 0, which have lenght N^β each, and after each excursion a period of length J_i in 0, with J_i iid Geo(ε). Hence coalescence can only happen in (random) time period of the form

$$\Big\{\sum_{i=0}^{l}(J_i + N^\beta), \ldots, \sum_{i=0}^{l}(J_i + N^\beta) + J_{l+1}\Big\}.$$

By the same reasoning as in the derivation of (4.10),

$$P_0\Big(T^N_{MRCA}(2) \geq \sum_{i=0}^{l}(J_i + N^\beta) + J_{l+1} \mid T^N_{MRCA}(2) > \sum_{i=0}^{l-1}(J_i + N^\beta) + J_l\Big) \geq 1 - \frac{1}{\varepsilon N}. \tag{4.11}$$

Hence the number of excursions away from 0 of one particle before coalescence dominates a geometric random variable with parameter $1/(\varepsilon N)$, and the expected number of excursions away from 0 is at least εN. Since each excursion has length at least N^β, we get

$$E_0[T^N_{MRCA}(2)] \geq \varepsilon N^{1+\beta} \vee N. \tag{4.12}$$

Obviously our argument was independent of the initial distribution γ, and hence (i) is proven.

Let now $\beta < 1/3$. Let $\tau_0 = 0$, and $\tau_k, k \in \mathbb{N}$, be such that $(\tau_k - \tau_{k-1}), k \in \mathbb{N}$ are independent geometric random variables with parameter $1/N$. Due to Lemma 4.1, $T_{MRCA}(2) \geq \tau_1$ in distribution (meaning that $P_{\gamma\otimes\gamma}(T_{MRCA}(2) \geq x) \geq P_{\gamma\otimes\gamma}(\tau_1 \geq x)$ for all $x \in \mathbb{R}$), and $\tau_1 \sim$ Geo$(1/N)$. Hence by (3.18), there exists $\delta' > 0$ such that

$$P_{\gamma\otimes\gamma}(T^N_{MRCA}(2) < N^{3\beta+\delta'}) \to 0$$

as $N \to \infty$. By Lemma 3.3, the Markov property, and Lemma 4.1, we get for all $\delta > 0$

$$\begin{aligned} P_{\gamma\otimes\gamma}\big(T^N_{MRCA}(2) \leq \varepsilon^2 N^{1+2\beta-\delta} \mid T^N_{MRCA}(2) > N^{3\beta+\delta'}\big) \\ \leq & P_{\nu_N\otimes\nu_N}\big(T^N_{MRCA}(2) \leq \varepsilon^2 N^{2\beta+1-\delta} - N^{3\beta+\delta'}\big) + o(1) \\ \leq & 1 - \Big(1 - \frac{1}{N}\cdot\nu_N(0)^2\Big)^{\varepsilon^2 N^{2\beta+1-\delta} - N^{3\beta+\delta'}} + o(1) \\ \to & 0, \end{aligned} \tag{4.13}$$

since

$$\frac{1}{N} \cdot \nu_N(0)^2 \sim \frac{1}{\varepsilon^2 N^{2\beta+1}} = o(\varepsilon^2 N^{2\beta+1-\delta} - N^{3\beta+\delta'})$$

as $N \to \infty$, and this finishes the proof. □

5 Simulations

Since our main results, Theorem 2.2 and Corollary 2.3, are valid only for $0 < \beta < 1/4$, it is natural to investigate the coalescence time $T^N_{MRCA}(2)$ defined in (2.8) for $\beta \geq 1/4$ via simulations. The results in Fig. 3, which show estimates (histograms) of the distribution of the scaled time $T^N_{MRCA}(2)/\left(\varepsilon^2 N^{1+2\beta}\right)$ when $\beta = 1/3$, suggest that $\varepsilon^2 N^{1+2\beta}$ is indeed still the correct scaling. The distribution of $T^N_{MRCA}(2)$ fits an exponential distribution with corresponding mean $\overline{T}$ quite well in all cases. The mean $\overline{T}$ should be 1 in all cases. We see that this is not the case for small ε, but the fit gets better as ε increases. This can be explained by noting that for small ε,

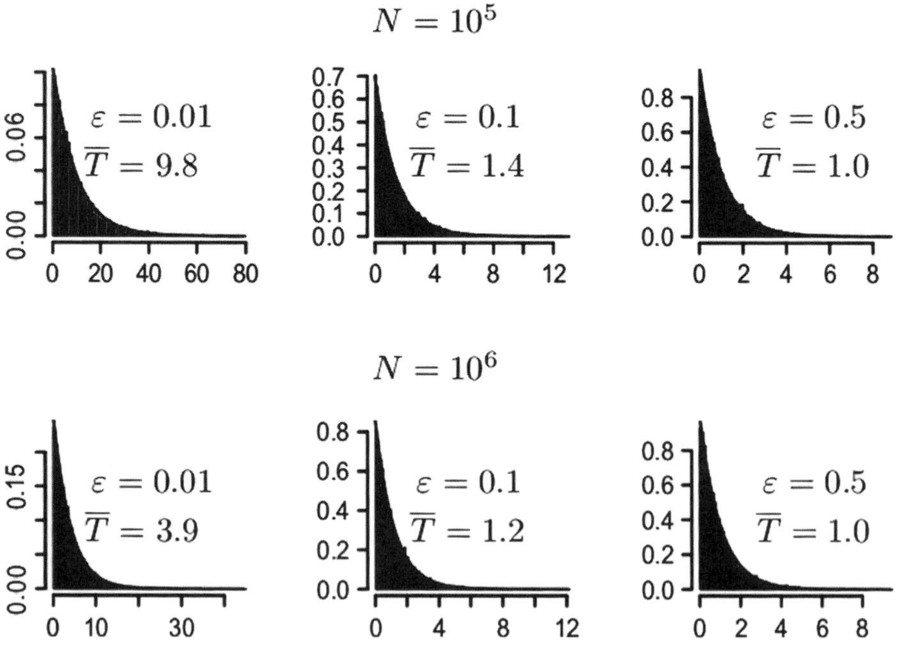

Fig. 3 Histograms of $T^N_{MRCA}(2)/\left(\varepsilon^2 N^{1+2\beta}\right)$ when $\beta = 1/3$, and ε, N as shown. The *solid fitted line* is the density of the exponential with mean the corresponding mean $(\overline{T})$ of the simulated datapoints. Each histogram is normalised to have unit mass one

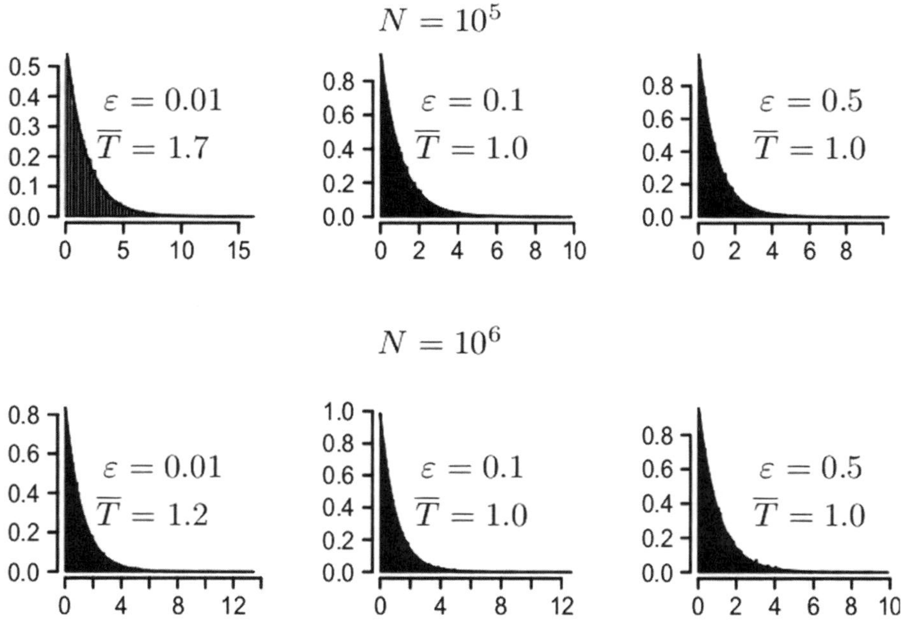

Fig. 4 Histograms of $T^N_{MRCA}(2)/\left(\varepsilon^2 N^{1+2\beta}\right)$ when $\beta = 1/2$, and ε, N as shown. The *solid fitted line* is the density of the exponential with mean the corresponding mean $(\overline{T})$ of the simulated datapoints. Each histogram is normalised to have unit mass one

the scales ε^{-2} and $N^{1+2\beta}$ may not differ very much, illustrating the fact that the asymptotic results hold only if ε is independent of N. Similar results and conclusions hold for $\beta = 1/2$ (Fig. 4). The estimate of the standard error is close to the mean in all cases. The simulation results for $\beta = 1/2$ are particularly interesting since the long ancestral jumps ($\lfloor N^\beta \rfloor$) are 10^3 generations when $N = 10^6$, which can be considered a significant extension of the ancestry of the lines. C code for the simulations is available at http://page.math.tu-berlin.de/~eldon/programs.html.

Acknowledgements JB, BE and NK acknowledge support by the DFG SPP 1590 "Probabilistic structures in evolution". AGC is supported by the DFG RTG 1845, the Berlin Mathematical School (BMS), and the Mexican Council of Science in collaboration with the German Academic Exchange Service (DAAD). The authors wish to thank Julien Berestycki and Dario Spanò for interesting discussions.

References

1. N. Berestycki, *Recent Progress in Coalescent Theory*. Ensaios Matemáticos, vol. 16 (Sociedade brasileira de matemática, Rio de Janeiro, 2009)
2. J. Blath, A. González Casanova, N. Kurt, D. Spanò, The ancestral process of long-range seedbank models. J. Appl. Probab. **50**(3), 741–759 (2013)

3. J. Blath, A. González Casanova, N. Kurt, M. Wilke Berenguer, A new coalescent for seed bank models. Ann. Appl. Probab. (Accepted for publication)
4. A. González Casanova, E. Aguirre-von Wobeser, G. Espín, L. Servín-González, N. Kurt, D. Spanò, J. Blath, G. Soberón-Chávez, Strong seed-bank effects in bacterial evolution. J. Theor. Biol. **356**, 62–70 (2014)
5. I. Kaj, S. Krone, M. Lascoux, Coalescent theory for seed bank models. J. Appl. Probab. **38**, 285–300 (2001)
6. D.A. Levin, The seed bank as a source of genetic novelty in plants. Am. Nat. **135**, 563–572 (1990)
7. D.A. Levin, Y. Peres, E.L. Wilmer, *Markov Chains and Mixing Times* (AMS, Providence, 2009)
8. M. Möhle, A convergence theorem for Markov chains arising in population genetics and the coalescent with selfing. Adv. Appl. Probab. **30**, 493–512 (1998)
9. M. Möhle, S. Sagitov, A classification of coalescent processes for haploid exchangeable population models. Ann. Probab. **29**(4), 1547–1562 (2001)
10. J. Pitman, *Combinatorial Stochastic Processes*. Lecture Notes in Mathematics, vol. 1875 (Springer, New York, 2002)
11. F. Spitzer, *Principles of Random Walk*, 2nd edn. (Springer, New York, 1976)
12. A. Tellier, S.J.Y. Laurent, H. Lainer, P. Pavlidis, W. Stephan, Inference of seed bank parameters in two wild tomato species using ecological and genetic data. PNAS **108**(41), 17052–17057 (2011)
13. R. Vitalis, S. Glémin, I. Oliviere, When genes got to sleep: the population genetic consequences of seed dormacy and monocarpic perenniality. Am. Nat. **163**(2), 295–311 (2004)
14. J. Wakeley, *Coalescent Theory: An Introduction* (Roberts and Company Publishers, Greenwood Village, Colorado, 2009)

Shifting Processes with Cyclically Exchangeable Increments at Random

Loïc Chaumont and Gerónimo Uribe Bravo

Abstract We propose a path transformation which applied to a cyclically exchangeable increment process conditions its minimum to belong to a given interval.

This path transformation is then applied to processes with start and end at 0. It is seen that, under simple conditions, the weak limit as $\varepsilon \to 0$ of the process conditioned on remaining above $-\varepsilon$ exists and has the law of the Vervaat transformation of the process.

We examine the consequences of this path transformation on processes with exchangeable increments, Lévy bridges, and the Brownian bridge.

Keywords Cyclic exchangeability • Vervaat transformation • Brownian bridge • Three dimensional Bessel bridge • Uniform law • Path transformation • Occupation time

Mathematics Subject Classification (2010). 60G09, 60F17, 60G17, 60J65.

Research supported by UNAM-DGAPA-PAPIIT grant no. IA101014.

L. Chaumont (✉)
LAREMA, Département de Mathématiques, Université d'Angers, 2, Bd Lavoisier-49045, Angers Cedex 01, France
e-mail: loic.chaumont@univ-angers.fr

G. Uribe Bravo
Instituto de Matemáticas, Universidad Nacional Autónoma de México, Área de la Investigación Científica, Ciudad Universitaria 04510, Ciudad de México, México
e-mail: geronimo@matem.unam.mx

R.H. Mena et al. (eds.), *XI Symposium on Probability and Stochastic Processes*, Progress in Probability 69, DOI 10.1007/978-3-319-13984-5_5

1 Introduction

In this paper, we use symmetries of the law of a stochastic process, which appear through its invariance under a group of transformations, to construct a version of the process conditioned on certain events. The objects of interest will be processes with cyclically exchangeable increments. These processes, denoted by $X = (X_t, t \in [0, 1])$, are defined by having a law which is invariant under the cyclic shift $\theta_t X$ that interchanges the pre and post-t part of the process X, preserving the same values at times 0 and 1. The precise definition of the shift θ_t is found in Eq. (2.1). The kind of transformations we will be concerned with are of the type $\theta_\nu X$ where ν is a random variable. Informally, we will chose ν to be uniform on the set of indices t such that the minimum of $\theta_t X$ belongs to a given interval I. Our conclusion, stated in Theorem 2.2, will be that $\theta_\nu X$ has the same law as X conditioned on its minimum belonging to the interval I, and that ν is uniform on $[0, 1]$ and independent of $\theta_\nu X$.

Our main motivation in performing such a construction is to show that classical results regarding the normalized Brownian excursion are in fact a direct consequence of the cyclical exchangeability property of the increments of the Brownian bridge. Indeed, Durrett, Iglehart and Miller proved in [11] that the law of the normalized Brownian excursion can be obtained as the weak limit of the standard Brownian bridge conditioned to stay above $\varepsilon > 0$, as $\varepsilon \to 0$. We will obtain this result by applying the above random shift when the interval I shrinks to a point. A pathwise relationship was then found by Vervaat in [28] who proved in his famous transformation that the path of the normalized Brownian excursion can be constructed by inverting the pre-minimum part and the post-minimum part of the standard Brownian bridge. Then in [4], Biane noticed that the latter process is independent of the position of the minimum of the initial Brownian bridge, and that this minimum time is uniformly distributed. He derived from this result and Vervaat transformation a path construction of the Brownian bridge from the normalized Brownian excursion. We will refer to both transformations as the Vervaat-Biane transformation. The above mentioned results can be stated as follows.

Theorem 1.1 ([28] and [4]) *Let X be the standard Brownian bridge. Then, the law of X conditioned to remain above $-\varepsilon$ converges weakly as $\varepsilon \to 0$ toward the law of the normalized Brownian excursion. If ρ is the unique instant at which X attains its minimum, then the process $\theta_\rho(X)_t$ has the law of a normalized Brownian excursion. Conversely, if Y is a normalized Brownian excursion and if U is a uniformly distributed random variable, independent of Y then the process $\theta_U(Y)_t$ has the law of a standard Brownian bridge.*

The aim of this paper is to show that Vervaat-Biane transformation is actually a direct consequence of the cyclical exchangeability property of the increments of the Brownian bridge. Therefore the same type of transformation can be obtained between any process with cyclically exchangeable increments and its version conditioned to stay positive, provided its minimum is attained at a unique instant. This will be done in Sect. 3 where we will also study the example of processes with exchangeable increments. Then in Sect. 4, we prove some refinements of Vervaat-Biane transformation for the Brownian bridge conditioned by its minimum value. Section 2, is devoted to the main theorem of this paper which provides the essential argument from which most of the other results will be derived.

Inverting Vervaat's path transformation, first considered by Biane in [4], leads naturally to relationships for the normalized Brownian excursion, sampled at an independent uniform time, and the Brownian bridge. Developments around this are found in [2, 22] and [24].

Other examples of Vervaat type transformations, almost always connected to Lévy processes, are found in [5–7, 12, 18, 19] and [27].

2 Conditioning the Minimum of a Process with Cyclically Exchangeable Increments

We now turn to our main theorem in the context of cyclically exchangeable increment processes.

We use the canonical setup: let **D** stand for the Skorohod space of càdlàg functions $f : [0,1] \to \mathbb{R}$ on which the canonical process $X = (X_t, t \in [0,1])$ is defined. Recall that $X_t : \mathbf{D} \to \mathbb{R}$ is given by

$$X_t(f) = f(t)\,.$$

Then, **D** is equipped with the σ- field $\sigma(X_t, t \in [0,1])$. Denote by $\{t\}$ and $\lfloor t \rfloor$ the fractional part and the lower integer part of t, respectively, and introduce the shift θ_u by means of

$$\theta_u f(t) = f(\{t+u\}) - f(u) + f(\lfloor t+u \rfloor)\,. \tag{2.1}$$

The transformation θ_u consists in inverting the paths $\{f(t), 0 \le t \le u\}$ and $\{f(t), u \le t \le 1\}$ in such a way that the new path $\theta_u(f)$ has the same values as f at times 0 and 1, i.e. $\theta_u f(0) = f(0)$ and $\theta_u f(1) = f(1)$. We call θ_u the **shift** at time u of X over the interval [0,1]. Note that we will always use the transformation θ_u with $f(0) = 0$ (Fig. 1).

Definition 2.1 (CEI process) A càdlàg stochastic process has **cyclically exchangeable increments (CEI)** if its law satisfies the following identities in law:

$$\theta_u X \overset{(d)}{=} X \text{ for every } u \in [0,1].$$

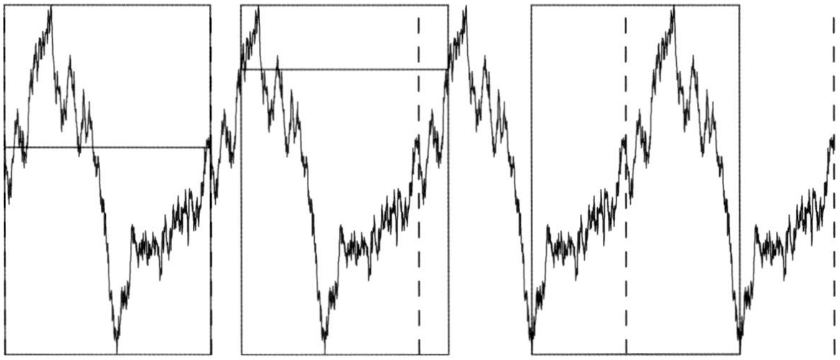

Fig. 1 Repeated trajectory of a Brownian bridge. The first frame shows the original trajectory. The second shows its shift at $u = .14634$. The third frame shows the shift at the location of the unique minimum, illustrating the Vervaat transformation

The overall minimum $\underline{X}$, which can be defined now as a functional on the Skorohod space, is given by

$$\underline{X} = \inf_{0 \le t \le 1} X_t.$$

Intuitively, to condition X on having a minimum on a given interval $I \subset (-\infty, 0]$, we choose t uniformly on the set in which $\underline{X} \circ \theta_t \in I$ by using the occupation time process

$$A_t^I = \int_0^t \mathbf{1}_{\{\underline{X} \circ \theta_s \in I\}} \, ds.$$

Here is the main result. It provides a way to construct CEI processes conditioned on their overall minimum.

Theorem 2.2 *Let $(X, \mathbb{P})$ be any non trivial CEI process such that $X_0 = 0$, $X_1 \ge 0$ and $\mathbb{P}(\underline{X} \in I) > 0$. Let U be an independent random time which is uniformly distributed over $[0, 1]$ and define:*

$$\nu = \inf\{t : A_t^I = U A_1^I\}. \tag{2.2}$$

Conditionally on $A_1^I > 0$, the process $\theta_\nu(X)$ is independent of ν and has the same law as X conditionally on $\underline{X} \in I$. Moreover the time ν is uniformly distributed over $[0, 1]$.

Conversely, if Y has the law of X conditioned on $\underline{X} \in I$ and U is uniform and independent of Y then $\theta_U(Y)$ has the same law as X conditioned on $A_1^I > 0$.

Remark 2.3 When $X_1 = 0$, the set $\{A_1^I > 0\}$ can be written in terms of the amplitude $H = \overline{X} - \underline{X}$ (where $\overline{X} = \sup_{t\in[0,1]} X_t$) as $\{H \geq -\inf I\}$.

Proof of Theorem 2.2 We first note that the law of $X\circ\theta_U$ conditionally on $\underline{X}\circ\theta_U \in I$ is equal to the law of X conditionally on $\underline{X} \in I$. Indeed, using the CEI property:

$$\begin{aligned}\mathbb{E}\big(f(U)\,F(\theta_U X)\,\mathbf{1}_{\{\underline{X}\circ\theta_U\in I\}}\big) &= \int_0^1 f(u)\,\mathbb{E}\big(F(\theta_u X)\,\mathbf{1}_{\{\underline{X}\circ\theta_u\in I\}}\big)\,du \\ &= \mathbb{E}\big(F\mathbf{1}_{\{\underline{X}\in I\}}\big)\int_0^1 f(u)\,du.\end{aligned}$$

Additionally, we conclude that the random variable U is uniform on $(0,1)$ and independent of $X\circ\theta_U$ conditionally on $\underline{X}\circ\theta_U \in I$.

Write U in the following way:

$$U = \inf\left\{t : A_t^I = \frac{A_U^I}{A_1^I}A_1^I\right\}\,. \tag{2.3}$$

Then it suffices to prove that conditionally on $\underline{X}\circ\theta_U \in I$, the random variable A_U^I/A_1^I is uniformly distributed over [0,1] and independent of X. Indeed from the conditional independence and (2.3), we deduce that conditionally on $\underline{X}\circ\theta_U \in I$, the law of $(\theta_U(X), U)$ is the same as that of $(\theta_\nu(X)\,,\,\nu)$.

Let F be any positive, measurable functional defined on D and f be any positive Borel function. From the change of variable $s = A_t^I/A_1^I$, we obtain

$$\begin{aligned}&\mathbb{E}\big(F(X)f(A_U^I/A_1^I)\mathbf{1}_{\{I\}}(\underline{X}\circ\theta_U)\big)\\ &= \mathbb{E}\left(\int_0^1 f(A_t^I/A_1^I)F(X)\mathbf{1}_{\{I\}}(\underline{X}\circ\theta_t)\;dt\right)\\ &= \mathbb{E}\left(\int_0^1 f(A_t^I/A_1^I)F(X)\,dA_t^I\right)\\ &= \mathbb{E}\big(F(X)A_1^I\big)\int_0^1 f(t)\,dt\,,\end{aligned}$$

which proves the conditional independence mentioned above.

The converse assertion is immediate using the independence of $\theta_\nu X$ and ν and the fact that the latter is uniform. □

We will now apply Sect. 2 to particular situations to get diverse generalizations of Theorem 2.2.

3 Exchangeable Increment Processes and the Vervaat Transformation

In Sect. 2 we shifted paths at random using θ_η to condition a given CEI process to have a minimum in a given interval I. When $I = (-\varepsilon, 0]$ and $X_1 = 0$, and under a simple technical condition, we now see that the limiting transformation of θ_η as $\varepsilon \to 0$ is the Vervaat transformation. Hence, we obtain an extension of Theorem 1.1.

Corollary 3.1 *Let $(X, \mathbb{P})$ be any non trivial CEI process such that $X_0 = 0 = X_1$. Assume that there exists a unique $\rho \in (0, 1)$ such that $X_\rho = \underline{X}$ and that $X_{\rho-} = X_\rho$. Then:*

1. *The law of X conditioned to remain above $-\varepsilon$ converges weakly in the Skorohod J_1 topology as $\varepsilon \to 0$. Furthermore, the weak limit is the law of $\theta_\rho X$.*
2. *Conversely, let Y be a process with the same law as $\theta_\rho X$ and let U be uniform on $(0, 1)$ and independent of Y. Then the process $\theta_U Y$ has the same law as X. In particular, ρ is uniformly distributed on $(0, 1)$.*

Note that we assume that the infimum of the process X is achieved at ρ. Actually if the infimum is only achieved as a limit (from the left) at ρ and $X_{\rho-} < X_\rho$ then the transformation θ_ν converges, as $\varepsilon \to 0$ pointwise to a process θ_ρ which satisfies $\theta_\rho(0) = 0$, $\theta_\rho(0+) = X_\rho - X_{\rho-}$. Hence, convergence cannot take place in the Skorohod space. A similar fact happens when $X_{\rho-} > X_\rho$. After the proof, we shall examine an example of applicability of Corollary 3.1 to exchangeable increment processes.

Proof We use the notation of Theorem 2.2. Recall the definition of ν, given in Eq. (2.2) of Theorem 2.2. Intuitively, $\nu = \nu(\varepsilon)$ is a uniform point on the set

$$\{t : X_t - \underline{X}_t < \varepsilon\}.$$

The uniqueness of the minimum implies that $\nu \to \rho$ as $\varepsilon \to 0$. Since X is continuous at ρ, by assumption, for any $\gamma > 0$ we can find $\delta > 0$ such that $|X_\rho - X_s| < \gamma$ if $s \in [\rho - \delta, \rho + \delta]$. On $[0, \rho - \delta]$ and $[\rho + \delta, 1]$, we use the càdlàg character of X to construct partitions $0 = t_0^1 < \cdots < t_{n_1}^1 = \rho - \delta$ and $\rho + \delta = t_0^2 < \cdots < t_{n_2}^2 = 1$ such that

$$|X_s - X_t| < \gamma \quad \text{if} \quad s, t \in [t_{j-1}^i, t_j^i) \text{ for } j \leq n_i.$$

We use these partitions to construct the piecewise linear increasing homeomorphism $\lambda : [0, 1] \to [0, 1]$ which satisfies $\|\theta_\nu \circ \lambda - \theta_\rho\|_{[0,1]} \leq \gamma$. Indeed, construct λ which scales the interval $[0, t_1^2 - \nu]$ to $[0, t_1^2 - \rho]$, shifts every interval $[t_{i-1}^2 - \nu, t_i^2 - \nu]$ to $[t_{i-1}^2 - \rho, t_i^2 - \rho]$ for $i \leq n_2$, also shifts $[1 - \nu + t_{i-1}^1, 1 - \nu + t_i^1]$ to $[1 - \rho + t_{i-1}^1, 1 - \rho + t_i^1]$, and finally scales $[1 - \nu + (\rho - \delta), 1]$ to $[1 - \delta, 1]$. Note that by choosing ν close enough to ρ, which amounts to choosing ε small enough, we can make $\|\lambda - \mathrm{Id}\,\|_{[0,1]} \leq \gamma$. Hence, $\theta_\nu \to \theta_\rho$ in the Skorohod J_1 topology as $\varepsilon \to 0$.

Also, from Theorem 2.2, we know that ν is uniform on $(0, 1)$ and independent of $\theta_\nu X$. Taking weak limits, we deduce that ρ is uniform and independent of $\theta_\rho X$, which finishes the proof. □

Our main example of the applicability of Corollary 3.1 is to exchangeable increment processes.

Definition 3.2 A càdlàg stochastic process has **exchangeable increments (EI)** if its law satisfies that for every $n \geq 1$, the random variables

$$X_{k/n} - X_{(k-1)/n}, 1 \leq k \leq n$$

are exchangeable.

Note that an EI process is also a CEI process.

According to [14], an EI process has the following canonical representation:

$$X_t = \alpha t + \sigma b_t + \sum_i \beta_i \left[\mathbf{1}_{\{U_i \leq t\}} - t \right]$$

where

1. α, σ and $\beta_i, i \geq 1$ are (possibly dependent) random variables such that $\sum_i \beta_i^2 < \infty$ almost surely.
2. b is a Brownian bridge
3. $(U_i, i \geq 1)$ are iid uniform random variables on $(0, 1)$.

Furthermore, the three groups of random variables are independent and the sum defining X_t converges uniformly in L_2 in the sense that

$$\lim_{m\to\infty} \sup_{n\geq m} \mathbb{E}\left(\sup_{t\in[0,1]} \left[\sum_{i=m+1}^{n} \beta_i^2 \left[\mathbf{1}_{\{U_i \leq t\}} - t \right]^2 \right] \right) = 0.$$

The above representation is called the canonical representation of X and the triple (α, β, σ) are its canonical parameters.

Our main example follows from the following result:

Proposition 3.3 *Let X be an EI process with canonical parameters (α, β, σ). On the set*

$$\left\{ \sum_i \beta_i^2 \left| \log |\beta_i| \right|^c < \infty \textit{ for some } c > 1 \textit{ or } \sigma \neq 0 \right\},$$

X reaches its minimum continuously at a unique $\rho \in (0, 1)$.

We need some preliminaries to prove Proposition 3.3. First, a criterion to decide whether X has infinite or finite variation in the case there is no Brownian component.

Proposition 3.4 *Let X be an EI process with canonical parameters $(\alpha, \beta, 0)$. Then, the sets*

$$\{X \text{ has infinite variation on any subinterval of } [0, 1]\}$$

and

$$\left\{\sum_i |\beta_i| = \infty\right\}$$

coincide almost surely. If $\sum_i |\beta_i| < \infty$ then X_t/t has a limit as $t \to 0$.

It is known that for finite-variation Lévy processes, X_t/t converges to the drift of X as $t \to 0$ as shown in [26].

Proof We work conditionally on α and (β_i); assume then that the canonical parameters are deterministic. If $\sum_i |\beta_i| < \infty$, we can define the following two increasing processes

$$X_t^p = \alpha^+ t + \sum_{i:\beta_i > 0} \beta_i \mathbf{1}_{\{U_i \le t\}} \quad \text{and} \quad X_t^n = \alpha^- t + \sum_{i:\beta_i < 0} -\beta_i \mathbf{1}_{\{U_i \le t\}}$$

and note that $X = X^p - X^n$. Hence X has bounded variation on $[0, 1]$ almost surely.

On the other hand, if $\sum_i |\beta_i| = \infty$ we first assert that the set

$$A_{k,n} = \left\{\sum_i |\beta_i| \, \mathbf{1}_{\{k/n \le U_i \le (k+1)/n\}} = \infty\right\}$$

has probability 1 for any $n \ge 1$ and any $k \in \{0, \dots, n\}$. Note that for fixed n, $\cup_{0 \le k \le n-1} A_{k,n} = \Omega$. Also, $\mathbb{P}(A_{k_1,n}) = \mathbb{P}(A_{k_2,n})$ since the U_i are uniform. Finally, note that $A_{k,n}$ belongs to the tail σ-field of the sequence of random variables (U_i). Hence, $\mathbb{P}(A_{k_1,n}) = 1$ by the Kolmogorov 0-1 law. Since

$$\sum_i |\beta_i| \, \mathbf{1}_{\{a \le U_i \le b\}} = \sum_{t:\Delta X_t \ne 0} |\Delta X_t| \, \mathbf{1}_{\{a \le t \le b\}}$$

and the sum of jumps of a càdlàg function is a lower bound for the variation, we see that X has infinite variation on any subinterval of $[0, 1]$.

Recall that $X_t \in L_2$ (since we assumed that the canonical parameters are constant). Using the EI property, it is easy to see that

$$\mathbb{E}(X_s \,|\, X_t, t \ge s) = \frac{s}{t} X_t.$$

Hence the process $M = (M_t, t \in [0, 1))$ given by $M_t = X_{1-t}/(1-t)$ is a martingale. If $\sum_i |\beta_i| < \infty$ then

$$X_t = \alpha t + \sum_i \beta_i \mathbf{1}_{\{U_i \le t\}} - t \sum_i \beta_i$$

so that $\mathbb{E}(|M_t|) \le |\alpha| + 2\sum_i |\beta_i|$. Hence, M is bounded in L_1 as $t \to 1$ and so it converges almost surely. □

Secondly, we give a version of a result originally found in [23] for Lévy processes.

Proposition 3.5 *The set*

$$\left\{ \sum_i |\beta_i| = \infty, \sum_i \beta_i^2 \, |\log |\beta_i||^{\,c} < \infty \, \textit{for some } c > 1 \textit{ or } \sigma \ne 0 \right\}$$

is almost surely contained in

$$\left\{ \limsup_{t\to\infty} \frac{X_t}{t} = \infty \textit{ and } \liminf_{t\to\infty} \frac{X_t}{t} = -\infty \right\}.$$

Proof By conditioning on the canonical parameters, we will assume they are constant.

If $\sigma \ne 0$, let $f : [0, 1] \to \mathbb{R}$ be a function such that $\sqrt{t} = o(f(t))$ and $f(t) = o\Big((t \log\log 1/t)^{1/2}\Big)$ as $t \to 0$. Then, since the law of the Brownian bridge is equivalent to the law of B on any interval $[0, t]$ for $t < 1$, the law of the iterated logarithm implies that $\limsup_{t\to 0} b_t/f(t) = \infty$ and $\liminf_{t\to 0} b_t/f(t) = -\infty$. On the other hand, if $Y = X - \sigma b$, then Y is an EI process with canonical parameters $(\alpha, \beta, 0)$ independent of b. Note that $\mathbb{E}\big(Y_t^2\big) \sim t\sum_i \beta_i^2$ as $t \to 0$ to see that $Y_t/f(t) \to 0$ in L_2 as $t \to 0$. If t_n is a (random and b-measurable) sequence decreasing to zero such that $b_{t_n}/f(t_n)$ goes to ∞, we can use the independence of Y and b to construct a subsequence s_n converging to zero such that $b_{s_n}/f(s_n) \to \infty$ and $Y_{s_n}/f(s_n) \to 0$. We conclude that $X_{t_n}/t_n \to \infty$ and so $\limsup_{t\to 0} X_t/t = \infty$. The same argument applies for the lower limit.

Let us now assume that $\sigma = 0$. If $\sum_i |\beta_i| = \infty$ then X necessarily has infinite variation on any subinterval of $[0, 1]$. If furthermore $\sum_i \beta_i^2 \, |\log |\beta_i||^{\,c} < \infty$ then Theorem 1.1 of [15] allows us to write X as $Y + Z$ where Z is of finite variation process with exchangeable increments and Y is a Lévy process. Since Y has infinite variation, then $\liminf_{t\to 0} Y_t/t = -\infty$ and $\limsup_{t\to 0} Y_t/t = \infty$ thanks to [23]. Finally, since $\lim_{t\to 0} Z_t/t$ exists in $\mathbb{R}$ by Proposition 3.4 since Z is a finite variation EI process, then $\liminf_{t\to 0} X_t/t = -\infty$ and $\limsup_{t\to 0} X_t/t = \infty$. □

Proof of Proposition 3.3 Since $\liminf_{t\to 0+} X_t/t = -\infty$, we see that $\rho > 0$. Using the exchangeability of the increments, we conclude from Proposition 3.5 that at any deterministic $t \geq 0$ we have that $\limsup_{h\to 0+}(X_{t+h} - X_t)/h = \infty$ and $\liminf_{h\to 0+}(X_{t+h} - X_t)/h = -\infty$ almost surely. If we write $X_t = \beta_i \left[\mathbf{1}_{\{U_i \leq t\}} - t\right] + X'_t$ and use the independence between U_i and X', we conclude that at any jump time U_i we have: $\limsup_{h\to 0+}(X_{U_i+h} - X_{U_i})/h = \infty$ and $\liminf_{h\to 0+}(X_{U_i+h} - X_t)/h = -\infty$ almost surely. We conclude from this that X cannot jump into its minimum. By applying the preceeding argument to $\left(X_1 - X_{(1-t)-}, t \in [0,1]\right)$, which is also EI with the same canonical parameters, we see that $\rho < 1$ and that X cannot jump into its minimum either. □

In contrast to the case of EI processes where we have only stated sufficient conditions for the achievement of the minimum, necessary and sufficient conditions are known for Lévy processes. Indeed, Theorem 3.1 of [20] tells us that if X is a Lévy process such that neither X nor $-X$ is a subordinator, then X achieves its minimum continuously if and only if 0 is regular for $(0,\infty)$ and $(-\infty,0)$. This happens always when X has infinite variation. In the finite-variation case, regularity of 0 for $(0,\infty)$ can be established through Rogozin's criterion: 0 is regular for $(-\infty,0)$ if and only if $\int_{0+} \mathbb{P}(X_t < 0)\,/t\,dt = \infty$. A criterion in terms of the characteristic triple of the Lévy process is available in [1]. We will therefore assume

H1: 0 is regular for $(-\infty,0)$ and $(0,\infty)$.

We now proceed then to give a statement of a Vervaat type transformation for Lévy processes, although actually we will use their bridges in order to force them to end at zero. Lévy bridges were first constructed in [16] (using the convergence criteria for processes with exchangeable increments of [14]) and then in [8] (via Markovian considerations) under the following hypothesis:

H2: For any $t > 0$, $\int \left|\mathbb{E}\left(e^{iuX_t}\right)\right| \, du < \infty$.

Under **H2**, the law of X_t is absolutely continuous with a continuous and bounded density f_t. Hence, X admits transition densities p_t given by $p_t(x,y) = f_t(y-x)$. If we additionally assume **H1** then the transition densities are everywhere positive as shown in [25].

Definition 3.6 The Lévy bridge from 0 to 0 of length 1 is the càdlàg process whose law $\mathbb{P}^1_{0,0}$ is determined by the local absolute continuity relationship: for every $A \in \mathscr{F}_s$

$$\mathbb{P}^1_{0,0}(A) = \mathbb{E}\left(\mathbf{1}_{\{A\}} \frac{p_{1-s}(X_s,0)}{p_t(0,0)}\right).$$

See [13, 16] or [8] for an interpretation of the above law as that of X conditioned on $X_t = 0$. Using time reversibility for Lévy processes, it is easy to see that the image of $\mathbb{P}^1_{0,0}$ under the time reversal map $\left(X_{(1-t)-}, t \in [0,1]\right)$ is the bridge of $-X$ from 0 to 0 of length 1 and that $X_1 = X_{1-} = 0$ under $\mathbb{P}^1_{0,0}$.

Proposition 3.7 *Under hypotheses **H1** and **H2**, the law $\mathbb{P}^1_{0,0}$ has the EI property. Under $\mathbb{P}^1_{0,0}$, the minimum is achieved at a unique place $\rho \in (0,1)$ and X is continuous at ρ.*

We conclude that Corollary 3.1 applies under $\mathbb{P}^1_{0,0}$. At this level of generality, this has been proved in [27]. In that work, the distribution of the image of $\mathbb{P}^1_{0,0}$ under the Vervaat transformation was identified with the (Markovian) bridge associated to the Lévy process conditioned to stay positive which was constructed there.

Under our hypotheses, the bridges of a Lévy process have exchangeable increments. Therefore it is natural to ask if Proposition 3.7 is not a particular case of Proposition 3.3. We did not address this particular point since under **H1** and **H2**, which are useful to construct weakly continuous versions of bridges, the minimum is attained at a unique place and continuously, as we now show.

Proof of Proposition 3.7 Using the local absolute continuity relationship and the regularity hypothesis **H1** we see that $\underline{X} < 0$ under $\mathbb{P}^1_{0,0}$. Let $\delta \in (0,1)$. On $[\delta, 1-\delta]$, the laws $\mathbb{P}^1_{0,0}$ and $\mathbb{P}$ are equivalent. Since the minimum of X on $[\delta, 1-\delta]$ is achieved at a unique place and continuously (because of regularity) under $\mathbb{P}$, the same holds under $\mathbb{P}^1_{0,0}$. We now let $\delta \to 0$ and use the fact that $\underline{X} < 0$ under $\mathbb{P}^1_{0,1}$ to conclude. □

4 Conditioning a Brownian Bridge on Its Minimum

In Corollary 3.1 we considered a limiting case of Theorem 2.2 by conditioning the minimum of a Brownian bridge to equal zero rather than to be close to zero when $X_1 = 0$. In this section, we will show that the limiting procedure is also valid when $X_1 > 0$ and for any value of the minimum. This will enable us to establish, in particular, a pathwise construction of the Brownian meander.

Theorem 4.1 *Let $\mathbb{P}_x$ be the law of the Brownian bridge from 0 to $x \geq 0$ of length 1. Consider the reflected process $R = X - J$ where*

$$J_t = \inf_{s \in [t,1]} X_s \vee \left[\underline{X}_t + X_1\right].$$

Then R admits a bicontinuous family of local times $\left(L^y_t, t \in [0,1], y \geq 0\right)$. Let $y \geq 0$ be fixed and U be a uniform random variable independent of X and define

$$\nu = \inf\left\{t \geq 0 : L^y_t = UL^y_1\right\}.$$

Let $\mathbb{P}^{y,x}$ be the law of $\theta_\nu(X)$ conditionally on $L^y_1 > 0$. Then $\mathbb{P}^{y,x}$ is a version of the law of X given $\underline{X} = y$ under $\mathbb{P}_x$ which is weakly continuous as a function of y.

Conversely, if $x = 0$ and Y has law $\mathbb{P}^{y,0}$, U is a uniform random variable independent of Y, and $H = \overline{X} - \underline{X}$ is the amplitude of the path X, then $\theta_U(Y)$ has the law of X conditionally on $H \geq -y$.

The process R is introduced in the preceding theorem for a very simple reason: when $X_1 \geq 0$, it is equal to $-\underline{X} \circ \theta_t$. See Fig. 2 for an illustration of its definition.

Proof of Theorem 4.1 To construct the local times, we first divide the trajectory of X in three parts. Let ρ be the unique instant at which the minimum is achieved and let $\underline{X}$ be the minimum. Using Denisov's decomposition of the Brownian bridge of [10], we can see that conditionally on $\rho = t$ and $\underline{X} = y$, the processes $X^{\leftarrow} = (X_{t-s} - y, s \leq t)$ and $X^{\rightarrow} = (X_{t+s} - y, s \leq 1-t)$ are three-dimensional Bessel bridges starting at 0, of lengths t and $1-t$, and ending at y and $y+x$ (see also Theorem 3 in [27], where the preceding result is stated for $x = 0$ for more general Lévy processes). Next, the trajectory of $X^{\rightarrow}$ will be further decomposed at

$$\Lambda_x = \sup\{r \leq 1-t : X^{\rightarrow}_r \leq x\}.$$

The backward strong Markov property (Theorem 2 in [8]) tells us that, conditionally on $\Lambda_x = s$, the process $X^{\rightarrow,1} = \left(X^{\rightarrow}_r, r \leq s\right)$ is a three-dimensional Bessel bridge from 0 to x of length s. Finally, the process $X^{\rightarrow,2}$ given by $X^{\rightarrow,2}_r = X^{\rightarrow}_{s+r} - x$ for $r \leq 1-t-s$ is a three-dimensional Bessel bridge from 0 to y of length $1-t-s$. Now, note that under the law of the three-dimensional Bessel process, one can construct a bicontinuous family of local times given as occupation densities. That is, if $\mathbb{P}^3_0$ is the law of the three-dimensional Bessel process, there exists a bicontinuous process $\left(L^y_t, t, y \geq 0\right)$ such that:

$$L^y_t = \lim_{\varepsilon \to 0} \frac{1}{\varepsilon} \int_0^t \mathbf{1}_{\{|X_s - y| \leq \varepsilon\}}\, ds$$

for any t, y almost surely. By Pitman's path transformation between $\mathbb{P}^3_0$ and the reflected Brownian motion found in [21], note that if $X_{\rightarrow t} = \inf_{s \geq t} X_s$ is the future infimum process of X, then $X - X_{\rightarrow}$ is a reflected Brownian motion for which one can also construct a bicontinuous family of local times. Therefore, the following limits exist and are continuous in t and y:

$$L^{r,y}_t = \lim_{\varepsilon \to 0} \int_0^t \mathbf{1}_{\{|X_r - X_{\rightarrow r}| \leq \varepsilon\}}\, dr.$$

Since the laws of the Bessel bridges are locally absolutely continuous with respect to the law of Bessel processes, we see that the following limits exist and are continuous

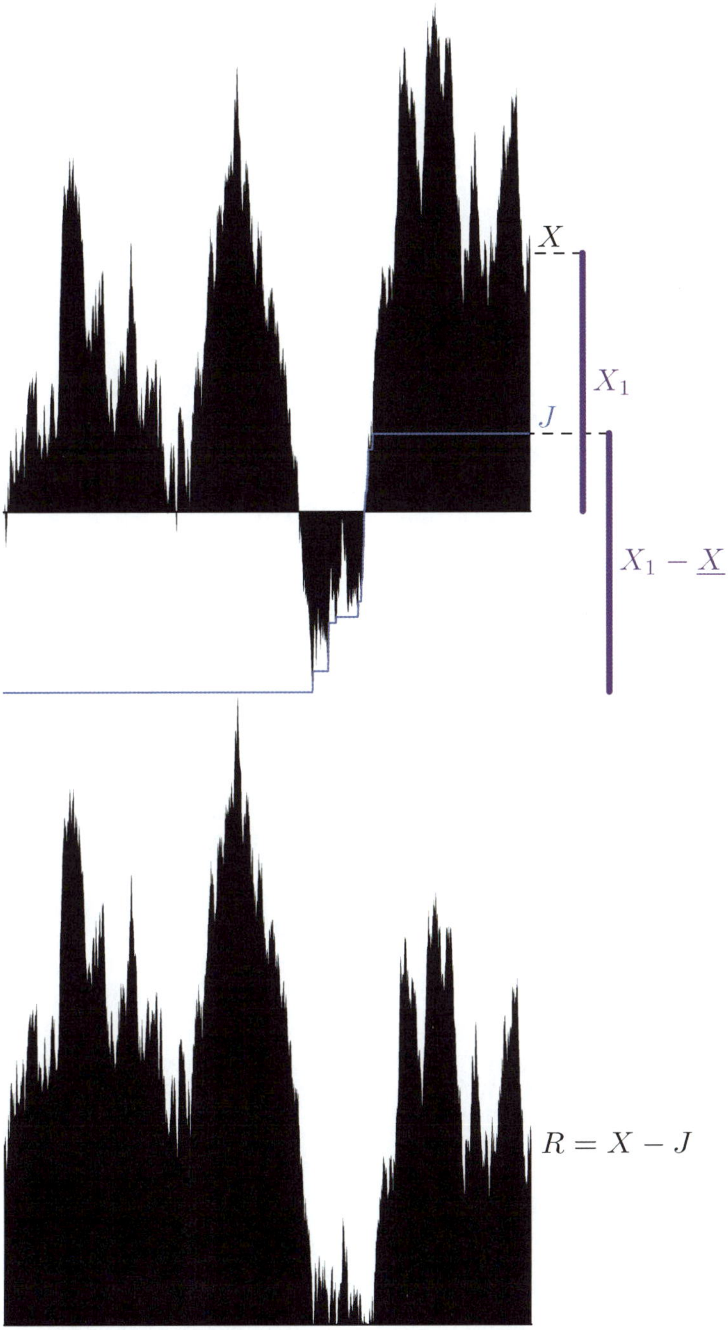

Fig. 2 Illustration of the reflected process R, given by $-\underline{X} \circ \theta_t$ when $X_1 \geq 0$

as functions of t and z under $\mathbb{P}_x$:

$$\begin{aligned} &L_r^z(R) \\ &= \lim_{\varepsilon\to 0} \frac{1}{\varepsilon} \int_{[0,r]} \mathbf{1}_{\{|R_u - z| \le \varepsilon\}} \, du \\ &= \lim_{\varepsilon\to 0} \frac{1}{\varepsilon} \left[\int_{[0,r]} \mathbf{1}_{\{|X_u - y - z| \le \varepsilon\}} \mathbf{1}_{\{u \in [0,t] \cup [t+s,1]\}} \, du \right. \\ &\qquad \left. + \int_{[0,r]} \mathbf{1}_{\left\{|X_u - X_{\rightarrow,u} - y - z| \le \varepsilon\right\}} \mathbf{1}_{\{u \in [t,t+s]\}} \, du \right] . \end{aligned}$$

(The bridge laws are not absolutely continuous with respect to the original law near the endpoint, but one can then argue by time-reversal.)

Note that R is cyclically exchangeable, so that the set $L_1^y > 0$ is invariant under θ_t for any $t \in [0, 1]$. Hence, X conditioned on $L_1^y > 0$ is cyclically exchangeable. Hence, by conditioning, we can assume that $L_1^y > 0$.

Define $I = (y - \varepsilon, y + \varepsilon)$ and let

$$\eta^I = \inf\left\{t \ge 0 : A_t^I = UA_1^I\right\}.$$

Note that the process L^y is strictly increasing at ν. Indeed, this happens because U is independent of L^y and therefore is different, almost surely, from any of the values achieved by L^y/L_1^y on any of its denumerable intervals of constancy. (The fact that $L_1^y > 0$ is used implicitly here.) Since A^I converges to L^y, it then follows that η^I converges to ν. Using the fact that X is continuous, it follows that $\theta_{\nu^I} X \to \theta_\nu X$. Since $L_1^y > 0$, then $\lim_{\varepsilon\to 0} \mathbb{P}_x\left(A_1^I > 0\right) = 1$. However, by Theorem 2.2, conditionally on $A_1^I > 0$, $\theta_{\eta^I} X$ has the law of X conditioned on $\underline{X} \in I$. Hence, the latter conditional law converges, as $\varepsilon \to 0$ to the law of $\theta_\nu X$. A similar argument applies to show that ν is continuous as a function of y and hence that the law $\mathbb{P}^{y,x}$ is weakly continuous as a function of y. But now, it is a simple exercise to show that $(\mathbb{P}^{y,x}, y \ge 0)$ disintegrates $\mathbb{P}_x$ with respect to $\underline{X}$.

Finally, suppose that $x = 0$. Since η^I is independent of X, then ν is independent of X also. Hence, the law of θ_U under $\mathbb{P}^{y,x}$ equals $\mathbb{P}_x$ conditioned on $L_1^y > 0$. However, note that $L_y^1 > 0$ implies that R (which equals $X - \underline{X}$ when $x = 0$) reaches level y. Conversely, if R reaches level y, then the local time at y must be positive. Hence the sets $\left\{L_1^y > 0\right\}$ and $\{H \ge y\}$ coincide. □

One could think of a more general result along the lines of Theorem 4.1 for processes with exchangeable increments. As our proof shows, it would involve technicalities regarding local times of discontinuous processes. We leave this direction of research open.

As a corollary (up to a time-reversal), we obtain the path transformation stated as Theorem 7 in [3].

Corollary 4.2 *Let $\mathbb{P}$ be the law of a Brownian bridge from* 0 *to* $x \geq 0$ *of length* 1 *and let* U *be uniform and independent of* X. *Let* $\nu = \sup\{t \leq 1 : X_t \leq \underline{X} + xU\}$. *Then* $\theta_\nu X$ *has the same law as the three-dimensional Bessel bridge from* 0 *to* x *of length* 1.

Proof We need only to note that, under the law of the three-dimensional Bessel process, the local time of $X - X_{\rightarrow}$ equals $X_{\rightarrow}$ (which can be thought of as a consequence of Pitman's construction of the three-dimensional Bessel process). Then, the local time at zero of R equals $J + \underline{X}$, its final value is $x + \underline{X}$, and then $\nu = \inf\left\{t \geq 0 : L_t^0 > UL_1^0\right\}$. □

By integrating with respect to x in the preceding corollary, we obtain a path construction of the Brownian meander in terms of Brownian motion. Indeed, consider first a Brownian motion B and define $X = B\,\mathrm{sgn}(B_1)$. Then X has the law of B conditionally on $B_1 > 0$ and it is cyclically exchangeable. Applying Theorem 4.1 to X, we deduce that if $\nu = \sup\{t \leq 1 : X_t \leq \underline{X} + xU\}$ then $X \circ \theta_\nu$ has the law of the weak limit as $\varepsilon \to 0$ of B conditioned on $\inf_{t \leq 1} B_t \geq -\varepsilon$, a process which is known as the Brownian meander.

Setting $x = 0$ in Theorem 4.1 gives us a novel path transformation to condition a Brownian bridge on achieving a minimum equal to y. In this case, we consider the local time process. This generalizes the Vervaat transformation, to which it reduces when $y = 0$.

Corollary 4.3 *Let* $\mathbb{P}$ *be the law of the Brownian bridge from* 0 *to* 0 *of length* 1, *let* $\left(L_t^y, y \in \mathbb{R}, t \in [0,1]\right)$ *be its continuous family of local times and let* U *be uniform and independent of* X. *For* $y \leq 0$, *let*

$$\eta_y = \inf\left\{t \geq 0 : L_t^{\underline{X}-y} > UL_1^{\underline{X}-y}\right\}.$$

Then the laws of $X \circ \theta_{\eta_y}$ *provide a weakly continuous disintegration of* $\mathbb{P}$ *given* $\underline{X} = y$.

The only difference with Theorem 4.1 is that the local times are defined directly in terms of the Brownian bridge since the reflected process R equals $X - \underline{X}$ when the ending point is zero. The equality between both notions follows from bicontinuity and the fact that local times were constructed as limits of occupation times. Also, note that since the minimum is achieved in a unique place $\rho \in (0,1)$, then $L_1^{\underline{X}} = 0$. Hence $\eta_y \to \rho$ as $y \to 0$ and the preceding path transformation converges to the Vervaat transformation.

Theorem 4.1 may be expressed in terms of the non conditioned process, that is, instead of considering the Bessel bridge, one may state the above transformation for the three dimensional Bessel process itself. More precisely, since path by path

$$\theta_u(X_t - xt\,, 0 \leq t \leq 1) = (\theta_u f(X)_t - xt\,, 0 \leq t \leq 1)\,,$$

and since the Brownian bridge b from 0 to x can be represented as

$$b_t = X_t - t(X_1 - x),\ 0 \le t \le 1)\,, \tag{4.1}$$

under the law of Brownian motion, then the process $(b_t - xt\,, 0 \le 1)$ is a Brownian bridge from 0 to 0 and then so is $(\theta_u(X)_t - xt\,, 0 \le t \le 1)$ under the law of the three dimensional Bessel bridge from 0 to x of length 1. In particular, the law of the latter process does not depend on x and we can state:

Corollary 4.4 *Under the law of the three-dimensional Bessel process on* $[0, 1]$*, if* U *is uniform and independent of* X*, then* $(\theta_U(X)_t - tX_1,\ 0 \le t \le 1)$ *is a Brownian bridge (from* 0 *to* 0 *of length* 1*) which is independent of* X_1.

Let us end by noting the following consequence of Theorem 4.1: if $\mathbb{P}$ is the law of the Brownian bridge from 0 to 0, then the law of $\overline{X} - y$ given $\underline{X} = y$ equals the law of A given $A \ge y$. When $y = 0$, we conclude that the law of the maximum of a normalized Brownian excursion equals the law of the range of a Brownian bridge. This equality was first proved in [9] and [17]. Providing a probabilistic explanation was the original motivation of Vervaat when proposing the path transformation θ_ρ in [28].

References

1. J. Bertoin, Regularity of the half-line for Lévy processes. Bull. Sci. Math. **121**(5), 345–354 (1997). MR 1465812
2. J. Bertoin, J. Pitman, J. Ruiz de Chávez, Constructions of a Brownian path with a given minimum. Electron. Commun. Probab. **4**, 31–37 (electronic) (1999). MR 1703609
3. J. Bertoin, L. Chaumont, J. Pitman, Path transformations of first passage bridges. Electron. Commun. Probab. **8**, 155–166 (2003). MR 2042754
4. Ph. Biane, Relations entre pont et excursion du mouvement brownien réel. Ann. Inst. H. Poincaré Probab. Stat. **22**(1), 1–7 (1986). MR 838369
5. P. Chassaing, S. Janson, A Vervaat-like path transformation for the reflected Brownian bridge conditioned on its local time at 0. Ann. Probab. **29**(4), 1755–1779 (2001). MR 1880241
6. L. Chaumont, Excursion normalisée, méandre et pont pour les processus de Lévy stables. Bull. Sci. Math. **121**(5), 377–403 (1997). MR 1465814
7. L. Chaumont, An extension of Vervaat's transformation and its consequences. J. Theor. Probab. **13**(1), 259–277 (2000). MR 1744984
8. L. Chaumont, G. Uribe Bravo, Markovian bridges: weak continuity and pathwise constructions. Ann. Probab. **39**(2), 609–647 (2011). MR 2789508
9. K.L. Chung, Excursions in Brownian motion. Ark. Mat. **14**(2), 155–177 (1976). MR 0467948
10. I.V. Denisov, Random walk and the Wiener process considered from a maximum point. Teor. Veroyatnost. i Primenen. **28**(4), 785–788 (1983). MR 726906
11. R.T. Durrett, D.L. Iglehart, D.R. Miller, Weak convergence to Brownian meander and Brownian excursion. Ann. Probab. **5**(1), 117–129 (1977). MR 0436353
12. S. Fourati, Vervaat et Lévy. Ann. Inst. H. Poincaré Probab. Stat. **41**(3), 461–478 (2005). MR MR2139029

13. P. Fitzsimmons, J. Pitman, M. Yor, Markovian bridges: construction, Palm interpretation, and splicing, in *Seminar on Stochastic Processes (Seattle, 1992)*, ed. by E. Çinlar, K.L. Chung, M. Sharpe. Progress in Probability, vol. 33 (Birkhäuser, Boston, 1993), pp. 101–134. MR 1278079
14. O. Kallenberg, Canonical representations and convergence criteria for processes with interchangeable increments. Z. Wahrscheinlichkeitstheorie und Verw. Gebiete **27**, 23–36 (1973). MR 0394842
15. O. Kallenberg, Path properties of processes with independent and interchangeable increments. Z. Wahrscheinlichkeitstheorie und Verw. Gebiete **28**, 257–271 (1973/1974). MR 0402901
16. O. Kallenberg, Splitting at backward times in regenerative sets. Ann. Probab. **9**(5), 781–799 (1981). MR 628873
17. D.P. Kennedy, The distribution of the maximum Brownian excursion. J. Appl. Probab. **13**(2), 371–376 (1976). MR 0402955
18. A. Lambert, P. Trapman, Splitting trees stopped when the first clock rings and Vervaat's transformation. J. Appl. Probab. **50**(1), 208–227 (2013). MR 3076782
19. G. Miermont, Ordered additive coalescent and fragmentations associated to Levy processes with no positive jumps. Electron. J. Probab. **6**(14), 33pp. (electronic) (2001). MR 1844511
20. P.W. Millar, Zero-one laws and the minimum of a Markov process. Trans. Am. Math. Soc. **226**, 365–391 (1977). MR 0433606
21. J.W. Pitman, One-dimensional Brownian motion and the three-dimensional Bessel process. Adv. Appl. Probab. **7**(3), 511–526 (1975). MR 0375485
22. J. Pitman, Brownian motion, bridge, excursion, and meander characterized by sampling at independent uniform times. Electron. J. Probab. **4**(11), 33pp. (electronic) (1999). MR 1690315
23. B.A. Rogozin, The local behavior of processes with independent increments. Teor. Verojatnost. i Primenen. **13**, 507–512 (1968). MR 0242261
24. M. Rosenbaum, M. Yor, Some explicit formulas for the brownian bridge, brownian meander and bessel process under uniform sampling. arXiv:1311.1900 (2013)
25. M. Sharpe, Zeroes of infinitely divisible densities. Ann. Math. Stat. **40**, 1503–1505 (1969). MR 0240850
26. E.S. Štatland, On local properties of processes with independent increments. Teor. Verojatnost. i Primenen. **10**, 344–350 (1965). MR 0183022
27. G. Uribe Bravo, Bridges of Lévy processes conditioned to stay positive. Bernoulli **20**(1), 190–206 (2014). MR 3160578
28. W. Vervaat, A relation between Brownian bridge and Brownian excursion. Ann. Probab. **7**(1), 143–149 (1979). MR 515820

Stochastic Integral and Covariation Representations for Rectangular Lévy Process Ensembles

J. Armando Domínguez-Molina and Alfonso Rocha-Arteaga

Abstract A Bercovici-Pata bijection Λ_c from the set of symmetric infinitely divisible distributions to the set of $\boxplus_c$-free infinitely divisible distributions, for certain free convolution $\boxplus_c$ is introduced in Benaych-Georges (Random matrices, related convolutions. Probab Theory Relat Fields 144:471–515, 2009. Revised version of F. Benaych-Georges: Random matrices, related convolutions. arXiv, 2005). This bijection is explained in terms of complex rectangular matrix ensembles whose singular distributions are $\boxplus_c$-free infinitely divisible. We investigate the rectangular matrix Lévy processes with jumps of rank one associated to these rectangular matrix ensembles. First as general result, a sample path representation by covariation processes for rectangular matrix Lévy processes of rank one jumps is obtained. Second, rectangular matrix ensembles for $\boxplus_c$-free infinitely divisible distributions are built consisting of matrix stochastic integrals when the corresponding symmetric infinitely divisible distributions under Λ_c admit stochastic integral representations. These models are realizations of stochastic integrals of nonrandom functions with respect to rectangular matrix Lévy processes. In particular, any $\boxplus_c$-free selfdecomposable infinitely divisible distribution has a random matrix model of Ornstein-Uhlenbeck type $\int_0^\infty e^{-t} \mathrm{d}\Psi(t)$, where $\{\Psi(t) : t \geq 0\}$ is a rectangular matrix Lévy process.

Keywords Random matrices • Rectangular random matrix model • Complex matrix semimartingales • Complex matrix Lévy processes • Lévy measures • Ornstein-Uhlenbeck rectangular type processes • Infinitely divisible distribution • Free infinitely divisible distribution • Bercovici-Pata bijection

Mathematics Subject Classification (2000). 46L54, 60E07, 60H05, 15A52, 60G51, 60G57.

J.A. Domínguez-Molina • A. Rocha-Arteaga (✉)
Facultad de Ciencias Físico-Matemáticas, Universidad Autónoma de Sinaloa, Culiacán, Sinaloa, México
e-mail: jadguez@uas.edu.mx; arteaga@uas.edu.mx

R.H. Mena et al. (eds.), *XI Symposium on Probability and Stochastic Processes*, Progress in Probability 69, DOI 10.1007/978-3-319-13984-5_6

1 Introduction

In the context of free probability, the Bercovici-Pata bijection Λ from the set of classical infinitely divisible distributions to the set of free infinitely divisible distributions is explained in [7] and [10] through random matrix ensembles. The Hermitian Lévy process ensembles associated to this class of infinitely divisible matrix ensembles are studied in [11] and [12]. In [11] a sample path covariation approximation for these Lévy process ensembles is proved; and in the bounded variation case a Wiener-Hopf factorization is obtained. In [12] stochastic integral realizations for several classes of these Lévy process ensembles are given.

An ensemble of square random matrices is a sequence $(M_d)_{d\geq 1}$ where M_d is a $d \times d$ complex random matrix. A random matrix model for a distribution μ is an ensemble for which the empirical spectral distribution of M_d converges weakly almost surely to μ as $d \to \infty$. In [7] and [10] for each one-dimensional infinitely divisible distribution μ a Hermitian random matrix model $(M_d)_{d\geq 1}$ for the free infinitely divisible distribution $\Lambda(\mu)$ is constructed.

A family $(M_{d,d'})_{d\geq 1,d'\geq 1}$ of $d \times d'$ complex random matrices $M_{d,d'}$ is called a rectangular random matrix model for a distribution μ if $(|M_{d,d'}|)_{d\geq 1}$ is a random matrix model for μ where $|M_{d,d'}|$ is the positive square root of $M_{d,d'}M^*_{d,d'}$.

In Benaych-Georges [8] is established a new Bercovici-Pata bijection Λ_c from the set of symmetric infinitely divisible distributions on $\mathbb{R}$ to the set of $\boxplus_c$-free infinitely divisible distributions on $\mathbb{R}$, for certain free convolution $\boxplus_c$, $0 \leq c \leq 1$, called rectangular convolution (see also [9]). This bijection plays a similar role as the one of Λ between $*$- and $\boxplus$-infinitely divisible distributions. Moreover, it is constructed, for each symmetric infinitely divisible distribution μ, a random matrix model $(M_{d,d'})_{d,d'\geq 1}$ of infinitely divisible rectangular matrices for the corresponding $\Lambda_c(\mu)$ when $\frac{d}{d'} \to c$, in a similar way as done in [7] and [10]. In the sequel these models $(M_{d,d'})_{d,d'\geq 1}$ will be called *BG* random matrix models.

Any one dimensional Lévy process $\{X^\mu(t) : t \geq 0\}$ with law μ at time $t = 1$ has associated a real-valued independently scattered random measure. A stochastic integral of a real-valued function h on $[0, \infty)$ with respect to $\{X^\mu(t) : t \geq 0\}$

$$\int_0^\infty h(t)\mathrm{d}X^\mu(t), \tag{1.1}$$

is a real-valued infinitely divisible random variable defined in the sense of integrals of non random functions with respect to scattered random measures, see [18] and [24]; and [22] for the $\mathbb{R}^d$ case. Several important classes of infinitely divisible distributions having this stochastic integral representation in law have been studied, see [1, 2] and [4].

Let $I_{\log}(\mathbb{R})$ denote the class of infinitely divisible distributions μ on $\mathbb{R}$ whose Lévy measures ν_μ satisfy the condition $\int_{|x|>2} \log|x|\, \nu_\mu(dx) < \infty$. It is shown in [14, 21] and [23], that the class of selfdecomposable distributions on $\mathbb{R}$ is characterized by the stochastic integrals (1.1) where $h(t) = e^{-t}$ and $\mu \in I_{\log}(\mathbb{R})$. That is, for any

$\mu \in I_{\log}(\mathbb{R})$ there exists a selfdecomposable distribution $\tilde{\mu}$ such that

$$\tilde{\mu} = \mathcal{L}\left(\int_0^\infty e^{-t}\mathrm{d}X^\mu(t)\right) \tag{1.2}$$

and viceversa, to any selfdecomposable distribution $\tilde{\mu}$ corresponds a distribution μ in $I_{\log}(\mathbb{R})$ such that (1.2) holds. This characterization of selfdecomposable distributions as stochastic integrals is related to Ornstein-Uhlenbeck type processes through the Langevin equation. The Langevin equation $\mathrm{d}Y(t) = \mathrm{d}X^\mu(t) - Y(t)\mathrm{d}t$ has stationary solution $\{Y(t) : t \geq 0\}$ if and only if $\mu \in I_{\log}(\mathbb{R})$. This stationary solution $\{Y(t) : t \geq 0\}$ is unique and $\mathcal{L}(Y(t)) = \mathcal{L}\left(\int_0^\infty e^{-t}\mathrm{d}X^\mu(t)\right)$ for each $t \geq 0$. The process $\{Y(t) : t \geq 0\}$ is called the stationary Ornstein-Uhlenbeck type process, see [19] and [21].

In this work we are concerned with rectangular matrix models for free infinitely divisible distributions corresponding to the image Λ_c of classical infinitely divisible distributions. A problem of interest is to understand the rectangular matrix Lévy processes $\{M_{d,d'}(t) : t \geq 0\}$ associated to the *BG* random matrix models. In Sect. 3 we prove that the Lévy measures of these models are supported in the subset of rectangular complex matrices of rank one. We find the Lévy triplet of these models giving a polar decomposition of their Lévy measures. Hence the rectangular Lévy processes associated to the *BG* random matrix models have rank one jumps, this means that the random matrix $M_{d,d'}$ is a realization, at time $t = 1$, of a rectangular Lévy process $\{M_{d,d'}(t) : t \geq 0\}$ with rank one jumps $\Delta M_{d,d'}(t) = M_{d,d'}(t) - M_{d,d'}(t-)$. In Sect. 4 we prove a representation for rectangular Lévy processes of bounded variation with jumps of rank one, as covariation realizations of a d-dimensional complex Lévy process and a d'-dimensional complex Lévy process, which are constructed via Lévy-Itô decomposition. In particular, a pathwise covariation representation of rectangular Lévy process ensembles of bounded variation associated to the *BG* random matrix models is obtained. In the case $d = d'$, it is proved in [11] that Hermitian Lévy process ensembles of bounded variation are realizations as difference of quadratic variations of d-dimensional Lévy processes. In this direction our result is of different nature since the square matrix case is not necessarily Hermitian. In Sect. 5 we prove that for every real symmetric infinitely divisible distribution with stochastic integral representation (1.1) there exists a rectangular matrix model for the corresponding $\boxplus_c$-free infinitely divisible distribution, these models are realizations of stochastic integrals of a non random function with respect to a rectangular matrix Lévy processes similar to (1.1). As a consequence, the rectangular Lévy processes given by the *BG* random matrix models are representable in law, as this type of rectangular stochastic integrals. Furthermore, the $\boxplus_c$-free selfdecomposable distribution $\Lambda_c(\tilde{\mu})$ corresponding to a symmetric selfdecomposable distribution $\tilde{\mu}$ with stochastic integral representation (1.2) where $\mu \in I_{\log}(\mathbb{R})$, has a realization as random matrix model of Ornstein-Uhlenbeck type $\left(\int_0^\infty e^{-t}\mathrm{d}\Psi^{d,d'}(t)\right)_{d\geq 1, d'\geq 1}$ where $\left\{\Psi^{d,d'}(t) : t \geq 0\right\}$ is a $d \times d'$ matrix-valued Lévy process satisfying an $I_{\log}$-condition. In [12] and [16] are studied representations as stochastic integrals of Hermitian matrix models for free infinitely divisible distributions under the bijection Λ, for several classes of classical infinitely divisible distributions having the stochastic integral representation (1.1).

2 Preliminaries

Let $\mathbb{M}_{p,q}$ be the linear space of $p \times q$ complex random matrices with scalar product $\langle A, B\rangle = \operatorname{tr}(AB^*)$ and the Frobenius norm $\|A\| = [\operatorname{tr}(AA^*)]^{1/2}$ where tr denotes the (non normalized) trace. Let denote $\tilde{\mathbb{M}}_{p,q}$ the set in $\mathbb{M}_{p,q}$ of matrices of rank one and $\tilde{\mathbb{S}}_{p,q} = \left\{V \in \tilde{\mathbb{M}}_{p,q} : \|V\| = 1\right\}$ the set of rank one matrices in the unit sphere of $\mathbb{M}_{p,q}$.

Remark 2.1 Given $V \in \tilde{\mathbb{M}}_{d,d'}$ let V_{ij} be the first entry (from left to right and up to down) such that $V_{ij} \neq 0$, that is $i = \inf\{l : V_{lk} \neq 0, k = 1, 2, \ldots, d'\}$ and $j = \inf\{k : V_{ik} \neq 0, k = 1, 2, \ldots, d'\}$. Hence $V = uv^*$ for some $u \in \mathbb{C}^d$ with $u_1 = \ldots = u_{i-1} = 0, u_i = 1$ and $v \in \mathbb{C}^{d'}$ with $v_1 = \ldots = v_{j-1} = 0, v_j \neq 0$ where $v_j = V_{ij}$. One can see that such a decomposition is unique, namely if $V = u'(v')^*$ where $u'_1 = \ldots = u'_{i-1} = 0, u'_i = 1$ and $v'_1 = \ldots v'_{j-1} = 0, v'_j \neq 0$ with $V_{ij} = v'_j$ then $u' = u$ and $v' = v$.

Covariation of complex matrix semimartingales An $\mathbb{M}_{p,q}$-valued process $X = \left\{(x_{ij})(t)\right\}_{t\geq 0}$ is a matrix semimartingale if $x_{ij}(t)$ is a complex semimartingale for each $i = 1, \ldots, p, j = 1, \ldots, q$. Let $X = \left\{(x_{ij})(t)\right\}_{t\geq 0}$ in $\mathbb{M}_{p,q}$ and $Y = \left\{(y_{ij})(t)\right\}_{t\geq 0}$ in $\mathbb{M}_{q,r}$ be semimartingales. Similar to the case of matrices with real entries in [5], we define the matrix covariation of X and Y as the $\mathbb{M}_{p,r}$-valued process $[X, Y] := \{[X, Y](t) : t \geq 0\}$ with entries

$$[X, Y]_{ij}(t) = \sum_{k=1}^{q} \left[x_{ik}, y_{kj}\right](t), \tag{2.1}$$

where $\left[x_{ik}, y_{kj}\right](t)$ is the covariation of the $\mathbb{C}$-valued semimartingales $\{x_{ik}(t)\}_{t\geq 0}$ and $\left\{y_{kj}(t)\right\}_{t\geq 0}$; see [17, pp 83]. One has the decomposition into a continuous part and a pure jump part

$$[X, Y](t) = [X^c, Y^c](t) + \sum_{s\leq t} (\Delta X(s))(\Delta Y(s)), \tag{2.2}$$

where $[X^c, Y^c]_{ij}(t) := \sum_{k=1}^{q} \left[x^c_{ik}, y^c_{kj}\right](t)$. We recall that for any semimartingale x, the process x^c is the *a.s.* unique continuous local martingale m such that $[x - m]$ is purely discontinuous.

The natural example of a continuous semimartingale is the standard complex $p \times q$ matrix Brownian motion $B = \{B(t)\}_{t\geq 0} = \{b_{jl}(t)\}_{t\geq 0}$ consisting of independent $\mathbb{C}$-valued Brownian motions $b_{jl}(t) = \mathrm{Re}(b_{jl}(t)) + \mathrm{i}\,\mathrm{Im}(b_{jl}(t))$ where $\mathrm{Re}(b_{jl}(t)), \mathrm{Im}(b_{jl}(t))$ are independent one-dimensional Brownian motions with common variance $t/2$. Then we have $[B, B^*]_{ij}(t) = \sum_{k=1}^{q} [b_{ik}, \overline{b}_{jk}](t) = qt\delta_{ij}$ and hence the matrix quadratic variation of B is given by the $d \times d$ matrix process:

$$[B, B^*](t) = qt\mathrm{I}_p. \tag{2.3}$$

The case $q = 1$ corresponds to the $\mathbb{C}^p$-valued standard Brownian motion B. We observe this corresponds to $[B, B^*](t) = t\mathrm{I}_p$ instead of the common $2t\mathrm{I}_p$ used in the literature.

Other examples of complex matrix semimartingales are Lévy processes considered next.

Complex matrix Lévy processes An infinitely divisible random matrix M in $\mathbb{M}_{p,q}$ is characterized by the Lévy-Khintchine representation of its Fourier transform $\mathbb{E}e^{\mathrm{i}\,\mathrm{tr}(\Theta^* M)} = \exp(\psi(\Theta))$ with Laplace exponent

$$\psi(\Theta) = \mathrm{i}\,\mathrm{tr}(\Theta^*\Psi) - \frac{1}{2}\mathrm{tr}\left(\Theta^*\mathcal{A}\Theta^*\right) + \int_{\mathbb{M}_{p\times q}} \left(e^{\mathrm{i}\,\mathrm{tr}(\Theta^*\xi)} - 1 - \mathrm{i}\frac{\mathrm{tr}(\Theta^*\xi)}{1 + \|\xi\|^2}\right) \nu(\mathrm{d}\xi), \tag{2.4}$$

$\Theta \in \mathbb{M}_{p\times q}$, where $\Psi \in \mathbb{M}_{p,q}$, $\mathcal{A} : \mathbb{M}_{q,p} \to \mathbb{M}_{p,q}$ is a positive symmetric linear operator (i.e. $\mathrm{tr}(\Phi^*\mathcal{A}\Phi^*) \geq 0$ for $\Phi \in \mathbb{M}_{p,q}$ and $\mathrm{tr}\left(\Theta_2^*\mathcal{A}\Theta_1^*\right) = \mathrm{tr}\left(\Theta_1^*\mathcal{A}\Theta_2^*\right)$ for $\Theta_1, \Theta_2 \in \mathbb{M}_{p,q}$) and ν is a measure on $\mathbb{M}_{p,q}$ (the Lévy measure) satisfying $\nu(\{0\}) = 0$ and $\int_{\mathbb{M}_{p,q}} (1 \wedge \|x\|^2)\nu(\mathrm{d}x) < \infty$. The triplet $(\mathcal{A}, \nu, \Psi)$ uniquely determines the distribution of M.

Let $\mathbb{S}_{p,q}$ be the unit sphere of $\mathbb{M}_{p,q}$. If ν is a Lévy measure on $\mathbb{M}_{p,q}$, then there are a measure τ on $\mathbb{S}_{p,q}$ with $\tau(\mathbb{S}_{p,q}) \geq 0$ and a measure ν_ξ on $(0, \infty)$ for each $\xi \in \mathbb{S}_{p,q}$ with $\nu_\xi((0, \infty)) > 0$ such that

$$\nu(E) = \int_{\mathbb{S}_{p,q}} \tau(\mathrm{d}\xi) \int_{(0,\infty)} 1_E(u\xi)\nu_\xi(\mathrm{d}u), \qquad E \in \mathcal{B}(\mathbb{M}_{p,q}\backslash\{0\}).$$

We call (τ, ν_ξ) a polar decomposition of ν. When $p = q = 1$, ν is a Lévy measure on $\mathbb{C}$ and τ is a measure in the unit sphere of $\mathbb{C}$.

Any $\mathbb{M}_{p,q}$-valued Lévy process $L = \{L(t)\}_{t\geq 0}$ with triplet $(\mathcal{A}, \nu, \Psi)$ is a semimartingale with the Lévy-Itô decomposition

$$L(t) = t\Psi + B_{\mathcal{A}}(t) + \int_{[0,t]} \int_{\|V\|\leq 1} V\tilde{J}_L(\mathrm{d}s, \mathrm{d}V) + \int_{[0,t]} \int_{\|V\|>1} VJ_L(\mathrm{d}s, \mathrm{d}V),\ t \geq 0, \tag{2.5}$$

where:

(a) $\{B_{\mathcal{A}}(t)\}_{t\geq 0}$ is a $\mathbb{M}_{p,q}$-valued Brownian motion with covariance $\mathcal{A}$, i.e. it is a Lévy process with continuous sample paths (a.s.) and each $B_{\mathcal{A}}(t)$ is centered Gaussian with

$$\mathbb{E}\left\{\mathrm{tr}(\Theta_1^* B_{\mathcal{A}}(t))\mathrm{tr}\left(\Theta_2^* B_{\mathcal{A}}(s)\right)\right\} = \min(s,t)\mathrm{tr}\left(\Theta_1^* \mathcal{A}\Theta_2^*\right) \text{ for each } \Theta_1, \Theta_2 \in \mathbb{M}_{p,q},$$

(b) $J_L(\cdot,\cdot)$ is the Poisson random measure of jumps on $[0,\infty) \times \mathbb{M}_{p,q}\backslash\{0\}$. That is, $J_L(t,E) = \#\{0 \leq s \leq t : \Delta L(s) \in E\}$, $E \subset \mathbb{M}_{p,q}\backslash\{0\}$, with intensity measure $Leb \otimes \nu$, and independent of $\{B_{\mathcal{A}}(t)\}_{t\geq 0}$,

(c) $\tilde{J}_L$ is the compensator measure of J_L, i.e.

$$\tilde{J}_L(\mathrm{d}t, \mathrm{d}V) = J_L(\mathrm{d}t, \mathrm{d}V) - \mathrm{d}t\nu(\mathrm{d}V);$$

see for example [3] for the most general case of Lévy processes with values in infinite dimensional Banach spaces.

An $\mathbb{M}_{p,q}$-valued Lévy process $L = \{L(t)\}_{t\geq 0}$ has bounded variation if and only if its Lévy-Itô decomposition takes the form

$$L(t) = t\Psi_0 + \int_{[0,t]}\int_{\mathbb{M}_{p,q}\backslash\{0\}} VJ_L(\mathrm{d}s, \mathrm{d}V) = t\Psi_0 + \sum_{s\leq t}\Delta L(s),\ t \geq 0, \tag{2.6}$$

where $\Psi_0 = \Psi - \int_{\|V\|\leq 1} V\nu(\mathrm{d}V)$.

3 The Lévy Triplet of *BG* Random Matrix Models

In [8] is constructed, for each symmetric infinitely divisible distribution μ on the real line a random matrix model $(M_{d,d'})_{d,d'\geq 1}$ for $\Lambda_c(\mu)$ when $\frac{d}{d'} \to c$. Moreover, the Fourier transform of $M_{d,d'}$, for each $d \geq 1, d' \geq 1$, is

$$\mathbb{E}[\exp(i\Re\mathrm{tr}(A^* M_{d,d'}))] = \exp\left[d\mathbb{E}_{u,v}\mathcal{C}_\mu\left(\Re\langle u, Av\rangle\right)\right]\ A \text{ is } d \times d', \tag{3.1}$$

where $\langle\cdot,\cdot\rangle$ denotes the canonical hermitian product of $\mathbb{C}^d$, u, v are independent random vectors uniformly distributed on the unit sphere of respectively $\mathbb{C}^d$, $\mathbb{C}^{d'}$ and $\mathcal{C}_\mu$ is the cumulant transform of μ. That is, $\mathcal{C}_\mu$ is the unique continuous function from $\mathbb{R}$ into $\mathbb{C}$ such that $\mathcal{C}_\mu(0) = 0$ and $\hat{\mu}(z) = \exp\left(\mathcal{C}_\mu(z)\right)$ for every $z \in \mathbb{R}$ where $\hat{\mu}$ denotes the Fourier transform of μ. Furthermore if μ has Lévy triplet (a^2, ν, ψ),

$$\mathcal{C}_\mu(z) = i\psi z - \frac{1}{2}a^2t^2 + \int_{\mathbb{R}}\left[e^{izx} - 1 - ixz1_{|x|\leq 1}(x)\right]\nu(\mathrm{d}x). \tag{3.2}$$

Let $\mathcal{C}_{M_{d,d'}}(A) = \log(\mathbb{E}[\exp(i\Re\mathrm{tr}(A^*M_{d,d'}))])$ for any $d \times d'$ complex matrix A, be the cumulant transform of $M_{d,d'}$. Then

$$\mathcal{C}_{M_{d,d'}}(A) = d\mathbb{E}_{u,v}\mathcal{C}_{\mu}(\Re\langle u, Av\rangle)$$
$$= i\psi d\mathbb{E}_{u,v}\Re\langle u, Av\rangle - \frac{1}{2}a^2 d\mathbb{E}_{u,v}(\Re\langle u, Av\rangle)^2$$
$$+ d\mathbb{E}_{u,v}\int_{\mathbb{R}}\left[e^{ix\Re\langle u,Av\rangle} - 1 - ix\Re\langle u, Av\rangle 1_{|x|\leq 1}(x)\right]\nu(\mathrm{d}x). \tag{3.3}$$

Henceforth we denote by $(\mathcal{A}_{d,d'}, \nu_{d,d'}, \Psi_{d,d'})$ the Lévy triplet of $M_{d,d'}$ for each $d \geq 1, d' \geq 1$. Theorem 3.1 gives a polar decomposition of the Lévy measures of the BG random matrix models, showing that they are supported in the subset of $d\times d'$ complex matrices of rank one, in a similar way as done in [12] for the Hermitian case.

Let us recall the polar decomposition of Lévy measures on $\mathbb{R}$, see [4] and [20]. The Lévy measure ν of an infinitely divisible distribution μ on $\mathbb{R}$ with $0 < \nu(\mathbb{R}) \leq \infty$, is expressed as

$$\nu(B) = \int_S \tau(\mathrm{d}\xi)\int_0^\infty 1_B(r\xi)\nu_\xi(\mathrm{d}r), \tag{3.4}$$

where τ is a measure on the unit sphere $S = \{-1, 1\}$ of $\mathbb{R}$ such that $0 < \tau(S) \leq \infty$ and ν_ξ is a measure on $(0, \infty)$ for each $\xi \in S$ such that $0 < \nu_\xi((0, \infty)) \leq \infty$. These measures τ and ν_ξ are called respectively the spherical and radial components of ν.

In the sequel we denote by $\omega_{d,d'}$ the probability measure on the set of $d \times d'$ complex random matrices of rank one induced by the mapping

$$(u, v) \to V = uv^*, \tag{3.5}$$

where u, v are independent random vectors, uniformly distributed on the unit sphere of respectively $\mathbb{C}^d$, $\mathbb{C}^{d'}$.

Theorem 3.1 *Let μ be an infinitely divisible distribution on $\mathbb{R}$ with Lévy measure ν and let $(M_{d,d'})_{d,d'\geq 1}$ be a random matrix model for $\Lambda_c(\mu)$ where $M_{d,d'}$ has the Fourier transform (3.1). Then the Lévy measure $\nu_{d,d'}$ of $M_{d,d'}$ is expressed as*

$$\nu_{d,d'}(B) = d\int_{\mathbb{S}_{d,d'}}\int_0^\infty 1_B(rV)\,\nu(\mathrm{d}r)\,\Pi(\mathrm{d}V) \quad B \in \mathcal{B}(\mathbb{M}_{d,d'}\backslash\{0\}),$$

where $\Pi\,(dV)$ *is a measure on* $\tilde{\mathbb{S}}_{d,d'}$ *such that*

$$\int_{\tilde{\mathbb{S}}_{d,d'}} 1_D\,(V)\,\Pi\,(\mathrm{d}V) = \int_{\tilde{\mathbb{S}}_{d,d'}} \int_{\{-1,1\}} 1_D\,(\xi V)\,\tau\,(\mathrm{d}\xi)\,\omega_{d,d'}\,(\mathrm{d}V) \qquad D \in \mathcal{B}\left(\tilde{\mathbb{S}}_{d,d'}\right),$$

τ *is the spherical measure of* ν *and* $\omega_{d,d'}$ *is the probability measure on* $\tilde{\mathbb{S}}_{d,d'}$ *given by (3.5).*

Proof Let $\tau(d\xi)$ and ν_ξ be the spherical and radial components of ν given by (3.4). Assume without loss of generality that the Gaussian term a^2 in (3.2) is zero. By Lemma 3.2 below $\mathbb{E}_{u,v}\Re\,\langle u, Av\rangle = 0$ for each $A \in \mathbb{M}_{d,d'}$, then (3.3) reduces to

$$\mathcal{C}_{M_{d,d'}}\,(A) = d\mathbb{E}_{u,v}\int_{\mathbb{R}}\left[e^{i\Re\langle u,Av\rangle x} - 1 - ix\Re\,\langle u, Av\rangle\,1_{|x|\le 1}\,(x)\right]\nu\,(\mathrm{d}x)$$

$$= d\int_{\tilde{\mathbb{S}}_{d,d'}}\int_{\mathbb{R}}\left[e^{i\Re\mathrm{tr}(Auv^*)x} - 1 - ix\Re\mathrm{tr}\left(Auv^*\right)1_{|x|\le 1}\,(x)\right]\nu\,(\mathrm{d}x)\,\omega_{d,d'}\,(\mathrm{d}V)$$

$$= d\int_{\tilde{\mathbb{S}}_{d,d'}}\int_{\mathbb{R}}\left[e^{i\Re\mathrm{tr}(AV)x} - 1 - ix\Re\mathrm{tr}\,(AV)\,1_{|x|\le 1}\,(x)\right]\nu\,(\mathrm{d}x)\,\omega_{d,d'}\,(\mathrm{d}V)$$

$$= d\int_{\tilde{\mathbb{S}}_{d,d'}}\int_{\{-1,1\}}\int_0^\infty\left[e^{i\Re\mathrm{tr}(AV)r\xi} - 1 - ir\xi\Re\mathrm{tr}\,(AV)\,1_{|r\xi|\le 1}\,(r\xi)\right]\nu_\xi\,(\mathrm{d}r)\,\tau\,(\mathrm{d}\xi)$$
$$\times\,\omega_{d,d'}\,(\mathrm{d}V)$$

$$= d\int_{\tilde{\mathbb{S}}_{d,d'}}\int_0^\infty\left[e^{i\Re\mathrm{tr}(A\tilde{V})r} - 1 - ir\Re\mathrm{tr}\left(A\tilde{V}\right)1_{|r|\le 1}\,(r)\right]\nu_\xi\,(\mathrm{d}r)\,\Pi\left(\mathrm{d}\tilde{V}\right),$$

and $\Pi\left(d\tilde{V}\right)$ is a measure on $\tilde{\mathbb{S}}_{d,d'}$ such that for any Borel set D of $\tilde{\mathbb{S}}_{d,d'}$

$$\int_{\{\tilde{V}:\,\mathrm{rank}(\tilde{V})=1,\|\tilde{V}\|=1\}} 1_D\left(\tilde{V}\right)\Pi\left(\mathrm{d}\tilde{V}\right)$$
$$= \int_{\{V:\,\mathrm{rank}(V)=1,\|V\|=1\}}\int_{\{-1,1\}} 1_D\,(\xi V)\,\tau\,(\mathrm{d}\xi)\,\omega_{d,d'}\,(\mathrm{d}V),$$

the equivalence of the regions of integration follows from the spectral representation theorem, since any V with $\mathrm{rank}\,(V) = 1$ and $\|V\| = 1$ can be written as $V = uv^*$ where u and v are orthonormal random vectors and noticing that $uv^* \stackrel{d}{=} -uv^*$. □

The following lemma, proved in Appendix, is useful to get the operator of the Gaussian part of BG random matrix models.

Lemma 3.2 *Let u, v be independent random vectors, uniformly distributed on the unit sphere of respectively $\mathbb{C}^d$, $\mathbb{C}^{d'}$. Then for any $A \in \mathbb{M}_{d,d'}$,*

(a)

$$\mathbb{E}_{u,v}\left(\langle u, Av\rangle\right) = 0,$$

(b)

$$\mathbb{E}_{u,v}\left(\Re\langle u, Av\rangle\right)^2 = \frac{1}{2dd'}\operatorname{tr}\left(AA^*\right).$$

Proposition 3.3 *Let μ be an infinitely divisible distribution in $\mathbb{R}$ with Lévy triplet (a^2, ν, ψ) and let $(M_{d,d'})_{d\geq 1}$ be the random matrix model for $\Lambda_c(\mu)$ where $M_{d,d'}$ has the Fourier transform (3.1). Then, for each $d \geq 1, d' \geq 1$, the Lévy triplet $(\mathcal{A}_{d,d'}$, $\nu_{d,d'}, \Psi_{d,d'})$ of $M_{d,d'}$ is given by*

(a) $\Psi_{d,d'} = 0.$

(b)

$$\mathcal{A}_{d,d'}\Theta = \frac{a^2}{2d'}\Theta^*, \qquad \Theta \in \mathbb{M}_{d,d'}.$$

(c)

$$\nu_{d,d'}(B) = d\int_{\tilde{\mathbb{S}}_{d,d'}}\int_0^\infty 1_B(rV)\,\nu(\mathrm{d}r)\,\Pi(\mathrm{d}V) \qquad B \in \mathcal{B}(\mathbb{M}_{d,d'}\setminus\{0\}),$$

where $\Pi(dV)$ is a measure on $\tilde{\mathbb{S}}_{d,d'}$ such that

$$\int_{\tilde{\mathbb{S}}_{\mathbb{M}_{d,d'}}} 1_D(V)\,\Pi(\mathrm{d}V) = \int_{\tilde{\mathbb{S}}_{d,d'}}\int_{\{-1,1\}} 1_D(\xi V)\,\tau(\mathrm{d}\xi)\,\omega_{d,d'}(\mathrm{d}V) \qquad D \in \mathcal{B}\left(\tilde{\mathbb{S}}_{d,d'}\right),$$

τ is the spherical measure of ν and $\omega_{d,d'}$ is the probability measure on $\tilde{\mathbb{S}}_{d,d'}$ given by (3.5).

Proof Assertion *(b)* follows from (3.3) and $\mathbb{E}_{u,v}\left(\Re\langle u, Av\rangle\right)^2 = \frac{1}{2dd'}\operatorname{tr}(A^*A)$ by Lemma 3.2. *(c)* follows from Theorem 3.1. It remains to prove (a). From (3.3) and (3.5) we get

$$i\psi d\mathbb{E}_{u,v}\Re\langle u, Av\rangle = i\psi \int_{\tilde{\mathbb{S}}_{d,d'}} \Re\left(\mathrm{tr}\left(A^*V\right)\right)\omega_{d,d'}(\mathrm{d}V)$$
$$= i\psi\Re\left(\mathrm{tr}\left(A^*\Psi_{d,d'}\right)\right),$$

where $\Psi_{d,d'} = \int_{\tilde{\mathbb{S}}_{d,d'}} V\omega_{d,d'}(\mathrm{d}V) = 0$ since $\mathbb{E}uv^* = 0$. □

4 Covariation Representation of Rectangular Lévy Processes

Theorem 4.1 *Let $L_{d,d'} = \{L_{d,d'}(t) : t \geq 0\}$ be a Lévy process in $\mathbb{M}_{d,d'}$ of bounded variation whose jumps are of rank one almost surely. Then there exist Lévy processes $X = \{X(t) : t \geq 0\}$ in $\mathbb{C}^d$ and $Y = \{Y(t) : t \geq 0\}$ in $\mathbb{C}^{d'}$ such that*

$$L_{d,d'}(t) = [X, Y](t).$$

Proof We construct X and Y as marginals of a Lévy-Itô decomposition realization. For each $d \geq 1, d' \geq 1$, $L_{d,d'}$ is an $\mathbb{M}_{d,d'}$-process of bounded variation with Lévy-Itô decomposition

$$L_d(t) = t\Psi_0 + \int_{[0,t]}\int_{\tilde{\mathbb{M}}_{d,d'}} VJ_L(\mathrm{d}s, \mathrm{d}V), t \geq 0,$$

where $\Psi_0 \in \mathbb{M}_{d,d'}$ and J_L is the Poisson random measure of $L_{d,d'}$. Let $Leb \otimes \nu_L$ denote the intensity measure of $L_{d,d'}$. Let $C^{d+d'}$ denote the set of $(u_1, \ldots, u_d, v_1, \ldots, v_{d'})^T \in \mathbb{C}^{d+d'}$ such that $u_1 = \ldots = u_{i-1} = 0, u_i = 1$ and $v_1 = \ldots = v_{j-1} = 0, v_j \neq 0$ for some $i = 1, \ldots, d$ and $j = 1, \ldots, d'$. Let $E \subset C^{d+d'}$ be defined by

$$E = \left\{\begin{pmatrix} u \\ v \end{pmatrix} \in C^{d+d'} : u \in \mathbb{C}^d, v \in \mathbb{C}^{d'}, |u|^2 + |v|^2 \leq |u|^2|v|^2\right\}.$$

Let $\varphi : \mathbb{R}_+ \times \tilde{\mathbb{M}}_{d,d'} \to \mathbb{R}_+ \times E$ be defined as $\varphi(t, V) = (t, \binom{u}{v})$ if $\binom{u}{v} \in E$ and $\varphi(t, V) = 0$ if $\binom{u}{v} \notin E$ where $V = uv^*$ with $\binom{u}{v} \in C^{d+d'}$. Let $\overline{\varphi} : \tilde{\mathbb{M}}_{d,d'} \to E$ be defined by $\overline{\varphi}(V) = \binom{u}{v}$ if $\binom{u}{v} \in E$ and $\overline{\varphi}(V) = 0$ if $\binom{u}{v} \notin E$ where $V = uv^*$ with $\binom{u}{v} \in C^{d+d'}$. The functions φ and $\overline{\varphi}$ are well defined by Remark 2.1. Let us define $J(\mathrm{d}s, \mathrm{d}u, \mathrm{d}v) = \left(J_L \circ \varphi^{-1}\right)(\mathrm{d}s, \mathrm{d}u, \mathrm{d}v)$ the random measure induced by the mapping φ which is a Poisson random measure on $\mathbb{R}_+ \times E$. Note that $\mathbb{E}[J(t, F)] = \mathbb{E}\left[J_L \circ \varphi^{-1}(\{t\} \times F)\right] = t\nu_L(\overline{\varphi}(F)) = t\left(\nu_L \circ \overline{\varphi}^{-1}\right)(F)$ for $F \in \mathcal{B}(E\backslash\{0\})$. Let us denote $\nu = \nu_L \circ \overline{\varphi}^{-1}$ which is a Lévy measure on $C^{d+d'}$ since

$$\int_{C^{d+d'}\setminus\{0\}} \left(1 \wedge |z|^2\right) \nu(\mathrm{d}z) = \int_{E\setminus\{0\}} \left(1 \wedge |z|^2\right) \nu_L \circ \overline{\varphi}^{-1}(\mathrm{d}z)$$

$$= \int_{E\setminus\{0\}} \left(1 \wedge \operatorname{tr}\left(zz^*\right)\right) \nu_L \circ \overline{\varphi}^{-1}(\mathrm{d}z) \leq \int_{\widetilde{\mathbb{M}}_{d,d'}} \left(1 \wedge \|V\|^2\right) \left(\nu_L \circ \overline{\varphi}^{-1}\right) \circ f^{-1}(\mathrm{d}V)$$

$$\leq \int_{\widetilde{\mathbb{M}}_{d,d'}} (1 \wedge \|V\|)\, \nu_L(\mathrm{d}V) < \infty,$$

where we have used $\operatorname{tr}(zz^*) = |u|^2 + |v|^2 \leq |u|^2 |v|^2 = \|V\|^2$ with $V = uv^*$ for $z = \binom{u}{v} \in E$ and $f : E \to \widetilde{\mathbb{M}}_{d,d'}$ is defined by $f(z) = uv^*$ which clearly satisfies $\left(\nu_L \circ \overline{\varphi}^{-1}\right) \circ f^{-1} = \nu$. Thus *Leb* $\otimes\, \nu$ is the intensity measure of the Poisson random measure J. Let us take the Lévy process in $\mathbb{C}^{d+d'}$,

$$Z(t) = B(t) + \int_{[0,t]} \int_{\mathbb{C}^{d+d'} \cap \{|z| \leq 1\}} z\widetilde{J}(\mathrm{d}s, \mathrm{d}u, \mathrm{d}v) + \int_{[0,t]} \int_{\mathbb{C}^{d+d'} \cap \{|z| > 1\}} zJ(\mathrm{d}s, \mathrm{d}u, \mathrm{d}v),$$

$t \geq 0$, where $B = (b_1^{(1)}, \ldots, b_d^{(1)}, b_1^{(2)}, \ldots, b_{d'}^{(2)})$ is a $\mathbb{C}^{d+d'}$-valued standard Brownian motion with quadratic variation

$$\begin{bmatrix} I_d & \Psi_0 \\ \Psi_0^* & I_{d'} \end{bmatrix}.$$

Let us take the marginal processes X in $\mathbb{C}^d$ and Y in $\mathbb{C}^{d'}$ of Z given by

$$X(t) = B^{(1)}(t) + \int_{[0,t]} \int_{\mathbb{C}^{d+d'} \cap \{|z| \leq 1\}} u\widetilde{J}(\mathrm{d}s, \mathrm{d}u, \mathrm{d}v)$$
$$+ \int_{[0,t]} \int_{\mathbb{C}^{d+d'} \cap \{|z| > 1\}} uJ(\mathrm{d}s, \mathrm{d}u, \mathrm{d}v),\ t \geq 0,$$
$$Y(t) = B^{(2)}(t) + \int_{[0,t]} \int_{\mathbb{C}^{d+d'} \cap \{|z| \leq 1\}} v\widetilde{J}(\mathrm{d}s, \mathrm{d}u, \mathrm{d}v)$$
$$+ \int_{[0,t]} \int_{\mathbb{C}^{d+d'} \cap \{|z| > 1\}} vJ(\mathrm{d}s, \mathrm{d}u, \mathrm{d}v),\ t \geq 0,$$

where $B^{(1)} = (b_1^{(1)}, \ldots, b_d^{(1)})$ and $B^{(2)} = (b_1^{(2)}, \ldots, b_{d'}^{(2)})$. Then

$$[X, Y](t) = \left[B^{(1)}, B^{(2)*}\right](t) + \int_{[0,t]} \int_{\mathbb{C}^{d+d'}\setminus\{0\}} uv^* J(\mathrm{d}s, \mathrm{d}u, \mathrm{d}v)$$

$$= \Psi_0 t + \int_{[0,t]} \int_{E\setminus\{0\}} uv^* J_L \circ \varphi^{-1}(\mathrm{d}s, \mathrm{d}u, \mathrm{d}v)$$

$$= \Psi_0 t + \int_{[0,t]} \int_{\tilde{\mathbb{M}}_{d,d'}} V J_L \circ \varphi^{-1} \circ g^{-1}(\mathrm{d}s, \mathrm{d}V)$$

$$= \Psi_0 t + \int_{[0,t]} \int_{\tilde{\mathbb{M}}_{d,d'}} V J_L(\mathrm{d}s, \mathrm{d}V) = L_d(t),$$

where $J_L \circ \varphi^{-1} \circ g^{-1} = J_L$ with $g : \mathbb{R}_+ \times E \to \mathbb{R}_+ \times \tilde{\mathbb{M}}_{d,d'}$ defined by $g\left(t, \binom{u}{v}\right) = (t, uv^*)$. □

Next we consider the rectangular matrix Lévy processes of bounded variation associated to the *BG* matrix ensembles $(M_{d,d'})_{d\geq 1, d'\geq 1}$. We have the following pathwise covariation representation.

Corollary 4.2 *Let $M_{d,d'} = \{M_{d,d'}(t) : t \geq 0\}$ be the matrix Lévy process associated to the BG random matrix ensembles.*

If $M_{d,d'}$ has bounded variation then there exist Lévy processes $X = \{X(t) : t \geq 0\}$ in $\mathbb{C}^d$ and $Y = \{Y(t) : t \geq 0\}$ in $\mathbb{C}^{d'}$ such that $M_{d,d'}(t) = [X, Y](t)$.

Remark 4.3 Let us consider $d = d'$. It is proved in [11] that for any Hermitian Lévy process $M_d = \{M_d(t) : t \geq 0\}$ of bounded variation of rank one jumps almost surely, there exist Lévy processes $X = \{X(t) : t \geq 0\}$ and $Y = \{Y(t) : t \geq 0\}$ in $\mathbb{C}^d$ such that $M_d(t) = [X](t) - [Y](t)$, where $\{[X](t) : t \geq 0\}$ and $\{[Y](t) : t \geq 0\}$ are independent. This result is of different nature than our result in Theorem 4.1, since the square case in Theorem 4.1 is not necessarily Hermitian. Neither of both results can be obtained from the other. We observe that there is not a natural cone in the rectangular case, as the one of nonnegative definite matrices in the square case, to allow a rectangular Wiener-Hopf factorization.

5 Stochastic Integral Representation of *BG* Random Matrix Models

The stochastic integral of a real function h on $[0, \infty)$ with respect to a real-valued Lévy process $\{X^\mu(t) : t \geq 0\}$ is a real-valued infinitely divisible random variable

$$w = \int_0^\infty h(t)\mathrm{d}X^\mu(t), \tag{5.1}$$

which is defined as the limit in probability of $\int_{[0,s]} h(t)\mathrm{d}X^\mu(t)$ as $s \to \infty$ when this limit exists, where the last integral is taken, for each $s \geq 0$, with respect to the unique real-valued independently scattered random measure on the bounded Borel sets of $[0, \infty)$ induced by this Lévy process, see [18] and [21]. Furthermore, its cumulant transform is given by

$$\mathcal{C}_w(x) = \int_0^\infty \mathcal{C}_\mu(h(t)x)\,\mathrm{d}t \quad x \in \mathbb{R}. \tag{5.2}$$

We have an analogous result for the complex $d \times d'$ matrix case; see [6] for the case of $d \times d$ real matrices. For any infinitely divisible matrix Ψ in $\mathbb{M}_{d,d'}$ with associated matrix Lévy process $\left\{\Psi^{d,d'}(t) : t \geq 0\right\}$, the infinitely divisible $d \times d'$ matrix valued stochastic integral

$$M = \int_0^\infty h(t)\mathrm{d}\Psi^{d,d'}(t), \tag{5.3}$$

whenever exists, has cumulant transform

$$\mathcal{C}_M(A) = \int_0^\infty \mathcal{C}_\Psi(h(t)A)\,\mathrm{d}t \quad A \in \mathbb{M}_{d,d'}. \tag{5.4}$$

In the sequel we consider matrix Lévy processes $\left\{\Psi^{d,d'}(t) : t \geq 0\right\}$ corresponding to Lévy measures of the form

$$\nu_\Psi^{d,d'}(B) = d\int_{\tilde{\mathbb{S}}_{d,d'}} \omega_{d,d'}(\mathrm{d}V) \int_{\mathbb{R}} 1_B(xV)\,\nu_\mu(\mathrm{d}x), \tag{5.5}$$

where $\omega_{d,d'}$ is the probability measure given by (3.5) and ν_μ is a Lévy measure of an infinitely divisible distribution μ on $\mathbb{R}$. Observe that $\nu_\Psi^{d,d'}$ is a Lévy measure supported in the subset of rank one matrices $\tilde{\mathbb{M}}_{d,d'}$.

The next result provides rectangular random matrix models for $\boxplus_c$-free infinitely divisible distributions given by matrix stochastic integrals of the form (5.3) whenever the corresponding symmetric infinitely divisible distributions under Λ_c are representable in law as real stochastic integrals.

Theorem 5.1 *Let μ_h be a symmetric infinitely divisible distribution on $\mathbb{R}$ with stochastic integral representation in law*

$$\mu_h = \mathcal{L}\left(\int_0^\infty h(t)\mathrm{d}X^\mu(t)\right), \tag{5.6}$$

where $\{X^\mu(t) : t \geq 0\}$ is a Lévy process on $\mathbb{R}$ with law μ at time $t = 1$ and Lévy measure ν_μ. Then the $\boxplus_c$-free infinitely divisible distribution $\Lambda_c(\mu_h)$ has a rectangular random matrix model $\left(M_h^{d,d'}\right)_{d\geq 1, d'\geq 1}$ given by infinitely divisible *matrix stochastic integrals*

$$M_h^{d,d'} := \int_0^\infty h(t)\mathrm{d}\Psi^{d,d'}(t), \tag{5.7}$$

where $\left\{\Psi^{d,d'}(t) : t \geq 0\right\}$ *is the* $\mathbb{M}_{d,d'}$*-valued Lévy process with Lévy measure* $\nu_\Psi^{d,d'}$ *given by (5.5) in terms of* $\omega_{d,d'}$ *and* ν_μ.

Proof We will prove that the random matrices defined by the ensemble $\left(M_h^{d,d'}\right)_{d\geq 1, d'\geq 1}$ in (5.7) and the random matrices of the *BG* random matrix model $(M_{d,d'})_{d\geq 1, d'\geq 1}$ for $\Lambda_c(\mu_h)$ given in [8] have the same law. From (3.1) the Fourier transform of $M_{d,d'}$ is given by

$$\mathbb{E}[\exp(i\mathrm{tr}(A^* M_{d,d'}))] = \exp\left[d\mathbb{E}_{u,v}\mathcal{C}_{\mu_h}(\Re\langle u, Av\rangle)\right], \quad A \in \mathbb{M}_{d,d'}.$$

Next we prove that this Fourier transform coincides with the Fourier transform of $M_h^{d,d'}$. First we calculate the cumulant transform of $\Psi^{d,d'}(t)$ at time $t = 1$ using (5.5),

$$\begin{aligned}
\mathcal{C}_{\Psi^{d,d'}(1)}(A) &= \int_{\mathbb{M}_{d,d'}} \left[e^{i\Re\mathrm{tr}(AX)} - 1 - i\Re\mathrm{tr}(AX)1_{\|X\|\leq 1}(X)\right]\nu_{\Psi(1)}^{d,d'}(\mathrm{d}X) \\
&= d\int_{\tilde{\mathbb{S}}_{d,d'}}\int_{\mathbb{R}} \left[e^{i\Re\mathrm{tr}(AV)x} - 1 - ix\Re\mathrm{tr}(AV)1_{|x|\leq 1}(x)\right]\omega_{d,d'}(\mathrm{d}V)\,\nu_\mu(\mathrm{d}x) \\
&= d\int_{\mathbb{R}} \mathbb{E}_V\left[e^{i\Re\mathrm{tr}(AV)x} - 1 - ix\Re\mathrm{tr}(AV)1_{|x|\leq 1}(x)\right]\nu_\mu(\mathrm{d}x) \\
&= d\mathbb{E}_{u,v}\int_{\mathbb{R}} \left[e^{i\Re\mathrm{tr}(Auv^*)x} - 1 - ix\Re\mathrm{tr}(Auv^*)1_{|x|\leq 1}(x)\right]\nu_\mu(\mathrm{d}x) \\
&= d\mathbb{E}_{u,v}\mathcal{C}_\mu(\Re\langle u, Av\rangle).
\end{aligned}$$

Using (5.4) the cumulant transform of $M_h^{d,d'}$ is

$$\begin{aligned}
\mathcal{C}_{M_h^{d,d'}}(A) &= \int_0^\infty \mathcal{C}_{\Psi^{d,d'}(1)}(h(t)A)\,\mathrm{d}t = \int_0^\infty d\mathbb{E}_{u,v}\mathcal{C}_\mu(\Re\langle u, h(t)Av\rangle)\,\mathrm{d}t \\
&= d\mathbb{E}_{u,v}\int_0^\infty \mathcal{C}_\mu(h(t)\Re\langle u, Av\rangle)\,\mathrm{d}t = d\mathbb{E}_{u,v}\mathcal{C}_{\mu_h}(\Re\langle u, Av\rangle),
\end{aligned}$$

where in the last equality we have used the relation (5.2) between the cumulant transforms of μ_h and μ corresponding to the stochastic integral representation (5.6). □

From the proof of Theorem 5.1 is obtained the following stochastic integral representation in law of the *BG* rectangular Lévy processes.

Corollary 5.2 *Let $M_{d,d'} = \{M_{d,d'}(t) : t \geq 0\}$ be the matrix Lévy process associated to the BG random matrix ensembles. If μ_h is a symmetric infinitely divisible distribution on $\mathbb{R}$ with stochastic integral representation (5.6) then*

$$M_{d,d'}(t) \stackrel{d}{=} \int_0^\infty h(t)\mathrm{d}\Psi^{d,d'}(t)$$

where $\left\{\Psi^{d,d'}(t) : t \geq 0\right\}$ is the $\mathbb{M}_{d,d'}$-valued Lévy process with Lévy measure $\nu_\Psi^{d,d'}$ given by (5.5).

5.1 Ornstein-Uhlenbeck Rectangular Type Processes

The class of selfdecomposable distributions denoted by $L(\mathbb{R})$ is characterized by the stochastic integral representation (5.6) where $\mu \in I_{\log}(\mathbb{R})$ and $h(t) = 1_{(0,\infty)}(t)e^{-t}$, see [14, 21] and [23].

Let $\mu_h \in L(\mathbb{R})$ with such an Ornstein-Uhlenbeck type integral representation. The random matrix models (5.7) for the corresponding $\boxplus_c$-free selfdecomposable distributions $\Lambda_c(\mu_h)$ are given by

$$\left(M_h^{d,d'} = \int_0^\infty e^{-t}\mathrm{d}\Psi^{d,d'}(t)\right)_{d\geq 1, d'\geq 1}$$

and satisfy the $I_{\log}$-condition, that is

$$\int_{\|X\|>2} \log\|X\| \, \nu_\Psi^{d,d'}(\mathrm{d}X) < \infty,$$

which follows from (5.5) and the fact that $\mu \in I_{\log}(\mathbb{R})$. According to (5.5) the Lévy measures of these Ornstein-Uhlenbeck type rectangular matrix integrals are supported on the subset of rank one matrices in $\mathbb{M}_{d,d'}$.

Next we get the Lévy measure $\nu_{M_h^{d,d'}}$ in terms of the Lévy measure ν_{μ_h}. From [22] the Lévy measures of μ and μ_h are related as

$$\nu_{\mu_h}(B) = \int_0^\infty dt \int_{\mathbb{R}} 1_B(h(t)x)\nu_\mu(\mathrm{d}x) \quad B \in \mathcal{B}(\mathbb{R}\backslash\{0\})$$

and from [6] it is obtained a similar relation between the Lévy measures of $M_h^{d,d'}$ and $\Psi^{d,d'}$,

$$\nu_{M_h^{d,d'}}(B) = \int_0^\infty dt \int_{\mathbb{M}_{d,d'}} 1_B(h(t)X)\nu_{\Psi^{d,d'}}(\mathrm{d}X) \quad B \in \mathcal{B}(\mathbb{M}_{d,d'}\backslash\{0\}).$$

Hence from (5.5),

$$\nu_{M_h^{d,d'}}(B) = d\int_{\tilde{\mathbb{S}}_{d,d'}} \omega_{d,d'}(dV) \int_{-\infty}^{\infty} 1_B(xV)\nu_{\mu_h}(\mathrm{d}x) \quad B \in \mathcal{B}\left(\tilde{\mathbb{M}}_{d,d'}\setminus\{0\}\right). \tag{5.8}$$

Let (τ, ν_ξ) and be (τ_h, ν_{h_ξ}) denote the polar decomposition (3.4) of the Lévy measures ν_μ and ν_{μ_h} respectively. Since μ_h is selfdecomposable, the radial component ν_{h_ξ} is expressed as

$$\nu_{h_\xi}(dr) = 1_{(0,\infty)}(r)\frac{k_{h_\xi}(r)}{r}dr,$$

where $k_{h_\xi}(r)$ is a nonnegative measurable function in $\xi \in S = \{1, -1\}$ and decreasing, right continuous in $r \in (0, \infty)$. It is proved in [1] that the k_{h_ξ}-function of ν_{μ_h} is given by

$$k_{h_\xi}(r) = \nu_\xi\left((r, \infty)\right).$$

Therefore, the Lévy measures of these random matrix models of Ornstein-Uhlenbeck type for free selfdecomposable distributions $\Lambda(\mu_h)$ are given by

$$\begin{aligned}\nu_{M_h^{d,d'}}(B) &= d\int_{\tilde{\mathbb{S}}_{d,d'}} \omega_{d,d'}(\mathrm{d}V) \int_{-\infty}^{\infty} 1_B(xV)\nu_{\mu_h}(\mathrm{d}x) \\ &= d\int_{\tilde{\mathbb{S}}_{d,d'}} \omega_{d,d'}(\mathrm{d}V) \int_S \tau_h(\mathrm{d}\xi) \int_0^\infty \frac{\mathrm{d}r}{r}\nu_\xi\left((r,\infty)\right) 1_B(r\xi V).\end{aligned}$$

Appendix

We prove Lemma 3.2 of Sect. 3.

Lemma 1 *Let u, v be independent random vectors, uniformly distributed on the unit sphere of respectively $\mathbb{C}^d$, $\mathbb{C}^{d'}$. Then for any $A \in \mathbb{M}_{d,d'}$,*

(a)

$$\mathbb{E}_{u,v}\left(\langle u, Av\rangle\right) = 0,$$

(b)

$$\mathbb{E}_{u,v}\left(\Re\langle u, Av\rangle\right)^2 = \frac{1}{2dd'}\mathrm{tr}\left(AA^*\right).$$

Proof The assertion *(a)* clearly follows from independence of u and v. Let us prove the assertion *(b)*. By noting that $(vu^*)_{ji} = v_j\bar{u}_i$ we get

$$\langle u, Av\rangle = \mathrm{tr}\left(Avu^*\right) = \sum_{i=1}^{d}\sum_{j=1}^{d'} a_{ij}\bar{u}_i v_j,$$

then

$$\Re\langle u, Av\rangle = \sum_{i=1}^{d}\sum_{j=1}^{d'} \Re\left(a_{ij}\bar{u}_i v_j\right).$$

If $a_{ij} = a_{ij1} + ia_{ij2}, u_i = u_{i1} + iu_{i2},\ v_j = v_{j1} + iv_{j2}$ then

$$\begin{aligned}\Re\left(a_{ij}\bar{u}_i v_j\right) &= \Re\left[\left(a_{ij1} + ia_{ij2}\right)\left(u_{i1} - iu_{i2}\right)\left(v_{j1} + iv_{j2}\right)\right] \\ &= \left(u_{i1}v_{j1} + u_{i2}v_{j2}\right)a_{ij1} + \left(u_{i2}v_{j1} - u_{i1}v_{j2}\right)a_{ij2}. \qquad (1)\end{aligned}$$

Now

$$\begin{aligned}(\Re\langle u, Av\rangle)^2 &= \left[\sum_{i=1}^{d}\sum_{j=1}^{d'} \Re\left(a_{ij}\bar{u}_i v_j\right)\right]\left[\sum_{i=1}^{d}\sum_{j=1}^{d'} \Re\left(a_{ij}\bar{u}_i v_j\right)\right] \\ &= \sum_{i=1}^{d}\sum_{j=1}^{d'}\sum_{k=1}^{d}\sum_{l=1}^{d'} \Re\left(a_{ij}\bar{u}_i v_j\right)\Re\left(a_{kl}\bar{u}_k v_l\right) \\ &= \sum_{i=1}^{d}\sum_{j=1}^{d'} \left(\Re\left(a_{ij}\bar{u}_i v_j\right)\right)^2 + \sum_{i=1}^{d}\sum_{j=1}^{d'}\sum_{k\neq i}^{d}\sum_{l\neq j}^{d'} \Re\left(a_{ij}\bar{u}_i v_j\right)\Re\left(a_{kl}\bar{u}_k v_l\right).\end{aligned}$$

Expanding $\left[\Re\left(a_{ij}\bar{u}_i v_j\right)\right]^2$ by using (1) we get

$$\begin{aligned}(\Re\langle u, Av\rangle)^2 &= \sum_{i=1}^{d}\sum_{j=1}^{d'} \left[\left(u_{i1}^2 v_{j1}^2 + u_{i2}^2 v_{j2}^2\right)a_{ij1}^2 + \left(u_{i2}^2 v_{j1}^2 + u_{i1}^2 v_{j2}^2\right)a_{ij2}^2\right] \\ &+ \sum_{i=1}^{d}\sum_{j=1}^{d'} [\left(2u_{i2}^2 v_{j1}v_{j2} - 2u_{i1}u_{i2}v_{j2}^2 + 2u_{i1}u_{i2}v_{j1}^2 - 2u_{i1}^2 v_{j1}v_{j2}\right)a_{ij1}a_{ij2} \\ &+ 2u_{i1}u_{i2}v_{j1}v_{j2}\left(a_{ij1}^2 - a_{ij2}^2\right)] \\ &+ \sum_{i=1}^{d}\sum_{j=1}^{d'}\sum_{k\neq i}^{d}\sum_{l\neq j}^{d'} \Re\left(a_{ij}\bar{u}_i v_j\right)\Re\left(a_{kl}\bar{u}_k v_l\right).\end{aligned}$$

Taking expectation of $(\Re\langle u, Av\rangle)^2$ we obtain that the expectation of the terms in the second and third summand are zero. Thus, using component-wise the Lemma 2 below,

$$\begin{aligned}
\mathbb{E}_{u,v}(\Re\langle u, Av\rangle)^2 &= \sum_{i=1}^{d}\sum_{j=1}^{d'}\mathbb{E}\left[\left(u_{i1}^2 v_{j1}^2 + u_{i2}^2 v_{j2}^2\right)a_{ij1}^2 + \left(u_{i2}^2 v_{j1}^2 + u_{i1}^2 v_{j2}^2\right)a_{ij2}^2\right] \\
&= \sum_{i=1}^{d}\sum_{j=1}^{d'}\left[\left(\frac{1}{2d}\frac{1}{2d'} + \frac{1}{2d}\frac{1}{2d'}\right)a_{ij1}^2 + \left(\frac{1}{2d}\frac{1}{2d'} + \frac{1}{2d}\frac{1}{2d'}\right)a_{ij2}^2\right] \\
&= \frac{1}{2dd'}\sum_{i=1}^{d}\sum_{j=1}^{d'}\left(a_{ij1}^2 + a_{ij2}^2\right) \\
&= \frac{1}{2dd'}\sum_{i=1}^{d}\sum_{j=1}^{d'}\left|a_{ij}\right|^2 \\
&= \frac{1}{2dd'}\operatorname{tr}\left(AA^*\right).
\end{aligned}$$

□

The following lemma is well known in the real vector case, see e.g. [15, eq. (2)]. We give a proof for the completeness of the paper.

Lemma 2 *Let $U = (U_1, \ldots, U_d)^T$ be a random vector uniformly distributed on the unit sphere of $\mathbb{C}^d$. Then their components U_k, $k = 1, 2, \ldots, d$ are identically distributed with symmetric density function*

$$\frac{d-1}{\pi}\left(1 - x^2 - y^2\right)^{d-2}, \quad x^2 + y^2 \leq 1. \tag{2}$$

The components X_k, Y_k of $U_k = X_k + iY_k$, are identically distributed with density

$$\frac{1}{\sqrt{\pi}}\frac{\Gamma(n)}{\Gamma\left(n - \frac{1}{2}\right)}\left(1 - t^2\right)^{n-3/2}, \ -1 \leq t \leq 1,$$

and their first two marginal moments are

$$EX_k = EY_k = 0$$

and

$$EX_k^2 = EY_k^2 = \frac{1}{2d}.$$

Proof Let $u = (u_1, \ldots, u_d)$ a random vector choosen uniformly on $\{u \in \mathbb{C}^d : \|u\| = 1\}$, observe that $\|u\|^2 = uu^* = \sum_{i=1}^d |u_i|^2$. By [13, pp 140] the distribution of u_k for $1 \le k \le d$ is

$$\frac{d-1}{\pi}\left(1-r^2\right)^{d-2} r\mathrm{d}r\mathrm{d}\theta \quad (u_k = re^{i\theta}, 0 \le r \le 1, 0 \le \theta \le 2\pi).$$

Using polar coordinates, $r = \sqrt{x^2+y^2}$ and $\theta = \arctan(y/x)$ wich implies $\frac{\partial(r,\theta)}{\partial(x,y)} = \frac{1}{\sqrt{x^2+y^2}}$. By the change of variable formula if we denote $u_k = x + iy$ then the distribution of u_k is

$$\frac{d-1}{\pi}\left(1-x^2-y^2\right)^{d-2},$$

where $x^2 + y^2 \le 1$. This proves (2). Hence X_k and Y_k are identically distributed symmetric distribution around zero and therefore $EX_k = EY_k = 0$. Next we compute the marginal density of X_k

$$\begin{aligned} f_{X_k}(x) &= \frac{n-1}{\pi}\int_{-\sqrt{1-x^2}}^{\sqrt{1-x^2}} \left(1-x^2-y^2\right)^{n-2}\mathrm{d}y \\ &= \frac{2(n-1)}{\pi}\int_0^{\sqrt{1-x^2}} \left(1-x^2-y^2\right)^{n-2}\mathrm{d}y, \end{aligned}$$

by change of variable $s = y^2/\left(1-x^2\right)$ we get

$$\begin{aligned} f_{X_k}(x) &= \frac{2(n-1)}{\pi}\int_0^{\sqrt{1-x^2}} \left(1-x^2-y^2\right)^{n-2}\mathrm{d}y \\ &= \frac{(n-1)\left(1-x^2\right)^{n-3/2}}{\pi}\int_0^1 s^{-\frac{1}{2}}(1-s)^{n-2}\mathrm{d}y \\ &= \frac{(n-1)\left(1-x^2\right)^{n-3/2}}{\pi}\mathrm{Beta}\left(\frac{1}{2}, n-1\right) \\ &= \frac{1}{\sqrt{\pi}}\frac{\Gamma(n)}{\Gamma\left(n-\frac{1}{2}\right)}\left(1-x^2\right)^{n-3/2}. \end{aligned}$$

Now

$$\mathrm{E}X_k^2 = \int\int_D x^2 f(x,y)\,\mathrm{d}x\mathrm{d}y = \frac{d-1}{\pi}\int_{-1}^1\int_{-\sqrt{1-y^2}}^{\sqrt{1-y^2}} x^2\left(1-x^2-y^2\right)^{d-2}\mathrm{d}x\mathrm{d}y$$

$$
\begin{aligned}
&= \frac{d-1}{\pi} \int_0^{2\pi} \int_0^1 r^2 \cos^2\theta \left(1-r^2\right)^{d-2} r\mathrm{d}r\mathrm{d}\theta \\
&= \frac{d-1}{\pi} \frac{\pi}{2d\,(d-1)} \\
&= \frac{1}{2d},
\end{aligned}
$$

where we have used the identity $\mathrm{Beta}\,(a,b) = \int_0^1 t^{a-1}\,(1-t)^{b-1}\,\mathrm{d}t = \frac{\Gamma(a)\Gamma(b)}{\Gamma(a+b)}$ to compute $\int_0^1 r^3\left(1-r^2\right)^{d-2}\mathrm{d}r = \frac{1}{2}\int_0^1 s\,(1-s)^{d-2}\,\mathrm{d}s = \frac{1}{2}\mathrm{Beta}\,(2,d-1) = \frac{1}{2d(d-1)}$, the change of variable $s = r^2$ with $dr = ds/\left(2\sqrt{s}\right)$ and $\int_0^{2\pi}\cos^2\theta\mathrm{d}\theta = \pi$. □

References

1. T. Aoyama, M. Maejima, Some classes of infinitely divisible distributions on $\mathbb{R}^d$ (a survey), unpublished note (2008). Revised version of the Research Report: T. Aoyama, M. Maejima, Some classes of infinitely divisible distributions on $\mathbb{R}^d$ (a survey), The Institute of Statistical Mathematics Cooperate Research Report, Tokyo, vol. 184 (2006), pp. 5–13
2. T. Aoyama, M. Maejima, Characterizations of subclasses of type G distributions on $\mathbb{R}^d$ by stochastic integral representations. Bermoulli **13**(1), 148–160 (2007)
3. D. Applebaum, Lévy processes and stochastic integrals in Banach spaces. Probab. Math. Stat. **27**, 75–88 (2007)
4. O.E. Barndorff-Nielsen, M. Maejima, K. Sato, Some classes of multivariate infinitely divisible distributions admitting stochastic integral representations. Bernoulli **12**, 1–33 (2006)
5. O.E. Barndorff-Nielsen, R. Stelzer, Positive-definite matrix processes of finite variation. Probab. Math. Stat. **27**, 3–43 (2007)
6. O.E. Barndorff-Nielsen, R. Stelzer, Multivariate supOU processes. Ann. Appl. Probab. **21**, 140–182 (2011)
7. F. Benaych-Georges, Classical and free infinitely divisible distributions and random matrices. Ann. Probab. **33**, 1134–1170 (2005)
8. F. Benaych-Georges, Infinitely divisible distributions for rectangular free convolution- classification and matricial interpretation. Probab. Theory Relat. Fields **139**, 143–189 (2007)
9. F. Benaych-Georges, Random matrices, related convolutions. Probab. Theory Relat. Fields **144**, 471–515 (2009). Revised version of F. Benaych-Georges, Random matrices, related convolutions. arXiv (2005)
10. T. Cabanal-Duvillard, A matrix representation of the Bercovici-Pata bijection. Electron. J. Probab. **10**, 632–661 (2005)
11. J.A. Domínguez-Molina, V. Pérez-Abreu, A. Rocha-Arteaga, Covariation representations for Hermitian Lévy process ensembles for free infinitely divisible distributions. Electron. Commun. Probab. **18**, 1–14 (2013)
12. J.A. Domínguez-Molina, A. Rocha-Arteaga, Random matrix models of stochastic integrals type for free infinitely divisible distributions. Periodica Mathematica Hungarica. **64**(2), 145–160 (2012)
13. F. Hiai, D. Petz, *The Semicircle Law, Free Random Variables and Entropy*. Mathematics Surveys and Monographs, vol. 77 (American Mathematical Society, Providence, 2000)
14. Z.J. Jurek, W. Vervaat, An integral representation for selfdecomposable Banach space valued random variables. Z. Wahrscheinlichkeitstheorie. Verw. Geb. **62**, 247–262 (1983)
15. J.F.C. Kingman, Random walks with spherical symmetry. Acta Math. **109**(1), 11–53 (1963)

16. V. Pérez-Abreu, N. Sakuma, Free generalized gamma convolutions. Electron. Commun. Probab. **13**, 526–539 (2008)
17. P. Protter, *Stochastic Integration and Differential Equations*. Stochastic Modelling and Applied Probability, vol. 21 (Springer, Berlin/New York, 2004)
18. B.S. Rajput, J. Rosiński, Spectral representations of infinitely divisible processes. Probab. Theory Relat. Fields **82**, 451–487 (1989)
19. A. Rocha-Arteaga, K. Sato, *Topics in Infinitely Divisible Distributions and Lévy Processes*. Aportaciones Matemáticas, Investigación, vol. 17 (Sociedad Matemática Mexicana, México, 2003)
20. J. Rosiński, On series representations of infinitely divisible random vectors. Ann. Probab. **18**, 405–430 (1990)
21. K. Sato, *Lévy Processes and Infinitely Divisible Distributions* (Cambridge University Press, Cambridge, 1999)
22. K. Sato, Additive processes and stochastic integrals. Ill. J. Math. **50**, 825–851 (2006)
23. K. Sato, M. Yamazato, Stationary processes of Ornstein-Uhlenbeck type, in *Probability Theory and Mathematical Statistics*, ed. by K. Ito, J.V. Prokhorov. Lecture Notes in Mathematics, vol. 1021 (Spriger, Berlin, 1983), pp. 541–551
24. K. Urbanik, W.A. Woyczynski, A random integral and Orlicz spaces. Bull. Acad. Polon. Sci. Math. Astro. e Phys. **15**, 161–168 (1967)

Asymptotic Behaviour of Poisson-Dirichlet Distribution and Random Energy Model

Shui Feng and Youzhou Zhou

Abstract The family of Poisson-Dirichlet distributions is a collection of two-parameter probability distributions $\{PD(\alpha, \theta) : 0 \leq \alpha < 1, \alpha + \theta > 0\}$ defined on the infinite-dimensional simplex. The parameters α and θ correspond to the stable and gamma component respectively. The distribution $PD(\alpha, 0)$ arises in the thermodynamic limit of the Gibbs measure of Derrida's Random Energy Model (REM) in the low temperature regime. In this setting α can be written as the ratio between the temperature T and a critical temperature T_c. In this paper, we study the asymptotic behaviour of $PD(\alpha, \theta)$ as α converges to one or equivalently when the temperature approaches the critical value T_c.

Keywords Dirichlet process • Large deviations • Phase transition • Poisson-Dirichlet distribution • Random energy model

Mathematics Subject Classification (2001). Primary 60F10; secondary 92D10.

1 Introduction

Derrida's Random Energy Model (henceforce REM) introduced in [2] and [3] is a very instructive toy model for disordered systems such as spin glasses. For any $N \geq 1$, let $S_N = \{-1, 1\}^N$ denote the configuration space. Then the REM is

---

This work was supported by the Natural Sciences and Engineering Research Council of Canada.

S. Feng (✉)
Department of Mathematics and Statistics, McMaster University, 1280 Main Street West, Hamilton, ON L8S 4K1, Canada
e-mail: shuifeng@mcmaster.ca

Y. Zhou
The School of Statistics and Mathematics, Zhongnan University of Economics and Law, 182 South Lake Avenue, East Lake New Technology Development Zone, Wuhan 430073, China
e-mail: youzhouzhou1984@gmail.com

R.H. Mena et al. (eds.), *XI Symposium on Probability and Stochastic Processes*, Progress in Probability 69, DOI 10.1007/978-3-319-13984-5_7

simply a family of i.i.d. random variables $\{H_N(\sigma) : \sigma \in S_N\}$ with common normal distribution of mean zero and variance $N/2$. Here $H_N(\sigma)$ is the Hamiltonian. Given the temperature T and $\beta = T^{-1}$, the Gibbs measure is a probability on S_N given by

$$Z_N^{-1}\exp\{-\beta H_N(\sigma)\}$$

where

$$Z_N = \sum_{\sigma \in S_N} \exp\{-\beta H_N(\sigma)\}$$

is the partition function. Let $T_c = \frac{1}{2\sqrt{2}}$. Then for $T < T_c$ or equivalently $\beta > 2\sqrt{2}$, and $\alpha = \frac{T}{T_c}$ the decreasing order statistic of the Gibbs measure is known (cf. [22]) to converge to a particular Poisson-Dirichlet distribution $PD(\alpha, 0)$ given below as N tends to infinity.

More generally, for any $0 \leq \alpha < 1, \theta + \alpha > 0$, let $U_1(\alpha, \theta), U_2(\alpha, \theta), \ldots$ be a sequence of independent random variables with $U_i(\alpha, \theta)$ having distribution $Beta(1 - \alpha, \theta + i\alpha)$ for $i \geq 1$. If we define

$$V_1(\alpha, \theta) = U_1(\alpha, \theta), V_n(\alpha, \theta) = (1 - U_1(\alpha, \theta)) \cdots (1 - U_{n-1}(\alpha, \theta))U_n(\alpha, \theta), \quad n \geq 2,$$

then the law of the decreasing order statistic of $(V_1(\alpha, \theta), V_2(\alpha, \theta), \ldots)$ is the two-parameter Poisson-Dirichlet distribution $PD(\alpha, \theta)$. It is clearly a probability on the infinite-dimensional simplex

$$\nabla = \{\mathbf{p} = (p_1, p_2, \ldots) : p_1 \geq p_2 \geq \cdots \geq 0, \sum_{i=1}^{\infty} p_i \leq 1\}.$$

The family $PD(0, \theta)$ was introduced by Kingman [15] as the law of relative jump sizes of a gamma subordinator over the interval $[0, \theta]$. It also arises in many other context most notably in population genetics. The family $PD(\alpha, 0)$ arising in the REM $(0 < \alpha < 1)$ was introduced in Kingman [15] through the stable subordinator. In [17] and [19], it was constructed from the ranked length of excursion intervals between zeros of a Brownian motion ($\alpha = 1/2$) or a recurrent Bessel process of order $2(1 - \alpha)$ for general α.

There has been an intensive study of the asymptotic behaviour for the Poisson-Dirichlet distribution in recent years with motivations from population genetics and Bayesian statistics. These include results for large θ [1, 5, 7, 8, 10–14], and results for small θ and α [6, 9].

In the context of REM, the large deviation result for small α and the structure of "energy ladder" obtained in [6] describe the microscopic transition from zero temperature to positive temperature. In this paper, we study the asymptotic behaviour of the Poisson-Dirichlet distribution when the parameter α converges to one. In terms of REM, we are dealing with the microscopic transition from the critical temperature to lower temperature.

The paper is organized as follows. In Sect. 2, we derive the law of large numbers of the two-parameter Poisson-Dirichlet distribution, and the two-parameter Dirichlet process defined in (2.1). The large deviation principle is established for the two-parameter Poisson-Dirichlet distribution. In Sect. 3, the result in Sect. 2 is applied to investigate models involving selection or external field and to study the asymptotic behaviour of homozygosity. Finally a detailed comparison with earlier work is carried out in Sect. 4.

2 Asymptotic Behaviour

Let S be a compact Polish space and $M_1(S)$ denote the space of probability measures on S equipped with the weak topology. Given a diffuse probability measure ν_0 in $M_1(S)$, let $\xi_1, \xi_2, \ldots$ be a sequence of i.i.d. random variables with common distribution ν_0. Independently for any $0 < \alpha < 1, \theta > -\alpha$, let $(P_1(\alpha,\theta), P_2(\alpha,\theta), \ldots)$ have the $PD(\alpha,\theta)$ distribution. Then the two-parameter Dirichlet process is defined to be the following random measure

$$\Xi_{\alpha,\theta,\nu_0} = \sum_{i=1}^{\infty} P_i(\alpha,\theta)\delta_{\xi_i}. \tag{2.1}$$

In this section, we first derive the limits for both $(P_1(\alpha,\theta), P_2(\alpha,\theta), \ldots)$ and $\Xi_{\alpha,\theta,\nu_0}$ when α converges to 1. The large deviation principle is then established for $(P_1(\alpha,\theta), P_2(\alpha,\theta), \ldots)$.

Theorem 2.1 *As α converges to 1, we have*

$$(P_1(\alpha,\theta), P_2(\alpha,\theta), \ldots) \to \mathbf{0} = (0, 0, \ldots) \text{ in probability in space } \nabla. \tag{2.2}$$

and

$$\Xi_{\alpha,\theta,\nu_0} \to \nu_0 \text{ in probabilty in space } M_1(S). \tag{2.3}$$

Proof To prove (2.2) it suffices to verify the convergence of $P_1(\alpha,\theta)$. For any $\epsilon > 0$ it follows from direct calculation that

$$\begin{aligned} \mathbb{P}\{P_1(\alpha,\theta) > \epsilon\} &\le \frac{\mathbb{E}[P_1^2(\alpha,\theta)]}{\epsilon^2} \\ &\le \frac{\mathbb{E}[\sum_{i=1}^{\infty} P_i^2(\alpha,\theta)]}{\epsilon^2} \\ &= \frac{1-\alpha}{\epsilon^2(\theta+1)} \to 0 \end{aligned}$$

where the last equality follows from the Pitman sampling formula [18, 20].

Let $C(S)$ be the space of continuous functions on S with uniform convergence topology. For any μ in $M_1(S)$ and f in $C(S)$, let $\langle \mu, f \rangle$ denote the integration of f with respect to μ. Since S is compact, one can find a countable dense subset $\{f_i : i = 1, 2, \ldots\}$ of $C(S)$ such that the weak topology can be generated by the metric

$$\rho(\mu, \nu) = \sum_{i=1}^{\infty} \frac{|\langle \mu - \nu, f_i \rangle| \wedge 1}{2^i}.$$

Hence it suffices to show that for any $\epsilon > 0$ and any $i \geq 1$

$$\mathbb{P}\{|\langle \Xi_{\alpha,\theta,\nu_0}, f_i \rangle - \langle \nu_0, f_i \rangle| > \epsilon\} \to 0 \text{ as } \alpha \to 1. \tag{2.4}$$

By direct calculation, one has

$$\begin{aligned}
\mathbb{E}[\langle \Xi_{\alpha,\theta,\nu_0}, f_i \rangle] &= \langle \nu_0, f_i \rangle \\
\mathbb{E}[(\langle \Xi_{\alpha,\theta,\nu_0}, f_i \rangle)^2] &= \langle \nu_0, f_i^2 \rangle \mathbb{E}[\sum_{i=1}^{\infty} P_i^2(\alpha, \theta)] \\
&\quad + (\langle \nu_0, f_i \rangle)^2 (1 - \mathbb{E}[\sum_{i=1}^{\infty} P_i^2(\alpha, \theta)]) \\
&= \frac{1-\alpha}{\theta+1} \langle \nu_0, f_i^2 \rangle + \frac{\theta+\alpha}{\theta+1} (\langle \nu_0, f_i \rangle)^2.
\end{aligned}$$

Thus

$$Var[\langle \Xi_{\alpha,\theta,\nu_0}, f_i \rangle] = \frac{1-\alpha}{\theta+1} [\langle \nu_0, f_i^2 \rangle - (\langle \nu_0, f_i \rangle)^2] \to 0$$

as α converges to 1. This combined with Chebyshev's inequality implies (2.4) and the theorem. □

Next we turn to the large deviation principle for $(P_1(\alpha, \theta), P_2(\alpha, \theta), \ldots)$. The terminologies and general results on large deviations can be found in [4].

The parameter $\theta > -1$ is fixed in the sequel. From the proof of Theorem 2.1 we can see that each $P_i(\alpha, \theta)$ stays away from zero at a probability comparable to $1-\alpha$ as α converges to 1. This suggests a large deviation speed $-\log(1-\alpha)$. Since

$$\Gamma(2-\alpha) = (1-\alpha)\Gamma(1-\alpha),$$

it follows that

$$\lim_{\alpha \to 1} \frac{-\log(1-\alpha)}{\log \Gamma(1-\alpha)} = 1. \tag{2.5}$$

Let $F_{\alpha,\theta}(p)$ denote the distribution function of $P_1(\alpha,\theta)$, and $f(p_1,\cdots,p_n;\alpha,\theta)$ the density function of $(P_1(\alpha,\theta),\ldots,P_n(\alpha,\theta))$. Then by Perman's formula [7, 16] one has for any $n \geq 1$

$$f(p_1,\ldots,p_n;\alpha,\theta) = C(\alpha,\theta,n)\frac{\hat{p}_n^{\theta+n\alpha-1}}{(\prod_{i=1}^n p_i)^{\alpha+1}}F_{\alpha,\theta+n\alpha}(\frac{p_n}{\hat{p}_n}) \tag{2.6}$$

where $\hat{p}_n = 1-\sum_{i=1}^n p_i$ and

$$C(\alpha,\theta,n) = \left(\frac{\alpha}{\Gamma(1-\alpha)}\right)^n \frac{\Gamma(\theta+1)\Gamma(\frac{\theta}{\alpha}+n+1)}{\Gamma(\frac{\theta}{\alpha}+1)\Gamma(\theta+n\alpha+1)}.$$

This representation will be the key in the establishment of the large deviation principle.

Theorem 2.2 *The family* $\{(P_1(\alpha,\theta),P_2(\alpha,\theta),\ldots) : \alpha \in (0\vee(-\theta),1)\}$ *satisfies a large deviation principle on space* ∇ *as* α *converges to* 1 *with speed* $-\log(1-\alpha)$ *and a good rate function*

$$I(\mathbf{p}) = \begin{cases} 0, & \textit{if } \mathbf{p}=\mathbf{0} \\ \max\{n\geq 1 : p_n > 0\}, & \textit{otherwise}, \end{cases} \tag{2.7}$$

We start the proof of the theorem with two lemmas.

Lemma 2.3 *For any* p *in* $[0,1]$

$$\begin{aligned}&\lim_{\delta\to 0}\liminf_{\alpha\to 1}\frac{1}{-\log(1-\alpha)}\log\mathbb{P}\{|P_1(\alpha,\theta)-p|<\delta\}\\ &=\lim_{\delta\to 0}\limsup_{\alpha\to 1}\frac{1}{-\log(1-\alpha)}\log\mathbb{P}\{|P_1(\alpha,\theta)-p|\leq\delta\}\\ &=-I_1(p).\end{aligned} \tag{2.8}$$

where

$$I_1(p) = \begin{cases} 0, \textit{ if } p=0 \\ 1, \textit{ if } 0<p\leq 1, \end{cases} \tag{2.9}$$

Proof It follows from Theorem 2.1 that

$$\begin{aligned}&\lim_{\delta\to 0}\liminf_{\alpha\to 1}\frac{1}{-\log(1-\alpha)}\log\mathbb{P}\{P_1(\alpha,\theta)<\delta\}\\ &=\lim_{\delta\to 0}\limsup_{\alpha\to 1}\frac{1}{-\log(1-\alpha)}\log\mathbb{P}\{P_1(\alpha,\theta)\leq\delta\}\\ &=0.\end{aligned} \tag{2.10}$$

For any p in $(0,1)$, one can choose δ small enough so that $(p-\delta, p+\delta)$ is contained in $(0,1)$. Applying (2.6) we obtain

$$\mathbb{P}\{|P_1(\alpha,\theta)-p|<\delta\} \geq 2\delta C(\alpha,\theta,1)\frac{(1-p-\delta)^{\theta+\alpha-1}}{(p+\delta)^{\alpha+1}}F_{\alpha,\theta+\alpha}(\frac{p-\delta}{1-p+\delta})$$
$$\mathbb{P}\{|P_1(\alpha,\theta)-p|\leq\delta\} \leq 2\delta C(\alpha,\theta,1)\frac{(1-p+\delta)^{\theta+\alpha-1}}{(p-\delta)^{\alpha+1}}F_{\alpha,\theta+\alpha}(\frac{p+\delta}{1-p-\delta})$$

which implies

$$\begin{aligned}
&\lim_{\delta\to 0}\liminf_{\alpha\to 1}\frac{1}{-\log(1-\alpha)}\log\mathbb{P}\{|P_1(\alpha,\theta)-p|<\delta\} \\
&= \lim_{\delta\to 0}\limsup_{\alpha\to 1}\frac{1}{-\log(1-\alpha)}\log\mathbb{P}\{|P_1(\alpha,\theta)-p|\leq\delta\} \\
&= -1.
\end{aligned} \tag{2.11}$$

For $p=1$, choose $\delta<1/2$. Then we have

$$\begin{aligned}
\mathbb{P}\{|P_1(\alpha,\theta)-1|<\delta\} &\geq \mathbb{P}\{1-\delta<P_1(\alpha,\theta)<1-\delta/2\} \\
&\geq \delta C(\alpha,\theta,1)\frac{(\delta/2)^{\theta+\alpha-1}}{(1-\delta/2)^{\alpha+1}}F_{\alpha,\theta+\alpha}(\frac{1-\delta}{\delta}) \\
\mathbb{P}\{|P_1(\alpha,\theta)-1|\leq\delta\} &\leq 2\delta C(\alpha,\theta,1)\frac{\delta^{\theta+\alpha-1}}{(1-\delta)^{\alpha+1}}
\end{aligned}$$

which implies

$$\begin{aligned}
&\lim_{\delta\to 0}\liminf_{\alpha\to 1}\frac{1}{-\log(1-\alpha)}\log\mathbb{P}\{|P_1(\alpha,\theta)-1|<\delta\} \\
&= \lim_{\delta\to 0}\limsup_{\alpha\to 1}\frac{1}{-\log(1-\alpha)}\log\mathbb{P}\{|P_1(\alpha,\theta)-1|\leq\delta\} \\
&= -1.
\end{aligned} \tag{2.12}$$

Putting together (2.10)–(2.12), we obtain the lemma. □

For any $n\geq 2$, set

$$\nabla_n = \{(p_1,\ldots,p_n) : (p_1,\ldots,p_n,0,0\ldots)\in\nabla\}.$$

For any $(p_1,\ldots,p_n)$ in ∇_n and $\delta>0$, set

$$B_n((p_1,\ldots,p_n),\delta) = \{(q_1,\ldots,q_n)\in\nabla_n : |q_i-p_i|<\delta, i=1,\ldots,n\}$$
$$\bar{B}_n((p_1,\ldots,p_n),\delta) = \{(q_1,\ldots,q_n)\in\nabla_n : |q_i-p_i|\leq\delta, i=1,\ldots,n\}.$$

Lemma 2.4 *Given $n \geq 2$ and $(p_1, \ldots, p_n)$ in ∇_n, we have*

$$\begin{aligned}
&\lim_{\delta \to 0} \liminf_{\alpha \to 1} \frac{1}{-\log(1-\alpha)} \log \mathbb{P}_{n,\alpha,\theta}\{B_n((p_1, \ldots, p_n), \delta)\} \\
&= \lim_{\delta \to 0} \limsup_{\alpha \to 1} \frac{1}{-\log(1-\alpha)} \log \mathbb{P}_{n,\alpha,\theta}\{\bar{B}_n((p_1, \ldots, p_n), \delta)\} \\
&= -I_n(p_1, \ldots, p_n).
\end{aligned} \tag{2.13}$$

where $\mathbb{P}_{n,\alpha,\theta}$ is the law of $(P_1(\alpha, \theta), \ldots, P_n(\alpha, \theta))$ and

$$I_n(p_1, \ldots, p_n) = \begin{cases} 0, & \text{if } p_1 = 0 \\ \max\{1 \leq i \leq n : p_i > 0\}, & \text{otherwise}, \end{cases} \tag{2.14}$$

Proof Fix $n \geq 2$ and a point $(p_1, \ldots, p_n)$ in ∇_n. If $p_1 = \cdots = p_n = 0$, then the result (2.13) follows directly from Theorem 2.1. Next we assume $p_1 > 0$ and set

$$r = \max\{1 \leq i \leq n : p_i > 0\}.$$

By definition, we have

$$\mathbb{P}_{n,\alpha,\theta}\{\bar{B}_n((p_1, \ldots, p_n), \delta)\} \leq \mathbb{P}_{r,\alpha,\theta}\{\bar{B}_r((p_1, \ldots, p_r), \delta)\}.$$

Choose δ small enough so that $p_i - \delta > 0$ for $1 \leq i \leq r$. Then on the set $\bar{B}_r((p_1, \ldots, p_r), \delta)$ the density function $f(q_1, \ldots, q_r; \alpha, \theta)$ is controlled from above by

$$C(\alpha, \theta, r) \frac{1}{(\prod_{i=1}^{r}(p_i - \delta))^{\alpha+1}}$$

which implies that

$$\lim_{\delta \to 0} \limsup_{\alpha \to 1} \frac{1}{-\log(1-\alpha)} \log \mathbb{P}_{n,\alpha,\theta}\{\bar{B}_n((p_1, \ldots, p_n), \delta)\} \leq -r \tag{2.15}$$

To prove the lemma, it suffices to verify that

$$\lim_{\delta \to 0} \liminf_{\alpha \to 1} \frac{1}{-\log(1-\alpha)} \log \mathbb{P}_{n,\alpha,\theta}\{B_n((p_1, \ldots, p_n), \delta)\} \geq -r \tag{2.16}$$

The proof of (2.16) is divided into two cases based on the value of $(p_1, \ldots, p_n)$.

Case 1 $r = n$.

In this case we have $p_i > 0$ for all $1 \leq i \leq n$. For any $\delta > 0$ choose $0 < \delta' < \delta$ such that

$$p_i - \delta' > 0, i = 1, \ldots, n.$$

Set

$$G((p_1,\ldots,p_n),\delta') = \{(q_1,\ldots,q_n) \in \nabla_n : p_i - \delta' < q_i < p_i - \frac{\delta'}{2}, i = 1,\ldots,n\}.$$

Clearly $G((p_1,\ldots,p_n),\delta')$ is a proper subset of $B_n((p_1,\ldots,p_n),\delta)$. For any $(q_1,\ldots,q_n)$ in $G((p_1,\ldots,p_n),\delta')$ one has

$$f(q_1,\ldots,q_n;\alpha,\theta) \geq C(\alpha,\theta,n)\frac{(n\delta'/2)^{\theta+n\alpha-1}}{\prod_{i=1}^n (p_i - \delta'/2)^{\alpha+1}} F_{\alpha,\theta+n\alpha}(\frac{p_n-\delta'}{1-\sum_{i=1}^n p_i + n\delta'})$$

which implies

$$\begin{aligned}
&\lim_{\delta\to 0}\liminf_{\alpha\to 1}\frac{1}{-\log(1-\alpha)}\log \mathbb{P}_{n,\alpha,\theta}\{B_n((p_1,\ldots,p_n),\delta)\}\\
&\geq \lim_{\delta\to 0}\liminf_{\alpha\to 1}\frac{1}{-\log(1-\alpha)}\log \mathbb{P}_{n,\alpha,\theta}\{G((p_1,\ldots,p_n),\delta')\}\\
&\geq \lim_{\delta\to 0}\liminf_{\alpha\to 1}\frac{1}{-\log(1-\alpha)}\log\Bigg(C(\alpha,\theta,n)\frac{(n\delta'/2)^{\theta+n\alpha-1}}{\prod_{i=1}^n (p_i-\delta'/2)^{\alpha+1}}\\
&\qquad \times F_{\alpha,\theta+n\alpha}(\frac{p_n-\delta'}{1-\sum_{i=1}^n p_i + n\delta'})\Bigg)\\
&= -n.
\end{aligned}$$

Case 2 $\quad 1 \leq r < n$.

In this case we have $p_i > 0$ for $i = 1,\ldots,r$ and $p_i = 0$ for $i = r+1,\ldots,n$. Choosing $0 < \delta' < \delta$ so that $p_i - \delta' > 0$ for $1 \leq i \leq r$ and $\delta'' = \frac{\delta'}{4(n-r)} < \delta$. Set

$$\begin{aligned}
\tilde{G}((p_1,\ldots,p_n),\delta') = \{&(q_1,\ldots,q_n) \in \nabla_n : p_i - \delta' < q_i < p_i - \frac{\delta'}{2}, i = 1,\ldots,r;\\
&0 < q_i < \delta'', r < i \leq n\}
\end{aligned}$$

It is clear that $\tilde{G}((p_1,\ldots,p_n),\delta')$ is a proper subset of $B((p_1,\ldots,p_n),\delta)$. Let $\chi_{\nabla_n}(\cdot)$ denote the indicator function of ∇_n. It follows from direct calculation that

$$\begin{aligned}
&\mathbb{P}_{n,\alpha,\theta}\{\tilde{G}((p_1,\ldots,p_n),\delta')\}\\
&= \int_{\tilde{G}((p_1,\ldots,p_n),\delta')} f(q_1,\ldots,q_n;\alpha,\theta)\,dq_1\cdots dq_n \qquad (2.17)\\
&= C(\alpha,\theta,n)\int_{p_1-\delta'}^{p_1-\delta'/2}\cdots\int_{p_r-\delta'}^{p_r-\delta'/2} dq_1\cdots dq_r \frac{\hat{q}_r^{\theta+r\alpha-1}}{(\prod_{i=1}^r q_i)^{\alpha+1}}\\
&\times \int_0^{\delta''}\cdots\int_0^{\delta''} dq_{r+1}\cdots dq_n\, \chi_{\nabla_n}(\mathbf{q})\frac{\hat{q}_n^{\theta+n\alpha-1}}{\hat{q}_r^{\theta+r\alpha-1}(\prod_{i=r+1}^n q_i)^{\alpha+1}} F_{\alpha,\theta+n\alpha}(\frac{q_n}{\hat{q}_n})
\end{aligned}$$

Given $q_1, \ldots, q_r$, set $q_i = \hat{q}_r u_{i-r}$ for $i = r+1, \ldots, n$ and

$$D = \{(u_1, \ldots, u_{n-r}) : \frac{q_r}{\hat{q}_r} \geq u_1 \geq \cdots \geq u_{n-r} \geq 0, \sum_{k=1}^{n-r} u_i \leq 1\}.$$

Then one obtains

$$\int_0^{\delta''} \cdots \int_0^{\delta''} d\,q_{r+1} \cdots d\,q_n\, \chi_{\nabla_n}(\mathbf{q}) \frac{\hat{q}_n^{\theta+n\alpha-1}}{\hat{q}_r^{\theta+r\alpha-1}(\prod_{i=r+1}^n q_i)^{\alpha+1}} F_{\alpha,\theta+n\alpha}(\frac{q_n}{\hat{q}_n})$$

$$= \int_0^{\delta''/\hat{q}_r} \cdots \int_0^{\delta''/\hat{q}_r} d\,u_1 \cdots d\,u_{n-r}$$

$$\times \chi_D \frac{\hat{u}_{n-r}^{\theta+r\alpha+(n-r)\alpha-1}}{(\prod_{i=1}^{n-r} u_i)^{\alpha+1}} F_{\alpha,\theta+r\alpha+(n-r)\alpha}(\frac{u_{n-r}}{\hat{u}_{n-r}}) \tag{2.18}$$

$$\geq (C(\alpha, \theta + r\alpha, n-r))^{-1} \mathbb{P}\{P_i(\alpha, \theta+r\alpha) < \frac{\delta'' \wedge (p_r - \delta')}{\hat{p}_r + r\delta'}, i = 1, \ldots, n-r\}.$$

Putting together (2.17) and (2.18) we obtain

$$\mathbb{P}_{n,\alpha,\theta}\{B((p_1, \ldots, p_n), \delta)\} \geq \mathbb{P}_{n,\alpha,\theta}\{\tilde{G}((p_1, \ldots, p_n), \delta')\}$$

$$\geq \frac{C(\alpha, \theta, n)}{C(\alpha, \theta + r\alpha, n-r)} \left(\frac{\delta'}{2}\right)^r \frac{(\delta'/2)^{\theta+r\alpha-1}}{(\prod_{i=1}^r (p_i - \delta'/2))^{\alpha+1}}$$

$$\times \mathbb{P}\{P_i(\alpha, \theta + r\alpha) < \frac{\delta'' \wedge (p_r - \delta')}{\hat{p}_r + r\delta'}, i = 1, \ldots, n-r\}$$

which implies

$$\lim_{\delta \to 0} \liminf_{\alpha \to 1} \frac{1}{-\log(1-\alpha)} \log \mathbb{P}_{n,\alpha,\theta}\{B_n((p_1, \ldots, p_n), \delta)\} \geq -r$$

and thus the lemma. □

The proof of Theorem 2.2 Introducing the metric

$$d(\mathbf{p}, \mathbf{q}) = \sum_{i=1}^{\infty} \frac{|p_i - q_i|}{2^i}$$

on space ∇. For any $\mathbf{p}$ in ∇ and $\delta > 0$, let $B(\mathbf{p}, \delta)$ and $\bar{B}(\mathbf{p}, \delta)$ denote the open and closed balls in ∇ centred at $\mathbf{p}$ with radius δ. Choosing m large enough so that

$1/2^m < \frac{\delta}{2}$. Then for any $0 < 2\delta'' < \delta < \delta'$, one has for any $n \geq 1$

$$\begin{aligned}&\{\mathbf{q} \in \nabla : |q_i - p_i| < \delta'', 1 \leq i \leq m\}\\ &\quad \subset B(\mathbf{p}, \delta) \subset \bar{B}(\mathbf{p}, \delta) \subset \{\mathbf{q} \in \nabla : |q_i - p_i| \leq 2^n\delta', 1 \leq i \leq n\}.\end{aligned}$$

This

$$\begin{aligned}&\mathbb{P}_{m,\alpha,\theta}\{B((p_1, \ldots, p_m), \delta'')\}\\ &= \mathbb{P}\{\{|P_i(\alpha, \theta) - p_i| < \delta'', 1 \leq i \leq m\}\}\\ &\leq \mathbb{P}\{(P_1(\alpha, \theta), \ldots) \in B(\mathbf{p}, \delta)\} \leq \mathbb{P}\{(P_1(\alpha, \theta), \ldots) \in \bar{B}(\mathbf{p}, \delta)\}\\ &\leq \mathbb{P}\{\{|P_i(\alpha, \theta) - p_i| \leq 2^n\delta', 1 \leq i \leq n\}\}\\ &= \mathbb{P}_{n,\alpha,\theta}\{\bar{B}((p_1, \ldots, p_n), 2^n\delta')\}.\end{aligned}$$

Taking limits following the order $\alpha \to 1, \delta'' \to 0, m \to \infty, \delta \to 0, \delta' \to 0$, and $n \to \infty$, we obtain

$$\begin{aligned}-I(\mathbf{p}) &\leq \lim_{m\to\infty} \lim_{\delta''\to 0} \liminf_{\alpha\to 1} \frac{1}{-\log(1-\alpha)} \log \mathbb{P}_{m,\alpha,\theta}\{B((p_1, \ldots, p_m), \delta'')\}\\ &\leq \lim_{\delta\to 0} \liminf_{\alpha\to 1} \frac{1}{-\log(1-\alpha)} \log \mathbb{P}\{(P_1(\alpha, \theta), \ldots) \in B(\mathbf{p}, \delta)\}\\ &\leq \lim_{\delta\to 0} \limsup_{\alpha\to 1} \frac{1}{-\log(1-\alpha)} \log \mathbb{P}\{(P_1(\alpha, \theta), \ldots) \in \bar{B}(\mathbf{p}, \delta)\} \qquad (2.19)\\ &\leq \lim_{n\to\infty} \lim_{\delta'\to 0} \limsup_{\alpha\to 1} \frac{1}{-\log(1-\alpha)} \log \mathbb{P}_{n,\alpha,\theta}\{\bar{B}(\mathbf{p}, 2^n\delta')\}\\ &\leq -I(\mathbf{p}).\end{aligned}$$

This combined with Theorem (P) in [21] or Theorem B.6 in [7] and the compactness of ∇ implies the theorem. □

3 Applications

For any $n \geq 1$ consider the following projection maps

$$\nabla \to \nabla_n, \ \mathbf{p} \mapsto (p_1, \ldots, p_n)$$

and

$$\nabla \to [0, 1/n], \ \mathbf{p} \mapsto p_n.$$

Since both maps are continuous, the next result follows from a direct application of the contraction principle.

Theorem 3.1 *For any fixed $n \geq 1$, the families*

$$\{(P_1(\alpha, \theta), \ldots, P_n(\alpha, \theta)) : \alpha \in (0 \vee (-\theta), 1)\}$$

and

$$\{P_n(\alpha, \theta) : \alpha \in (0 \vee (-\theta), 1)\}$$

satisfy large deviation principles on spaces ∇_n and $[0, 1/n]$ as α converges to 1 with the same speed $-\log(1-\alpha)$ and respective good rate functions $I_n(p_1, \ldots, p_n)$ defined in (2.14), and

$$S_n(p) = \begin{cases} 0, \text{ if } p = 0 \\ n, \text{ otherwise,} \end{cases} \tag{3.1}$$

For any $m \geq 2$, the function

$$\varphi_m(\mathbf{p}) = \sum_{i=1}^{\infty} p_i^m$$

plays an important role in both population genetics and statistical physics. Specifically $\varphi_2(\mathbf{p})$ is called the homozygosity in genetics representing the probability of two randomly selected genes having the same type. In spin glass models, $\varphi_2(\mathbf{p})$ is the probability of two randomly selected states or configurations falling in the same valley in the energy landscape. Applying the large deviation result in Theorem 2.2 we obtain

Theorem 3.2 *For any $m \geq 2$, the family $\{\varphi_m(P_1(\alpha, \theta), \ldots) : \alpha \in (0 \vee (-\theta), 1)\}$ satisfies a large deviation principles on space $[0, 1]$ as α converges to 1 with speed $-\log(1-\alpha)$ and the good rate function $I_1(p)$ defined in (2.9).*

Proof This follows from contraction principle and the fact that

$$\inf\{I(\mathbf{p}) : \mathbf{p} \in \nabla, \varphi_m(\mathbf{p}) = p\} = I(p^{1/m}, 0, \ldots) = I_1(p).$$

□

Let $h(\alpha)$ be a real-valued function such that $\frac{h(\alpha)}{-\log(1-\alpha)}$ has a limit as α converges to 1. Introducing the probability

$$PD(\alpha, \theta; h)(d\,\mathbf{p}) = C_{\alpha,\theta,h} \exp\{h(\alpha)\varphi_m(\mathbf{p})\} PD(\alpha, \theta)(d\,\mathbf{p})$$

on space ∇, where $C_{\alpha,\theta,h}$ is the normalizing constant.

In genetics, the function $\exp\{h(\alpha)\varphi_m(\mathbf{p})\}$ for $h(\alpha) \neq 0$ can be viewed as a selection force where the homozygotes are favoured over heterozygotes if $h(\alpha) > 0$ and vice versa if $h(\alpha) < 0$. In the context of statistical physics, one can write the two-parameter Poisson-Dirichlet distribution $PD(\alpha, \theta)$ in "Gibbs" form using a formal "Hamiltonian". The factor $h(\alpha)\varphi_m(\mathbf{p})$ can then correspond to an external field where the states in deep energy valley are favoured or disfavoured depending on whether $h(\alpha) > 0$ or < 0. The following result describes the impact of $h(\alpha)\varphi_m(\mathbf{p})$ in terms of large deviations.

Theorem 3.3 *Let c denote the limit of $\frac{h(\alpha)}{-\log(1-\alpha)}$ as α converges to 1. Then the family $\{PD(\alpha, \theta; h) : 0 < \alpha < 1, \theta + \alpha > 0\}$ satisfies a large deviation principle on ∇ as α converges to 1 with speed $-\log(1-\alpha)$ and a good rate function*

$$J_c(\mathbf{p}) = I(\mathbf{p}) - c\varphi_m(\mathbf{p}) + (c-1)^+ \tag{3.2}$$

Proof Rewrite $h(\alpha)\varphi_m(\mathbf{p})$ as $(-\log(1-\alpha))\frac{h(\alpha)}{-\log(1-\alpha)}\varphi_m(\mathbf{p})$. Clearly

$$\lim_{\alpha\to 1} \frac{h(\alpha)}{-\log(1-\alpha)}\varphi_m(\mathbf{p}) = c\varphi_m(\mathbf{p})$$

is bounded continuous on ∇. By Varadhan's lemma and Theorem 2.2, we obtain that the family $\{PD(\alpha, \theta; h) : 0 < \alpha < 1, \theta + \alpha > 0\}$ satisfies a large deviation principle on ∇ as α converges to 1 with speed $-\log(1-\alpha)$ and a good rate function

$$\sup\{c\varphi_m(\mathbf{q}) - I(\mathbf{q}) : \mathbf{q} \in \nabla\} - [c\varphi_m(\mathbf{p}) - I(\mathbf{p})].$$

A direct calculation leads to

$$\sup\{c\varphi_m(\mathbf{q}) - I(\mathbf{q}) : \mathbf{q} \in \nabla\} = (c-1)^+.$$

□

Remark The equation $J_c(\mathbf{p}) = 0$ has a unique solution $\mathbf{p} = \mathbf{0}$ if $c < 1$, and $\mathbf{p} = (1, 0, \ldots)$ if $c > 1$. But for $c = 1$, both $\mathbf{0}$ and $(1, 0, \ldots)$ are solutions of the equation. This implies that for $c < 1$

$$PD(\alpha, \theta; h) \Rightarrow \delta_{\mathbf{0}}, \quad \alpha \to 1$$

and for $c > 1$

$$PD(\alpha, \theta; h) \Rightarrow \delta_{(1,0,\ldots)}, \quad \alpha \to 1.$$

It is not clear what the limit is when $c = 1$. The impact of the external field is reflected from the magnitude of the rate function. When deep valley states are disfavoured ($c < 0$), the rate function is bigger than $I(\mathbf{p})$ indicating it is more difficult to move away from $\mathbf{0}$. On the other hand, when deep valley states are favoured ($0 < c < 1$), the rate function becomes smaller than $I(\mathbf{p})$ and it is relatively easier to deviate from $\mathbf{0}$. When the advantage of deep valley states becomes more strong ($c > 1$), the whole system converges to a new limit $(1, 0, \ldots)$.

4 Concluding Remarks

In [6], a large deviation principle is established for $PD(\alpha, 0)$ on ∇ when α converges to zero and the corresponding rate function is given by

$$I_{small}(\mathbf{p}) = \max\{i \geq 1 : \mathbf{p} \in \nabla, \sum_{k=1}^{\infty} p_k = 1, p_i > 0\} - 1. \tag{4.1}$$

As a by product, one obtains that

$$PD(\alpha, 0) \Rightarrow \delta_{(1,0,\ldots)}, \ \alpha \to 0.$$

By contrast, as α converges to 1 the result in this paper shows that

$$PD(\alpha, 0) \Rightarrow \delta_{(0,0,\ldots)}.$$

In the context of random energy model, the limits of $\alpha \to 0$ and $\alpha \to 1$ correspond to temperatures going down to zero and up to the critical temperature T_c, respectively. At zero temperature, the energy is concentrated on one deep valley while at the critical temperature the energy levels are relatively uniform. The respective effective domains of $I_{small}(\cdot)$ and $I(\cdot)$ are

$$\{\mathbf{p} \in \nabla : I_{small}(\mathbf{p}) < \infty\}$$
$$= \{(p_1, \ldots, p_n, 0, \ldots) : n \geq 1, p_1 \geq \cdots \geq p_n > 0, \sum_{i=1}^{n} p_i = 1\}$$

and

$$\{\mathbf{p} \in \nabla : I(\mathbf{p}) < \infty\}$$
$$= \{(p_1, \ldots, p_n, 0, \ldots) : n \geq 1, p_1 \geq \cdots \geq p_n > 0, \sum_{i=1}^{n} p_i \leq 1\}.$$

The large deviations near zero temperature describe the microscopic movement of the system at the instant when temperature increases from zero. One can see from the effective domain of $I_{small}(\cdot)$ that all states will move into finite number of valleys and there is a finite step "energy ladder" for the system to climb. The large deviations near critical temperature describe the microscopic movement of the system at the instant when temperature decreases from T_c. The effective domain of I shows that a proportion of the states will move into finite number of energy valleys and the energy landscape could be a mixture of valleys and "flat" regions.

A different comparison between the limits of $\alpha \to 0$ and $\alpha \to 1$ is through the following subordinator representation. More specifically, for any α in $(0, 1)$, let τ_t be a stable subordinator with index α and Lévy measure $\frac{\alpha}{\Gamma(1-\alpha)} x^{-(1+\alpha)} dx$. Let $J_1 \geq J_2 \geq \ldots \geq 0$ denote the decreasing order jump sizes of τ_t over the interval $[0, 1]$. Then $PD(\alpha, 0)$ is the law of

$$(\frac{J_1}{\tau_1}, \frac{J_2}{\tau_1}, \ldots).$$

The Laplace transform of τ_1 has the form

$$\mathbb{E}[\exp\{-\lambda \tau_t\}] = \exp\{-t\lambda^\alpha\}, \ \lambda \geq 0.$$

When α converges to 1, τ_t converges to t and there is no jumps anymore. Hence the limit of $(\frac{J_1}{\tau_1}, \frac{J_2}{\tau_1}, \ldots)$ is $\mathbf{0}$. On the other hand, if α converges to zero, then the subordinator τ_t becomes a pure killed subordinator with killing rate one. This explains why the limit is $(1, 0, \ldots)$.

Acknowledgements We wish to thank an anonymous referee for very helpful comments and suggestions.

References

1. D.A. Dawson, S. Feng, Asymptotic behavior of Poisson-Dirichlet distribution for large mutation rate. Ann. Appl. Probab. **16**, 562–582 (2006)
2. B. Derrida, Random-energy model: limit of a family of disordered models. Phys. Rev. Lett. **45**, 79–82 (1980)
3. B. Derrida, Random-energy model: an exactly solvable model of disordered systems. Phy. Rev. B **24**, 2613–2626 (1981)
4. A. Dembo, O. Zeitouni, *Large Deviations Techniques and Applications*. Applications of Mathematics, vol. 38, 2nd edn. (Springer, New York, 1998)
5. S. Feng, Large deviations for Dirichlet processes and Poisson–Dirichlet distribution with two parameters. Electron. J. Probab. **12**, 787–807 (2007)
6. S. Feng, Poisson–Dirichlet distribution with small mutation rate. Stoch. Proc. Appl. **119**, 2082–2094 (2009)
7. S. Feng, *The Poisson-Dirichlet Distribution and Related Topics*. Probability and Its Applications (New York) (Springer, Heidelberg, 2010)
8. S. Feng, F.Q. Gao, Moderate deviations for Poisson–Dirichlet distribution. Ann. Appl. Probab. **18**, 1794–1824 (2008)
9. S. Feng, F.Q. Gao, Asymptotic results for the two-parameter Poisson-Dirichlet distribution. Stoch. Proc. Appl. **120**, 1159–1177 (2010)

10. R.C. Griffiths, On the distribution of allele frequencies in a diffusion model. Theor. Pop. Biol. **15**, 140–158 (1990)
11. K. Handa, The two-parameter Poisson-Dirichlet point process. Bernoulli **15**, 1082–1116 (2009)
12. L.F. James, Large sample asymptotics for the two-parameter Poisson Dirichlet processes, in *Pushing the Limits of Contemporary Statistics: Contributions in Honor of Jayanta K. Ghosh*, ed. by B. Clarke, S. Ghosal. Institute of Mathematical Statistics Collections, vol. 3 (Institute of Mathematical Statistics, Beachwood, 2008), pp. 187–199
13. P. Joyce, S.M. Krone, T.G. Kurtz, Gaussian limits associated with the Poisson–Dirichlet distribution and the Ewens sampling formula. Ann. Appl. Probab. **12**, 101–124 (2002)
14. P. Joyce, S.M. Krone, T.G. Kurtz, When can one detect overdominant selection in the infinite-alleles model? Ann. Appl. Probab. **13**, 181–212 (2003)
15. J.C.F. Kingman, Random discrete distributions. J. R. Stat. Soc. B **37**, 1–22 (1975)
16. M. Perman, Order statistics for jumps of normalised subordinators. Stoch. Proc. Appl. **46**, 267–281 (1991)
17. M. Perman, J. Pitman, M. Yor, Size-biased sampling of Poisson point processes and excursions. Probab. Theory Relat. Fields **92**, 21–39 (1992)
18. J. Pitman, The two-parameter generalization of Ewens' random partition structure. Technical report **345**, Department of Statistics, University of California, Berkeley (1992)
19. J. Pitman, M. Yor, Arcsine laws and interval partitions derived from a stable subordinator. Proc. Lond. Math. Soc. (3) **65**, 326–356 (1992)
20. J. Pitman, M. Yor, The two-parameter Poisson-Dirichlet distribution derived from a stable subordinator. Ann. Probab. **25**, 855–900 (1997)
21. A.A. Puhalskii, On functional principle of large deviations, in *New Trends in Probability and Statistics*, ed. by V. Sazonov, T. Shervashidze (VSP Moks'las, Moskva, 1991), pp. 198–218
22. M. Talagrand, *Spin Glasses: A Challenge for Mathematicians*. Ergebnisse der Mathematik und ihrer Grenzgebiete 3. Folge A Series of Modern Surveys in Mathematics, vol. 46 Springer, Berlin, Heidelberg, 2003

Stability Estimation of Transient Markov Decision Processes

Evgueni Gordienko, Jaime Martinez, and Juan Ruiz de Chávez

Abstract We consider transient or absorbing discrete-time Markov decision processes with expected total rewards. We prove inequalities to estimate the stability of optimal control policies with respect to the total variation norm and the Prokhorov metric. Some application examples are given.

Keywords Discrete-time Markov control process • Expected total reward • Stability index • Stability inequalities • Total variation • Prokhorov metrics

Mathematics Subject Classification (2000). 90C40.

1 Motivation, Notation and Problem Setting

In this paper we study the stability of transition discrete-time Markov decision processes (MDPs). Also such MDPs could be considered as "absorbing". Indeed, as shown in [13] (see also [14]), for each stationary control policy that induces a transient process, the dynamics is given by a pseudo probability kernel, which can be expanded to a probability kernel by adding a special absorbing state. We use the term "transient" since it is the more common in the literature on MDPs [9, 10, 12–14, 16, 19]. Hindered et al. [11] use the term "absorbing" for countable state MDPs relevant to those considered here. (However Bertsekas et al. [1], have investigated similar countable MDPs under different terminology.) Discounted Markov control models can be transformed into transient MDPs [9, Ch.5].

We consider MDPs that include the following components:

- The state space, X, is a Borel space with metric ρ;
- The action space, A, is a Borel space with metric δ;
- Admissible action sets, $A(x)$, $x \in X$;

E. Gordienko • J. Martinez • J. Ruiz de Chávez (✉)
Departmento de Matematicas, Universidad Autonoma Metropolitana-Iztapalapa, Av. San Rafael Atlixco 186, col. Vicentina, C.P. 09340, Mexico City, Mexico
e-mail: gord@xanum.uam.mx; j_edum@yahoo.com; jrch@xanum.uam.mx

R.H. Mena et al. (eds.), *XI Symposium on Probability and Stochastic Processes*, Progress in Probability 69, DOI 10.1007/978-3-319-13984-5_8

- The set of admissible state-actions pairs

$$\mathbb{K} = \{(x, a) \in X \times A : a \in A(x), x \in X\},$$

which is assumed to be a Borel subset of $X \times A$, and has the metric:

$$d := \max\{\rho, \delta\};$$

- The metric space, (S, ϱ), of values of random vectors determining the process under consideration. We assume S to be separable metric space.

Suppose we have two discrete-time MDPs

$$x_t = F(x_{t-1}, a_t, \xi_t), \ t = 1, 2, \ldots; \tag{1.1}$$

$$\tilde{x}_t = F(\tilde{x}_{t-1}, \tilde{a}_t, \tilde{\xi}_t), \ t = 1, 2, \ldots; \tag{1.2}$$

where $F : \mathbb{K} \times S \to X$ is a measurable function; $x_t, \ \tilde{x}_t \in X$ are states of the processes; $a_t \in A(x_{t-1}), \ \tilde{a}_t \in A(\tilde{x}_{t-1})$ are actions in the corresponding states; and $\{\xi_t\}, \ \{\tilde{\xi}_t\}$ are two sequences of independent and identically distributed (i.i.d.) random vectors taking values in S.

The only difference between the MDPs in (1.1) and (1.2) is the possibly different distributions μ and $\tilde{\mu}$ of the random vectors ξ_1 and $\tilde{\xi}_1$, respectively. For notational simplicity, we may refer to these vectors as ξ and $\tilde{\xi}$, and their distributions as μ and $\tilde{\mu}$, respectively. Distributions of random vectors with values in S or X are defined on the corresponding Borel σ-algebras $\mathfrak{B}_S$ and $\mathfrak{B}_X$.

We will refer to (1.1) as the original MDP, and (1.2) as its "approximation".

Let $r : \mathbb{K} \to \mathbb{R}$ be a one-period reward function, where r is assumed to be measurable and *bounded* on $\mathbb{K}$.

A control policy $\pi = (\pi_1, \pi_2, \ldots)$ is defined in a standard manner [4]. We denote by a_t a realization of π given the whole history up to time $t - 1$ (as in (1.1), (1.2)).

Let Π be the set of all policies, and $\mathbb{F}$ the subset of all stationary deterministic policies. Each *stationary policy* $f = \{f, f, \ldots\} \in \mathbb{F}$ is defined by a measurable function $f : X \to A$ such that $f(x) \in A(x)$ for every $x \in X$. The application of f at a state x_{t-1} means applying the action $a_t = f(x_{t-1})$.

For each *initial state* $x \in X$ of the processes and every policy $\pi \in \Pi$ *the expected total rewards* for the processes (1.1) and (1.2) are

$$V(x, \pi) := \lim_{n\to\infty} \inf E_x^{\pi}[\sum_{t=1}^{n} r(x_{t-1}, a_t)], \tag{1.3}$$

$$\tilde{V}(x, \pi) := \lim_{n\to\infty} \inf \tilde{E}_x^{\pi}[\sum_{t=1}^{n} r(\tilde{x}_{t-1}, \tilde{a}_t)], \tag{1.4}$$

respectively, where E_x^π and $\tilde{E}_x^\pi$ are expectations with respect to probability measures P_x^π and $\tilde{P}_x^\pi$, generated by processes (1.1) and (1.2) on the space of trajectories when policy π is applied with initial state x.

The corresponding value functions (for processes (1.1) and (1.2)) are

$$V_*(x) := \sup_{\pi \in \Pi} V(x, \pi), \quad x \in X, \tag{1.5}$$

$$\tilde{V}_*(x) := \sup_{\pi \in \Pi} \tilde{V}(x, \pi), \quad x \in X. \tag{1.6}$$

In the next sections we introduce assumptions under which the functions V_* and $\tilde{V}_*$ are bounded, and there exist *optimal stationary policies* f_* and $\tilde{f}_*$ (for processes (1.1) and (1.2) respectively):

$$V(x, f_*) = V_*(x); \qquad \tilde{V}(x, \tilde{f}_*) = \tilde{V}_*(x), \; x \in X.$$

The problem of stability estimation of optimal control is set, in the same way as [5–8]. We are searching for a continuous function $g : [0, \infty) \to [0, \infty)$ with $g(0) = 0$, such that

$$\sup_{x \in X} \Delta(x) \le g[\nu(\mu, \tilde{\mu})], \tag{1.7}$$

where ν is a particular probability metric, and

$$\Delta(x) := V_*(x) - V(x, \tilde{f}_*) \equiv V(x, f_*) - V(x, \tilde{f}_*) \ge 0 \tag{1.8}$$

is *the stability index*, which expresses a cut-back of the expected total reward when the policy $\tilde{f}_*$ is applied to control the original process (1.1) in place of the optimal policy f_*. Policy $\tilde{f}_*$ is optimal for the approximate control process (1.2). We assume that the distribution $\tilde{\mu}$ of $\tilde{\xi}$ in (1.2) is completely known, and the distribution μ of ξ in (1.1) can be (at least partly) unknown.

We offer the sets of conditions under which we prove (1.7) in the following cases:

(a) In (1.7) $\nu = V$ is the total variation metric and $g(y) = Ky$.
(b) In (1.7) $\nu = \pi_r$ is the Prokhorov metric (metrizying the weak convergence), and $g(y) = \overline{K} y \max\{1, \log(1/y)\}$.

Remark 1.1 How can we estimate (or bound) the distance $\nu(\mu, \tilde{\mu})$ in (1.7) if μ is unknown?

The upper bounds of $\nu(\mu, \tilde{\mu})$ can be derived in at least the two following situations:

(a) When μ is actually known, but it is replaced by some approximation $\tilde{\mu}$, which produces a simpler optimal control problem.

(b) When either μ or some of its parameters are unknown, but are estimated using statistical procedures with available observations $\xi_1, \xi_2, \ldots, \xi_n$. In such cases, the distance $\nu(\mu, \tilde{\mu})$ becomes random, and its expectation can frequently be bounded from above.

2 First Assumptions and Existence of Optimal Stationary Policies

For each stationary policy, $f \in \mathbb{F}$, let P_f and $\tilde{P}_f$ be transition probabilities of processes (1.1) and (1.2), respectively. Thus,

$$P_f(B|x) = P(F(x,f(x),\xi) \in B), \ x \in X, \ B \in \mathfrak{B}_X, \tag{2.1}$$

$$\tilde{P}_f(B|x) = P(F(x,f(x),\tilde{\xi}) \in B), \ x \in X, \ B \in \mathfrak{B}_X. \tag{2.2}$$

The following supposition is a variant of Assumption 2 in [13]

Assumption 2.1 ("Transitory" or absorbing conditions) There exists a set $\Theta \in \mathfrak{B}_X$ such that:

(a) If $f \in \mathbb{F}$ then either

$$\Theta \text{ is absorbing under } P_f, \text{ and } \sup_{x \in X} E_x^f \tau_{x,f}(\Theta) \le M_f < \infty, \tag{2.3}$$

or

$$\inf_{x \in X} V(x,f) = -\infty.$$

(b) If $f \in \mathbb{F}$ then either

$$\Theta \text{ is absorbing under } \tilde{P}_f, \text{ and } \sup_{x \in X} \tilde{E}_x^f \tilde{\tau}_{x,f}(\Theta) \le \tilde{M}_f < \infty, \tag{2.4}$$

or

$$\inf_{x \in X} \tilde{V}(x,f) = -\infty.$$

(c) The set of stationary policies satisfying (2.3), is nonempty, and the set of stationary policies satisfying (2.4), is nonempty.

In (2.3), (2.4) $\tau_{x,f}(\Theta)$, $\tilde{\tau}_{x,f}(\Theta)$, are the times of first entrance (for $t \ge 0$!) in the set Θ of processes (1.1) and (1.2), respectively, when policy f is applied and the initial state is x.

(d) The value functions V_* and $\tilde{V}_*$ are bounded from above.

Let $\mathbb{B}$ be the space of all measurable *bounded* functions $u : X \to \mathbb{R}$ such that $u(x) = 0, x \in \Theta$. The linear space $\mathbb{B}$ is equipped with the supremum norm $||\cdot||$.

Assumption 2.2 (Boundedness and continuity conditions)

(a) The one-period return function, r, is *bounded* on $\mathbb{K}$; $r(x,a) = 0$ for every $x \in \Theta$, $a \in A(x)$; and for each $x \in X$, the function $a \to r(x,a)$ is upper semicontinuous on $A(x)$.
(b) For each $x \in X, A(x)$ is compact.
(c) For every $u \in \mathbb{B}$ and $x \in X_0 := X\backslash\Theta$, the functions

$$a \to Eu[F(x,a,\xi)],$$
$$a \to Eu[F(x,a,\tilde{\xi})]$$

are continuous on $A(x)$.

Lemma 1 given in Appendix shows that a stationary policy f satisfying (2.3), is transient as described in [13]. That is

$$||\sum_{t=0}^{\infty} Q_f^t||_0 < \infty, \tag{2.5}$$

where Q_f is the restriction of the kernel P_f in (2.1) to $(X_0, \mathfrak{B}_{X_0})$, $||\cdot||_0$ is the operator norm, and again $X_0 = X\backslash\Theta$. The same is true for any stationary policy f satisfying (2.4), (replacing P_f by $\tilde{P}_f$ and Q_f by $\tilde{Q}_f$). Hence, the following assertion is a direct consequence of Theorem 1 in [13].

Proposition 2.3 *Let Assumptions* 2.1 *and* 2.2 *hold. Then:*

(a) *There exist stationary policies* f_* *and* $\tilde{f}_*$ *that are optimal for processes* (1.1) *and* (1.2), *respectively;*
(b) f_* *satisfies* (2.3) *and* $\tilde{f}_*$ *satisfies* (2.4).
(c) $V_*, \tilde{V}_* \in \mathbb{B}$, *and in particular equal zero on* Θ.

Remark 2.4 If a stationary policy, f, satisfies (2.3), then from (2.5) it follows that the total reward in (1.3) is

$$V_f(x) \equiv V(x,f) = E_x^f \sum_{t=1}^{\infty} r(x_{t-1}, f(x_{t-1})), \tag{2.6}$$

and

$$V_f \in \mathbb{B}. \tag{2.7}$$

Similarly, if f satisfies (2.4), then

$$\tilde{V}_f(x) \equiv \tilde{V}(x,f) = \tilde{E}_x^f \sum_{t=1}^{\infty} r(\tilde{x}_{t-1}, f(\tilde{x}_{t-1})), \tag{2.8}$$

and

$$\tilde{V}_f \in \mathbb{B}. \tag{2.9}$$

3 Estimation of Stability with Respect to the Total Variation Metric

For probability measures μ, $\tilde{\mu}$ on $(S, \mathfrak{B}_S)$ *the total variation distance* between μ and $\tilde{\mu}$ is

$$\mathbb{V}(\mu, \tilde{\mu}) := \sup\{|\int_S v d\mu - \int_S v d\tilde{\mu}| \, : \, v : S \to \mathbb{R}, \text{ with } \sup_{s \in S} |v(s)| \le 1\}. \tag{3.1}$$

We need the next assumption to make the constant K in the "stability inequality" (3.3) to be completely determined by characteristics of the original MDP (1.1) and the constants in Assumption 3.1. In such cases, fixing μ and considering $\tilde{\mu} \equiv \tilde{\mu}_n$ such that $\mathbb{V}(\mu, \tilde{\mu}_n) \to 0$, the constant K in (3.3) will not depend on $\mathbb{V}(\mu, \tilde{\mu}_n)$.

Assumption 3.1

(a) The set Θ from Assumption 2.1 is absorbing under $P_{\tilde{f}_*}$ and $\tilde{P}_{f_*}$
(b) Constants $M < \infty$, $\gamma < 1$ and integer $m \ge 1$ are known that

$$\sup_{x \in X_0} \tilde{E}_x^{f_*} [\tilde{\tau}_{x, f_*}(\Theta)] \le M;$$

$$\sup_{x \in X_0} \tilde{E}_x^{\tilde{f}_*} [\tilde{\tau}_{x, \tilde{f}_*}(\Theta)] \le M, \text{ and}$$

$$\sup_{x \in X_0} P_x^{\tilde{f}_*} (\tau_{x, \tilde{f}_*}(\Theta) > m) \le \gamma. \tag{3.2}$$

Theorem 3.2 *Let Assumptions* 2.1, 2.2 *and* 3.1 *hold. Then there exist a constant* $K < \infty$ *uniquely determined by the properties of MDP* (1.1) *and the constants* M, γ, m *in Assumption* 3.1, *such that*

$$\sup_{x \in X} \Delta(x) \le K \mathbb{V}(\mu, \tilde{\mu}). \tag{3.3}$$

Remark 3.3

(a) The proof of Theorem 3.2 (given in the Appendix) uses the method of contractive operators, as in [8, 20]. If we omit the phrase "uniquely determined by...", then inequality (3.3) holds true without Assumption 3.1.
(b) Following [8], one can obtain (using a similar proof) a version of (3.3) in a more general setting of the stability estimation problem: In the definitions of MDP (1.2) the function F (denoted now, say, by $\tilde{F}$) could be different from the function F in (1.1) (but in a certain sense closed).

Example (Optimal stopping problem) Consider the usual problem (see, e.g. [18]) of optimal stopping of discrete-time Markov process:

$$y_t = \Phi(y_{t-1}, \xi_t), \ \ t \geq 1. \tag{3.4}$$

The process (3.4) and its approximation

$$\tilde{y}_t = \Phi(\tilde{y}_{t-1}, \tilde{\xi}_t), \ \ t \geq 1, \tag{3.5}$$

both take values in a Borel phase space Y. Note, that if Y is a separable metric space then each homogeneous discrete-time Markov process can be represented in the form (3.4) with i.i.d. random vectors $\xi_1, \xi_2 \ldots$ in some Borel space S (see e.g. [2]).

Let us consider a standard extension (see [17, 18]) of (3.4), (3.5) to MDPs on $X = Y \cup *$, where $*$ is the absorbing state where the process "lives" after being stopped. The action sets are $A(x) \equiv A = \{0, 1\}$. Action 0 means continuation of observations, and action 1 stops the process. Hence, we can define the function F to obtain two MDPs:

$$x_t = F(x_{t-1}, a_t, \xi_t), \ t \geq 1, \tag{3.6}$$

$$\tilde{x}_t = F(\tilde{x}_{t-1}, \tilde{a}_t, \tilde{\xi}_t), \ t \geq 1, \tag{3.7}$$

and the optimal stopping problem is reduced to maximization of the expected total reward for processes (3.6), (3.7) with the following one-step return function:

$$r(x, a) = \begin{cases} R(x) & \text{if } a = 1, \ x \in Y; \\ -c(x) & \text{if } a = 0, \ x \in Y; \\ 0 & \text{if } x = *, \ a \in \{0, 1\}. \end{cases} \tag{3.8}$$

In (3.8), c, R are given bounded functions $Y \to \mathbb{R}$, representing payments for continuation of observations and a reward for stopping, respectively. By the definition of transition probabilities corresponding to processes (3.6), (3.7), the set $\Theta = \{*\}$ is absorbing for every policy $\pi \in \Pi$. The simplest situation, where we are able to meet Assumptions 2.1, 2.2 and 3.1, in this example, is the case where the following condition holds:

$$\inf_{y \in Y} c(y) \geq \alpha > 0. \tag{3.9}$$

Assumption 2.2 is satisfied because A is finite. The conditions (2.3), (2.4) in Assumption 2.1 hold, for instance for the stationary policy: "to stop at moment $t = 0$ (in an initial state)", and due to (3.9), for every stationary policy f with infinite average time till stopping $V(x,f) \equiv -\infty$. Assumption 3.1 is satisfied, and inequality (3.2) holds with $\gamma_n = 0$ for all

$$n \geq \left[\frac{\sup_{y \in Y} R(y)}{\alpha} \right] + 1.$$

Note that under (3.9) it is very simple to provide an upper bound of K in (3.3), and this bound could be chosen to be dependent only on $\sup_{y \in Y} R(y)$ and α.

Remark 3.4 Condition (3.9) makes the problem of stability estimation "almost trivial" (since all things are reduced to a finite time horizon). While condition (3.9) can be relaxed in several ways, it cannot be skipped completely in the general case. Indeed, examples of *unstable* stopping problems are given in [20]. In those examples $\mathbb{V}(\mu, \mu_\varepsilon) \to 0$ as $\varepsilon \to 0$, but for a certain $y \in Y$ the stability indices $\Delta_\varepsilon(y)$ are greater than a positive constant. Notably that in such examples $\inf_{y \in Y} c(y) = 0$. In [7, 20, 21], to obtain the inequality as in (3.3) (the optimal stopping problem without condition (3.9)), specific strong ergodicity conditions on processes (3.4), (3.5) were imposed.

Example (Simplest investment-consumption model with "very risky" investments) Let $X = [0, \infty)$, $A(x) = [0, x]$, $x \in X$, $S = \mathbb{R}$, and

$$x_t = [(x_{t-1} - a_t)\xi_t]^+, \; t \geq 1; \tag{3.10}$$

$$\tilde{x}_t = [(\tilde{x}_{t-1} - \tilde{a}_t)\tilde{\xi}_t]^+, \; t \geq 1. \tag{3.11}$$

In this model if $x_{t-1} = x$ is a current capital, then the quantity $(x - a)$ is used for investments, and the rest of money a is consumed.

Let $\Theta = \{0\}$ and r be a return function satisfying Assumption 2.2 (a). The following supposition means that in this model investments are "really risky".

There is a constant $\delta > 0$ such that

$$P(\xi \leq 0) \geq \delta \text{ and } P(\tilde{\xi} \leq 0) \geq \delta. \tag{3.12}$$

If we suppose that in (3.10) and (3.11) the random variables ξ and $\tilde{\xi}$ have continuous densities g and $\tilde{g}$, which vanish fast enough in $\pm\infty$, then Assumption 2.2 (c) can be met. It is clear from (3.10), (3.11) that the state $x = 0$ is absorbing for every $f \in \mathbb{F}$. Also $P_x^f(\tau_{x,f}(\{0\}) > n) \leq P(\xi_1 > 0, \xi_2 > 0, \ldots, \xi_n > 0) \leq (1 - \delta)^n$ due to (3.12).

Therefore, Assumptions 2.1 and 3.1 are satisfied. We can apply inequality (3.3), which is converted in this example into

$$\sup_{x\geq 0} \Delta(x) \leq K \int_{-\infty}^{\infty} |g(s) - \tilde{g}(s)| ds.$$

4 Stability Estimating with Respect to the Prokhorov Metric

Since the total variation metric is too strong, the stability inequality (3.3) is not always applicable.

In this section we strengthen Assumption 2.1 and assume that the return function, r, and the transition probabilities of MDPs (1.1) and (1.2) satisfy certain Lipschitz conditions. This allows us to prove a version of inequality (3.3) with the Prokhorov metric, π_r, in place of $\mathbb{V}$ in its right-hand side.

Rather than Assumptions 2.1 and 3.1 we introduce a further assumption as follows.

Assumption 4.1 (Uniform transitory or absorbing conditions) There exists a set $\Theta \in \mathfrak{B}_X$ such that:

(a) For every $f \in \mathbb{F}$, the set Θ is absorbing under P_f and $\tilde{P}_f$;
(b) There exists a constant $M < \infty$ such that:

$$\sup_{f\in\mathbb{F}} \sup_{x\in X} E_x^f \tau_{x,f}(\Theta) \leq M, \tag{4.1}$$

$$\sup_{f\in\mathbb{F}} \sup_{x\in X} \tilde{E}_x^f \tilde{\tau}_{x,f}(\Theta) \leq M. \tag{4.2}$$

The next assumption substitutes for Assumption 2.2.

Assumption 4.2 (Lipschitz conditions)

(a) The function r is bounded on $\mathbb{K}$; $r(x,a) = 0$ for every (x,a) with $x \in \Theta$; and there exists a finite constant L_0 such that

$$|r(k) - r(k^{'})| \leq L_0 d(k, k^{'});\ k, k^{'} \in \mathbb{K}. \tag{4.3}$$

(b) For each $x \in X$, the set $A(x)$ is compact, and there exists a finite constant L_1 such that

$$h[A(x), A(x^{'})] \leq L_1 \rho(x, x^{'});\ x, x^{'} \in X_0, \tag{4.4}$$

where h is the Hausdorff metric.

(c) There exist finite constants L, $\tilde{L}$, such that for every $u \in \mathbb{B}$ with $||u|| \leq 1$,

$$|Eu[F(k, \xi)] - Eu[F(k', \xi)]| \leq Ld(k, k');$$

$$|Eu[F(k, \tilde{\xi})] - Eu[F(k', \tilde{\xi})]| \leq \tilde{L}d(k, k'), \ k, k' \in \mathbb{K}. \tag{4.5}$$

(d) There exists a finite constant L_* such that for all $k \in \mathbb{K}$,

$$\rho[F(k, s), F(k, s')] \leq L_* \varrho(s, s'); \ s, s' \in S. \tag{4.6}$$

We recall the definition of the Prokhorov metric, π_r, on the space of probability distributions on $(S, \mathfrak{B}_S)$:

$$\pi_r(\mu, \tilde{\mu}) = \inf\{\varepsilon \geq 0 : \mu(A) \leq \tilde{\mu}(A^\varepsilon) + \varepsilon, \ \tilde{\mu}(A) \leq \mu(A^\varepsilon) + \varepsilon, \ A \in \mathfrak{B}_S\},$$

where $A^\varepsilon = \{s \in S : \varrho(s, s') < \varepsilon$ for some $s' \in A\}$. It is well-known that π_r metrizes for the weak convergence.

Theorem 4.3 *Let Assumptions* 4.1 *and* 4.2 *hold. Then there exists a finite constant,* $\bar{K}$*, depending only on the properties of MDP* (1.1) *and the constants involved in Assumptions* 4.1 *and* 4.2*, such that*

$$\sup_{x \in X} \Delta(x) \leq \bar{K}\pi_r(\mu, \tilde{\mu}) \max\left\{1, \log\left(\frac{1}{\pi_r(\mu, \tilde{\mu})}\right)\right\}. \tag{4.7}$$

The proof of this theorem is given in Appendix.

Example Let $X = [0, H]$, where $H > 0$ is a given number, $A(x) = [0, x]$, $x \in X$, and $S = \mathbb{R}$. Let us consider the following pair of MDPs:

$$x_t = \min\left\{H, \left[x_{t-1} - a_t - \min\left\{\frac{1}{x_{t-1} - a_t}, \gamma\right\} + \xi_t\right]^+\right\}, \tag{4.8}$$

$$\tilde{x}_t = \min\left\{H, \left[\tilde{x}_{t-1} - \tilde{a}_t - \min\left\{\frac{1}{\tilde{x}_{t-1} - \tilde{a}_t}, \gamma\right\} + \tilde{\xi}_t\right]^+\right\}, \tag{4.9}$$

$$t = 1, 2, \ldots.$$

In (4.8), (4.9) γ is a given positive number involved in Assumption 4.4, below.

Let $\Theta = \{0\}$, and $r(k)$, $k \in \mathbb{K}$ be a one-step reward function satisfying Assumption 4.2 (a).

Assumption 4.4 (Only for the current example)

(a) Random variables ξ and $\tilde{\xi}$ have bounded continuous densities g and $\tilde{g}$, and $g(x) = \tilde{g}(x) = 0$ for all $x \geq \gamma$.
(b) The densities $g, \tilde{g}$ are almost everywhere differentiable, and their derivatives are bounded (where they exist).
(c) $E\xi < 0$; $E\tilde{\xi} < 0$.

First let us show that for every $f \in \mathbb{F}$ the set $\Theta = \{0\}$ is absorbing under P_f and $\tilde{P}_f$. Indeed, if $x_{t-1} = 0$, then $a_t = 0$, $\min\left\{\dfrac{1}{x_{t-1} - a_t}, \gamma\right\} = \gamma$. But in (4.8) the random variable $\xi_t - \gamma$ is non-positive, and therefore $x_t = 0$. Thus Assumption 4.1 (a) is satisfied.

On the other hand Assumption 4.4 (c) yields (4.1) and (4.2). From (4.8), for every $f \in \mathbb{F}$ with probability 1 $x_t \leq x_t^{'}$, $t = 0, 1, \ldots$, where $x_t^{'} = [x_{t-1}^{'} + \xi_t]^+$, $t \geq 1$ is the random walk absorbed at $x = 0$, and for this walk $E\xi_t < 0$. It is well-known [15] that for such random walk the average time to enter in $\{0\}$ is uniformly bounded over initial states from a bounded interval. Therefore Assumption 4.1 holds for this example.

For $A(x) = [0, x]$ we can verify (4.4) with $L_1 = 1$, and from (4.8) inequality (4.6) holds with $L_* = 1$.

Let us, finally, verify Assumption 4.2 (c). For $y = x - a$, the function $\psi(y) := y - \min\{\frac{1}{y}, \gamma\}$ is Lipschitzian function on $[0, \infty)$. Let u be an arbitrary function from $\mathbb{B}$ with $||u|| \leq 1$. For any $y, y^{'} \in (0, H]$, we have

$$\begin{aligned} \mathrm{I}(y) :=& Eu[\min\{H, [\psi(y) + \xi]^+\}] \\ =& u(H)P(\psi(y) + \xi \geq H) + \int_0^H u(\psi(y) + t)g(t)dt \\ =& u(H)\int_{H-\psi(y)}^{\infty} g(z)dz + \int_{\psi(y)}^{\psi(y)+H} u(z)g(z - \psi(y))dz. \end{aligned} \tag{4.10}$$

To see that $|\mathrm{I}(y) - \mathrm{I}(y^{'})| \leq L|y - y^{'}|$, note that the first term on the right-hand side of (4.10) is Lipschitzian function because g is bounded and its support is bounded from the right. Let, for instance, $y > y^{'}$. Then $\psi(y) > \psi(y^{'})$ and the difference of the second terms on the right in (4.10) is less than

$$\begin{aligned} &\int_{\psi(y^{'})+H}^{\psi(y)+H} |u(z)|g(z - \psi(y))dz + \int_{\psi(y^{'})}^{\psi(y)} |u(z)|g(z - \psi(y^{'}))dz \\ &\quad + \int_{\psi(y)}^{\psi(y^{'})+H} |u(z)| \sup_z |g^{'}(z)||\psi(y) - \psi(y^{'})|dz. \end{aligned} \tag{4.11}$$

Since $\psi(y^{'})$, $\psi(y) \in [-\gamma, H]$, the last summand in (4.11) is less than $(2M + \gamma)||u|| \sup_z |g^{'}(z)||\psi(y) - \psi(y^{'})|$, and also ψ is Lipschitzian. From the boundedness of g, it follows that the first and second terms in (4.11) are less than $L^{'}|y - y^{'}|$.

Now let $y^{'} = 0$. From (4.10) and Assumption 4.4 (a), we have that $\mathrm{I}(y^{'}) = u(0) = 0$. For small enough y, $\psi(y) = y - \gamma$, and since $\xi - \gamma \leq 0$, in (4.10),

$$|\mathrm{I}(y)| \leq ||u||P(y + \xi - \gamma > 0) = \int_{\gamma - y}^{\gamma} g(z)dz \leq \text{const} \cdot y.$$

Remark 4.5 In [8] the example of MDPs unstable with respect to π_r was discussed. In that example the discount criteria of optimization was considered (with unbounded one-step cost function). The example can be modified for the case of a bounded cost function. On the other hand [9], a discounted MDP can be transformed in a transient MDP.

Appendix: Proofs of the Results

The following lemma establishes a connection between Assumption 2.1 and the definitions of transient policies given in [13, 16].

Lemma 1 *Let Θ be as in Assumption 2.1, $X_0 = X \backslash \Theta$ and for $f \in \mathbb{F}$, Q_f is the restriction of the kernel P_f in (2.1) to $(X_0, \mathfrak{B}_{X_0})$. If (2.3) holds, then*

$$\left\| \sum_{t=0}^{\infty} Q_f^t \right\|_0 \leq M_f. \tag{1}$$

In (1) $\|\cdot\|_0$ is the operator norm corresponding to the supremum norm $\|\cdot\|$ in $\mathbb{B}$.

Proof Since Q_f^t is a monotone operator,

$$\left\| \sum_{t=0}^{\infty} Q_f^t \right\|_0 = \left\| \sum_{t=0}^{\infty} Q_f^t \mathrm{I} \right\| = \sup_{x \in X_0} \left| \sum_{t=0}^{\infty} Q_f^t \mathrm{I}(x) \right|. \tag{2}$$

For every $t \geq 0$

$$Q_f^t \mathrm{I}(x) = P_x^f(x_t \in X_0) = P_x^f(\tau_{x,f}(\Theta) > t) \tag{3}$$

because Θ is absorbing set for P_f.

Thus, from (2), (3) we find

$$\left\| \sum_{t=0}^{\infty} Q_f^t \right\|_0 = \sup_{x \in X_0} \sum_{t=0}^{\infty} P_x^f(\tau_{x,f}(\Theta) > t) \leq M_f.$$

□

1. Proof of Theorem 3.2

Proof Let $f_*, \tilde{f}_*$ be the optimal stationary policies introduced in Proposition 2.3, and $F_* := \{f_*, \tilde{f}_*\}$.

Under Assumptions 2.1 and 3.1, for every $f \in \mathbb{F}_*$ the corresponding rewards $V_f \equiv V(x,f)$, $\tilde{V}_f(x) \equiv \tilde{V}(x,f)$ are bounded functions, and can be rewritten as

$$V_f(x) = E_x^f\left[\sum_{t=1}^{\infty} r(x_{t-1}, f(x_{t-1}))\right], \tag{4}$$

$$\tilde{V}_f(x) = \tilde{E}_x^f\left[\sum_{t=1}^{\infty} r(\tilde{x}_{t-1}, f(\tilde{x}_{t-1}))\right]. \tag{5}$$

From Proposition 2.3 and Assumption 3.1, the following operators G_f, $\tilde{G}_f$ ($f \in F_*$)

$$G_f u(x) := r(x, f(x)) + Eu[F(x, f(x), \xi)], \tag{6}$$

$$\tilde{G}_f u(x) := r(x, f(x)) + Eu(F(x, f(x), \tilde{\xi})) \tag{7}$$

act from $\mathbb{B}$ to $\mathbb{B}$.

Using (4), (5) and standard arguments (Markov property [17]) we find that, for $f \in \mathbb{F}_*$

$$V_f = G_f V_f \quad \text{and} \quad \tilde{V}_f = \tilde{G}_f \tilde{V}_f. \tag{8}$$

For the stability index in (1.8) we have [8, 20]:

$$\Delta(x) \le 2 \max_{f \in \mathbb{F}_*} \left| V(x,f) - \tilde{V}(x,f) \right|. \tag{9}$$

First, let $f = f_*$ (omitting subindex $*$). Then, by (8), for every $n \ge 1$,

$$\begin{aligned} \left| V(x,f) - \tilde{V}(x,f) \right| &\le \left\| V_f - \tilde{V}_f \right\| = \left\| G_f^n V_f - \tilde{G}_f^n \tilde{V}_f \right\| \\ &\le \left\| G_f^n V_f - G_f^n \tilde{V}_f \right\| + \left\| G_f^n \tilde{V}_f - \tilde{G}_f^n \tilde{V}_f \right\|. \end{aligned} \tag{10}$$

From Proposition 2.3 (b), the policy $f = f_*$ satisfies (2.3). Therefore, by Lemma 1, and the corresponding result in [13], there exists an integer $n \ge 1$ such that the operator G_f^n is contractive in $\mathbb{B}$ with some module $\beta < 1$. Thus, from (10),

$$\left\| V_f - \tilde{V}_f \right\| \leq \frac{1}{(1-\beta)} \left\| G_f^n \tilde{V}_f - \tilde{G}_f^n \tilde{V}_f \right\|. \tag{11}$$

Taking into account (6), (7) and applying the arguments used, for example, in [20], we obtain

$$\left\| G_f^n \tilde{V}_f - \tilde{G}_f^n \tilde{V}_f \right\| \leq$$
$$n \left\| \tilde{V}_f \right\| \sup_{x \in X_0, a \in A(x)} \sup_{B \in B_{X_0}} \left| P(F(x,a,\xi) \in B) - P(F(x,a,\tilde{\xi}) \in B) \right|. \tag{12}$$

The last term on the right-hand side of (12) is less than $\frac{1}{2}\mathbb{V}(\xi, \tilde{\xi})$.

On the other hand, since $r \equiv 0$ on Θ, from Assumption 3.1 we have:

$$\left\| \tilde{V}_f \right\| \equiv \left\| \tilde{V}_{f_*} \right\| = \sup_{x \in X_0} \left| \tilde{E}_x^f \sum_{t=1}^{\infty} r(\tilde{x}_{t-1}, f_*(\tilde{x}_{t-1})) \mathrm{I}_{\{\tilde{\tau}_{x,f_*}(\Theta) > t-1\}} \right| \leq bM, \tag{13}$$

where $b = \sup_{k \in \mathbb{K}} |r(k)|$, and M is the constant from Assumption 3.1.

Second, in (9) let $f = \tilde{f}_*$. Now we have the inequality (10) with $f = \tilde{f}_*$. Let $m \geq 1$ and $\gamma < 1$ be the constants from Assumption 3.1. From (3) and (3.2) in Assumption 3.1, $\left\| Q_f^m \right\| \leq \gamma$.

Since the set Θ is absorbing under $P_f \equiv P_{\tilde{f}_*}$ (see Assumption 3.1), and iterating (6), for each $u, \upsilon \in \mathbb{B}$, $\left\| Q_f^m u - Q_f^m \upsilon \right\| \leq \gamma \left\| u - \upsilon \right\|$. Thus, from (10) it follows that

$$\left\| V_f - \tilde{V}_f \right\| \leq \frac{1}{(1-\gamma)} \left\| G_f^m \tilde{V}_f - \tilde{G}_f^m \tilde{V}_f \right\|.$$

Proceeding as in (12) and (13) (with $f \equiv \tilde{f}_*$ rather than $f = f_*$), and applying Assumption 3.1 (b), we get that for a given constant $\tilde{K}$:

$$\left\| V_{\tilde{f}_*} - \tilde{V}_{\tilde{f}_*} \right\| \leq \tilde{K} \mathbb{V}(\xi, \tilde{\xi}). \tag{14}$$

To conclude the proof of (3.3) it suffices to gather the inequalities (9) and (11)–(14). □

2. *Proof of Theorem 4.3*

Proof Let $f_*, \tilde{f}_* \in \mathbb{F}$ be the stationary policies optimal for MDPs (1.1), (1.2), respectively, and $V_* = V_{f_*}$, $\tilde{V}_* = V_{\tilde{f}_*}$ be the corresponding value functions. The existence of f_* and $\tilde{f}_*$ was ensured in Proposition 2.3. From Assumption 4.1 (a), for every $f \in \mathbb{F}$ the corresponding rewards V_f and $\tilde{V}_f$ (see (2.6)–(2.9)) are zero on Θ. Particularly $V_*(x) = \tilde{V}_*(x) = 0$, for $x \in \Theta$. Hence, we can consider all functions V_f and $\tilde{V}_f, f \in \mathbb{F}$ as elements of the space $\mathbb{B}$ (taking into account their boundedness which follows from Assumption 4.1).

In the usual manner, we introduce the dynamic programming operators $T, \tilde{T} : \mathbb{B} \to \mathbb{B}$:

$$Tu(x) := \sup_{a \in A(x)} \{r(x,a) + Eu(F(x,a,\xi))\}, \; x \in X, \tag{15}$$

$$\tilde{T}u(x) := \sup_{a \in A(x)} \left\{r(x,a) + Eu(F(x,a,\tilde{\xi}))\right\}, \; x \in X. \tag{16}$$

From Assumption 4.2 (a) (b), it follows that for each $u \in \mathbb{B}$ there exists a stationary policy (selector), f_u, such that

$$\begin{aligned}\sup_{a \in A(x)} \{r(x,a) + Eu[F(x,a,\xi)]\} &= r(x, f_u(x)) + Eu[F(x, f_u(x), \xi)] \\ &= r(x, f_u(x)) + E_x^{f_u} u(x_1), \;\; x \in X.\end{aligned}$$

Thus for $x \in \Theta$ by Assumption 4.1 (a) $Tu(x) = 0$, and $T\mathbb{B} \subseteq \mathbb{B}$. (Similarly $\tilde{T}\mathbb{B} \subseteq \mathbb{B}$.)

As it was proven in [13, 16] the fulfilment of Assumptions 4.1 and 4.2 is sufficient for validity of the following assertions.

Proposition 2

(a) $V_* = TV_*, \;\; \tilde{V}_* = \tilde{T}\tilde{V}_*$.
(b) *The optimal policy f_* is a selector in the right-hand side of (15) with $u = V_*$; and the optimal policy $\tilde{f}_*$ is a selector in the right-hand side of (16) with $u = \tilde{V}_*$;*
(c) *There exists an integer $m \geq 1$ such that the operator T^m is contractive in $\mathbb{B}_0$ with some module $\beta < 1$.*

For any $(x,a) \in \mathbb{K}$ let

$$H(x,a) := r(x,a) + EV_*[F(x,a,\xi)], \tag{17}$$

$$\tilde{H}(x,a) := r(x,a) + E\tilde{V}_*[F(x,a,\tilde{\xi})]. \tag{18}$$

To simplify notation let $f = \tilde{f}_*$. Similarly to [5], let $\Gamma_t = \{x, a_1, x_1, a_2, \ldots, x_{t-1}, a_t\}$, $(t \geq 1)$, be the part of a trajectory of process (1.1) under the control policy $f = \{f, f, \ldots, \}$ (with the initial state $x \in X_0$). By the Markov property, we have

$$\begin{aligned}\zeta_t &:= E^f[V_*(x_t)|\Gamma_t] \\ &= H(x_{t-1}, a_t) - r(x_{t-1}, a_t) - \sup_{a\in A(x_{t-1})} H(x_{t-1}, a) \\ &\quad + \sup_{a\in A(x_{t-1})} H(x_{t-1}, a).\end{aligned} \tag{19}$$

By (15), (17) and Proposition 2 (a) we obtain:

$$\begin{aligned}\zeta_t &= H(x_{t-1}, a_t) - \sup_{a\in A(x_{t-1})} H(x_{t-1}, a) - r(x_{t-1}, a_t) + V_*(x_{t-1}) \\ &= \Lambda_t - r(x_{t-1}, a_t) + V_*(x_{t-1}),\end{aligned} \tag{20}$$

where

$$\Lambda_t := \sup_{a\in A(x_{t-1})} H(x_{t-1}, a) - H(x_{t-1}, a_t) \geq 0. \tag{21}$$

From (19) and (20) we get:

$$E_x^f V_*(x_t) = E_x^f \Lambda_t - E_x^f r(x_{t-1}, a_t) + E_x^f V_*(x_{t-1}).$$

Summing the last equality over $t \in [1, n]$, we obtain that

$$E_x^f \sum_{t=1}^n r(x_{t-1}, a_t) = V_*(x) - E_x^f V_*(x_n) - \sum_{t=1}^n E_x^f \Lambda_t. \tag{22}$$

Since $r, V_* \in \mathbb{B}$, under Assumption 4.1 (b) as $n \to \infty$, $E_x^f \sum_{t=1}^n r(x_{t-1}, a_t) \to E_x^f \sum_{t=1}^\infty r(x_{t-1}, a_t) = V(x, f(x))$ (see (2.6)), and $E_x^f V_*(x_n) = [Q_f^n V_*](x) \to 0$, where Q_f is the kernel defined in Lemma 1.

Thus we can pass to the limit in (22) to find

$$\Delta(x) = V_*(x) - V(x, f) \leq \limsup_{n\to\infty} \sum_{t=1}^n E_x^f \Lambda_t. \tag{23}$$

Similarly to Lemma 1 it is proven that (4.1) yields that

$$\left\| \sum_{t=0}^{\infty} Q_f^t \right\|_0 \leq M, \quad \text{for every } f \in \mathbb{F}. \tag{24}$$

On the other hand, in [13] it was shown that

$$V_*(x) = V(x, f_*) = \left[\sum_{t=0}^{\infty} Q_{f_*}^t r \right](x).$$

Thus, from (24) we see that $\|V_*\| \leq M \|r\|$, and, similarly, $\|\tilde{V}_*\| \leq M \|r\|$. From the first of these inequalities it follows that (see (17), (18)) in (23) Λ_t is a function of x_{t-1} (a state under the policy f) bounded by

$$2 \|r\| (1 + M) =: b. \tag{25}$$

From Proposition 2 (a), (17) and (21):

$$\Lambda_t(x_{t-1}) = V_*(x_{t-1}) - r(x_{t-1}, f(x_{t-1})) - EV_*(x_t),$$

and by Assumption 4.1, if $x_{t-1} \in \Theta$, then $x_t \in \Theta$, and therefore, (since r and V_* are zero on Θ) $\Lambda_t(x_{t-1}) = 0$ when $x_{t-1} \in \Theta$. Hence, $\Lambda_t = \Lambda(x_{t-1})$, where Λ is a function from $\mathbb{B}$.

In [16] was proven that under Assumption 4.1 that there exist constants $c < \infty$ and $\alpha < 1$ such that for every $f \in \mathbb{F}$,

$$\left\| Q_f^n \right\|_0 \leq c\alpha^n, \quad n = 1, 2, \ldots. \tag{26}$$

On the other hand, in view of the above properties of Λ, the right-hand side of (23) can be rewritten as follows. Let $N \geq 1$ be an arbitrary (for now) fixed integer. Then,

$$\begin{aligned} I(x) := \limsup_{n\to\infty} \sum_{t=1}^{n} E_x^f \Lambda_t &= \sum_{t=1}^{\infty} E_x^f \Lambda_t \\ &= \sum_{t=1}^{N} E_x^f \Lambda_t + \sum_{t>N} E_x^f \Lambda_t, \end{aligned} \tag{27}$$

And from (26),

$$\sup_{x\in X_0} \left| \sum_{t>N} E_x^f \Lambda_t \right| = \left\| \sum_{t>N} Q_f^t \Lambda_t \right\| \leq \sum_{t>N} \|Q_f^t \Lambda_t\| \leq \frac{bc}{1-\alpha} \alpha^{N+1}. \tag{28}$$

Combining (23), (27) and (28), we obtain the inequality

$$\Delta(x) \leq \sum_{t=1}^{N} E_x^f \Lambda_t + \frac{bc}{1-\alpha}\alpha^{N+1}. \tag{29}$$

Let us bound Λ_t in the last inequality. From the definition of Λ_t in (21) and from (16)–(18), Proposition 2 (a), we have:

$$\begin{aligned}
\Lambda_t &= \sup_{a\in A(x_{t-1})} H(x_{t-1},a) - \sup_{a\in A(x_{t-1})} \tilde{H}(x_{t-1},a) + \tilde{H}(x_{t-1},a_t) - H(x_{t-1},a_t) \\
&\leq 2 \sup_{a\in A(x_{t-1})} \left|H(x_{t-1},a) - \tilde{H}(x_{t-1},a)\right| \\
&\leq 2 \sup_{a\in A(x_{t-1})} \left|EV_*[F(x_{t-1},a,\xi)] - E\tilde{V}_*[F(x_{t-1},a,\tilde{\xi})]\right|,
\end{aligned} \tag{30}$$

where expectations are interpreted as conditional expectations with x_{t-1} being fixed.

From (30) we get:

$$\begin{aligned}
\Lambda_t &\leq 2 \sup_{a\in A(x_{t-1})} \left|EV_*[F(x_{t-1},a,\xi)] - EV_*[F(x_{t-1},a,\tilde{\xi})]\right| \\
&\quad + 2 \sup_{a\in A(x_{t-1})} \left|EV_*[F(x_{t-1},a,\tilde{\xi})] - E\tilde{V}_*[F(x_{t-1},a,\tilde{\xi})]\right| \\
&\leq 2 \sup_{k\in\mathbb{K}} \left|EV_*[F(k,\xi)] - EV_*[F(k,\tilde{\xi})]\right| + 2\left\|V_* - \tilde{V}_*\right\|.
\end{aligned} \tag{31}$$

From Proposition 2 (c), there exists integers $m \geq 1$ and $\beta < 1$ such that the operator T^m is contractive with module $\beta < 1$. Thus, again using Proposition 2,

$$\left\|V_* - \tilde{V}_*\right\| = \left\|T^m V_* - \tilde{T}^m \tilde{V}_*\right\| \leq \left\|T^m V_* - T^m \tilde{V}_*\right\| + \left\|T^m \tilde{V}_* - \tilde{T}^m \tilde{V}_*\right\|,$$

or

$$\left\|V_* - \tilde{V}_*\right\| \leq \frac{1}{1-\beta}\left\|T^m \tilde{V}_* - \tilde{T}^m \tilde{V}_*\right\|. \tag{32}$$

Now, since T is a nonexpansive operator, by induction we have

$$\begin{aligned}
\left\|T^m \tilde{V}_* - \tilde{T}^m \tilde{V}_*\right\| &\leq \left\|TT^{m-1}\tilde{V}_* - T\tilde{T}^{m-1}\tilde{V}_*\right\| + \left\|T\tilde{T}^{m-1}\tilde{V}_* - \tilde{T}\tilde{T}^{m-1}\tilde{V}_*\right\| \\
&\leq \left\|T^{m-1}\tilde{V}_* - \tilde{T}^{m-1}\tilde{V}_*\right\| + \left\|T\tilde{V}_* - \tilde{T}\tilde{V}_*\right\| \\
&\leq m\left\|T\tilde{V}_* - \tilde{T}\tilde{V}_*\right\| \\
&\leq m \sup_{k\in K}\left|E\tilde{V}_*[F(k,\xi)] - E\tilde{V}_*[F(k,\tilde{\xi})]\right|.
\end{aligned} \tag{33}$$

From (16) and Proposition 2 (a),

$$\tilde{V}_*(x) = \sup_{a \in A(x)} \left\{ r(x,a) + E\tilde{V}_*[F(x,a,\tilde{\xi})] \right\}. \tag{34}$$

Since $\tilde{V}_*$ is bounded by $M\,\|r\|$, from Assumption 4.2 (a) and (c), in (34) the function under supremum is Lipschitzian with respect to $k = (x,a)$. Then, as it was shown in [6], this fact and Assumption 4.2 (b), proves that the function $\tilde{V}_*$ in (34) is Lipschitzian. Therefore applying (4.6) in Assumption 4.2 (d), to the function $s \to \tilde{V}_*[F(k,s)]$ in (33) we obtain that this function satisfies the Lipschitz condition with a constant not depending on k.

In the same way (using Assumption 4.2 (c)) we can confirm that the function $s \to \tilde{V}_*[F(k,s)]$ is Lipschitzian.

Finally, combining inequalities (31)–(33), Λ_t in (29) is less than $\sup \left| E\varphi(\xi) - E\varphi(\tilde{\xi}) \right|$ over a certain class of functions φ, which are bounded by the same constant $\bar{b}$ and satisfy the Lipschitz conditions with the same constant $\bar{L}$ (and these constants depend only on m, α, and the constant involved in Assumptions 4.1 and 4.2).

Therefore,

$$\begin{aligned} \Lambda_t &\le (\bar{b} + \bar{L}) Dud(\xi, \tilde{\xi}) \\ &\le 2(\bar{b} + \bar{L}) \pi_r(\xi, \tilde{\xi}), \end{aligned} \tag{35}$$

where $Dud(\xi,\tilde{\xi})$ denotes the Dudley distance between the distributions of random vectors ξ and $\tilde{\xi}$. (See [3] for the definition of the Dudley metric, and the inequality between Dudley and Prokhorov metrics.)

If $\tilde{b} = 2(\bar{b} + \bar{L})$, then from (35) and (29)

$$\sup_{x \in X_0} \Delta(x) \le N\tilde{b}\pi_r(\xi,\tilde{\xi}) + \frac{bc}{1-\alpha}\alpha^{N+1} \equiv N\tilde{b}\pi_r(\mu,\tilde{\mu}) + \frac{bc}{1-\alpha}\alpha^{N+1}. \tag{36}$$

Finally, the desired inequality (4.7) follows from (36) if we choose

$$N = \left[\max\left\{ 1, \log_\alpha \left(\frac{1}{\pi_r(\mu,\tilde{\mu})} \right) \right\} \right] + 1.$$

□

References

1. D.P. Bertsekas, J.N. Tsitsiklis, An analysis of stochastic shortest path problems. Math. Oper. Res. **16**(3), 580–595 (1991)
2. A.A. Borovkov, S.G. Foss, Stochastically recursive sequences and their generalization. Sib. Adv. Math. **2**, 16–81 (1992)

3. R.M. Dudley, *Real Analysis and Probability*. Volume 74 of Cambridge Studies in Advanced Mathematics (Cambridge University Press, Cambridge, 2002). Revised reprint of the 1989 original
4. E.B. Dynkin, A.A. Yushkevich, *Controlled Markov Processes* (Springer, New York, 1979)
5. E.I. Gordienko, E. Lemus-Rodriguez, R. Montes-de Oca, Discounted cost optimality problem: stability with respect to weak metrics. Math. Methods Oper. Res. **68**, 77–96 (2008)
6. E.I. Gordienko, E. Lemus-Rodriguez, R. Montes-de Oca, Average cost Markov control processes: stability with respect to the Kantorovich metric. Math. Methods Oper. Res. **70**, 13–33 (2009)
7. E.I. Gordienko, A. Novikov, Characterization of optimal policies in a general stopping problem and stability estimating. Probab. Eng. Inf. Sci. **28**(3), 335–352 (2014)
8. E.I. Gordienko, F. Salem, Estimates of stability of Markov control processes with unbounded cost. Kybernetika **36**, 195–210 (2000)
9. O Hernández-Lerma, J.B. Lasserre, *Further Topics on Discrete-Time Markov Control Processes* (Springer, New York, 1999)
10. O. Hernández-Lerma, G. Carrasco, R. Pérez-Hernández, Markov control processes with the expected total cost criterion: optimality, stability, and transient models. Acta Appl. Math. **59**(3), 229–269 (1999)
11. K. Hinderer, K.H. Waldmann, Algorithms for countable state Markov decision models with an absorbing set. SIAM J. Control Optim. **43**(6), 2109–2131 (electronic) (2005)
12. A. Hordijk, *Dynamic Programming and Markov Potential Theory*. Volume No. 51 of Mathematical Centre Tracts (Mathematisch Centrum, Amsterdam, 1974)
13. H.W. James, E.J. Collins, An analysis of transient Markov decision processes. J. Appl. Probab. **43**(3), 603–621 (2006)
14. L.C.M. Kallenberg, *Linear Programming and Finite Markovian Control Problems*. Volume 148 of Mathematical Centre Tracts (Mathematisch Centrum, Amsterdam, 1983)
15. S.P. Meyn, R.L. Tweedie, *Markov Chains and Stochastic Stability*. Communications and Control Engineering Series (Springer, London, 1993)
16. S.R. Pliska, On the transient case for Markov decision chains with general state spaces, in *Dynamic Programming and Its Applications (Proc. Conf., University of British Columbia, Vancouver, 1977)* (Academic, New York/London, 1978), pp. 335–349
17. S.M. Ross, *Applied Probability Models with Optimization Applications* (Dover Publications, New York, 1992). Reprint of the 1970 original
18. A.N. Shiryayev, *Optimal Stopping Rules* (Springer, New York/Heidelberg, 1978). Translated from the Russian by A.B. Aries, *Applications of Mathematics*, vol. 8
19. A.F. Veinott, Discrete dynamic programming with sensitive discount optimality criteria. Ann. Math. Stat. **40**, 1635–1660 (1969)
20. E. Zaitseva, Stability estimating in optimal stopping problem. Kybernetika (Prague) **44**(3), 400–415 (2008)
21. E. Zaitseva, Robustness estimating of optimal stopping problem with unbounded revenue and cost functions. Int. J. Pure Appl. Math. **59**(3), 291–306 (2010)

Solution of the HJB Equations Involved in Utility-Based Pricing

Daniel Hernández–Hernández and Shuenn-Jyi Sheu

Abstract In this paper the connection between the utility pricing methodology and risk sensitive control is explored for stochastic volatility models. It is proved that the utility based price of a European option can be written as the difference of the value functions of two different stochastic optimal control problems. The smoothness of those value functions and gradient estimates are proved, to give a complete solution to these problems. As a consequence of these results, the relation with quadratic BSDEs, as well as the description of a risk neutral measure associated with this pricing approach are formalized.

Keywords Utility-based pricing • Portfolio optimization • Risk sensitive control • Stochastic volatility

Mathematics Subject Classification (2000). Primary 91G10, 91G80; Secondary 93E20.

The research od D. Hernández was partially supported by Conacyt, through the Laboratory LEMME. The research of S.-J. Sheu was supported by the grants from the Ministry of Science and Technology (No. NSC 102-2115-M-008 -002 -MY3), NCTS (MOST), NCTS (NCU), and the Ministry of Education (No. 103G906-9).

D. Hernández–Hernández (✉)
Centro de Investigación en Matemáticas, Apartado Postal 402, Guanajuato, Gto. 36000, México
e-mail: dher@cimat.mx

S.-J. Sheu
National Central University, Jongli, Taiwan

Academia Sinica, Taipei, Taiwan
e-mail: sheusj@math.sinica.edu.tw

R.H. Mena et al. (eds.), *XI Symposium on Probability and Stochastic Processes*, Progress in Probability 69, DOI 10.1007/978-3-319-13984-5_9

1 Introduction

In complete markets arbitrage-free arguments lead to a unique way to determine the price of contingent claims written in terms of the underlying assets, being the most famous example the Black-Scholes formula to obtain the price when the asset is modeled as a geometric Brownian motion with drift. In that context, the problem of pricing and hedging a contingent claim has a well understood solution, which can be written in terms of the unique martingale measure. However, in incomplete markets there is not a unique way to determine the price and build dynamic strategies that hedge the risk involved in the transaction, and both are very important problems in mathematical finance. Different methodologies have been proposed to deal with these problems so far, and in this note we follow the original idea proposed by Hodges and Neuberger [19], based in utility theory.

In recent years this pricing method has received a lot of attention, and has been analyzed from different perspectives using a large variety of frameworks; see, for instance, [3, 4, 17, 18, 23] and [24]. In order to define the utility-based price the preferences of the investor are assumed to be of exponential type, which allows explicit formulas. In this paper we present a different perspective to this problem, putting together ideas from risk-sensitive control and PDEs. The price formula is written in terms of the solutions of the HJB equations and, in this sense, the closest to our work is the paper by Sircar and Zariphopoulou [27], where they study the HJB equation involved in the definition of the utility-based price using the theory of viscosity solutions for PDEs. Our main results can be summarized as follows: having proven existence of classical solutions of the HJB equation associated with the optimal investment problem with a contingent claim, useful estimates of the gradient are obtained to derive the form of the optimal solutions. This results turn out to be very powerful to derive important results, as the uniqueness of solution, through a verification theorem, the existence of solution of forward-backward stochastic differential equations (FBSDE) with quadratic growth, as well as the form of an equivalent martingale measure associated with this pricing approach. Finally, it is worth mentioning that the area of risk sensitive control is very active, in particular the analysis of the solution of the HJB equations as well as their connections with the long time behaviour of controlled processes; see, for instance, [5, 13, 20, 25].

This paper is organized as follows. In Sect. 2 we present the factors model for the dynamics of the risky asset. Then, the optimal investment problems when there is a contingent claim in the portfolio, and without it, are presented, to give the proper definition of the utility-based price of the contingent claim in terms of the value functions of two different risk-sensitive control problems. Next, in Sect. 3, the HJB equations of the control problems involved are solved, obtaining very useful bounds on the gradient of the solutions; this result is the main contribution of this paper; see Theorem 3.4. Section 4 presents a verification theorem for the value functions of the risk-sensitive control problem. The gradient estimates obtained in Theorem 3.4 have important consequences, and are useful to prove this verification result, but also to give a straightforward response to the existence of solution to the quadratic BSDE associated with utility-based pricing, in Sect. 5. Finally, in Sect. 6 we present the form of an equivalent martingale measure that can be read directly from the PDE solved by the buyer's utility-based price.

2 The Model

Throughout we fix a probability space $(\Omega, \mathcal{F}, \mathbf{P})$, where a standard two dimensional Brownian motion (W^1, W^2) is defined; the augmented filtration generated by this Brownian motion is denoted by $\{\mathcal{F}_t;\ t \geq 0\}$. The securities market we shall consider consist of a riskless bond, which can be used to lending or borrowing, paying a zero interest rate, as well as a risky asset, with dynamics

$$dS_t = S_t[\mu(Y_t)dt + \sigma(Y_t)dW_t^1].$$

The stochastic process Y_t appearing in the coefficients of the above equation represents the stochastic volatility in the market, and it is assumed that it satisfies the following SDE:

$$dY_t = g(Y_t)dt + \beta(Y_t)[\rho dW_t^1 + \sqrt{1-\rho^2}dW_t^2],$$

with initial condition $Y_0 = y$. The number $\rho \in (-1, 1)$ represents the correlation between the noises driving the risky asset and the volatility.

Remark 2.1 There are different ways to model the volatility of the risky asset S_t. For instance, introducing a local volatility function $\sigma_t = \sigma(S_t)$ or using external factor processes Y_t, with correlated noises. In this note we use a diffusion model for Y_t, which provides flexibility to obtain qualitative properties of the distribution of the risky asset observed in the market; see, for instance, [14].

Specific assumptions about the functions μ, σ, β and g will be given below. The assumption that the interest rate paid by the bond is zero can be dispensed by discounting in an appropriate way.

The previous market model is incomplete and hence, given a European contingent claim $h(S_T)$ with expiration date $T > 0$, an arbitrage pricing method analogous to the Black-Scholes is not available any more. In order to determine an arbitrage free price for derivatives in incomplete markets, different approaches have been proposed in recent years. As it was explained in the Introduction, in this paper we are interested in the utility-based price, which takes into account the risk preferences of the investor, who is willing to buy the contingent claim $h(S_T)$.

Following the original idea of Hodges and Neuberger in [19] based in utility theory, an arbitrage free *utility-based price* of option $h(\cdot)$ at time $t < T$ is introduced next. Consider an investor with initial capital $x > 0$ at time t and risk preferences defined by the exponential utility function $U : I\!R \to I\!R$, with

$$U(w) = -\exp(-\gamma w), \tag{2.1}$$

where $\gamma > 0$ is fixed throughout. Let α_t be an $\mathcal{F}_t$-adapted process representing the amount of money invested in the risky asset at time t such that

$$\mathbf{E}\int_0^T \alpha_t^2 dt < \infty,$$

the class of such processes satisfying the following additional condition is denoted by $\mathcal{A}$.

The process M_t defined by

$$M_t = \exp\left\{\int_0^t [-\gamma\alpha_u\sigma(Y_u)dW_u^1 - \frac{1}{2}\gamma^2\alpha_u^2\sigma^2(Y_u)du]\right\}, \tag{2.2}$$

is a martingale.

Then, the dynamics for the wealth process are given by

$$dX_t = \alpha_t(\mu(Y_t)dt + \sigma(Y_t)dW_t^1), \quad X_0 = x.$$

Throughout the following assumptions on the coefficients of the model will be assumed. The first two guarantee the existence of strong solutions to the SDEs describing the dynamics of S_t and Y_t. Notice that when Y_t is the mean reverting Ornstein-Uhlenbeck process, it satisfies Assumption A.2 below. The rest of the assumptions are technical and are needed to guarantee the existence of classical solutions to the HJB equations studied in the next sections; observe that smooth approximations of Put options satisfy Assumptions A.4 and A.5.

Assumption A

1. The functions $\mu(\cdot)$, $\sigma(\cdot)$ are bounded and of class $C_b^1(I\!R)$, where $C_b^1(I\!R)$ is the space of functions which are bounded together with their first derivative.
2. The functions $g(\cdot)$ and $\beta(\cdot)$ belong to $C^1(I\!R)$, and are Lipschitz continuous.
3. There exists a constant c such that $\sigma(\cdot)$, $\beta(\cdot) \geq c > 0$.
4. The function $h(\cdot)$ is nonnegative, continuous differentiable up to the second order and bounded.
5. The derivatives of $\tilde{h}(z) := h(e^z)$ up to the second order are bounded continuous.

Now consider the following two optimal investment problems, with value functions

$$W(t,x,y) = \sup_{\alpha\in\mathcal{A}} \mathbf{E}[U(X_T) \mid X_t = x,\ Y_t = y] \tag{2.3}$$

$$V(t,x,y,s) = \sup_{\alpha\in\mathcal{A}} \mathbf{E}[U(X_T + h(S_T)) \mid X_t = x, Y_t = y, S_t = s], \tag{2.4}$$

with U as in (2.1).

In the first case, it corresponds to the classical maximum expected utility problem studied originally by Merton, while in the second case the investor receives at time T the value of his portfolio and the value of the contingent claim $h(S_T)$. So, at time $t \in [0, T)$ the investor has two alternatives, one consists in investing his money in the market and the second is to buy the option for price p, and to invest in the market the rest of his money; at time T he receives his capital and the value of the option at the exercise time T. For ease of notation, we shall write $\mathbf{E}_{t,x,y,s}$ for the expectation conditioned on $X_t = x,\ Y_t = y$ and $S_t = s$; the same simplification applies when we write $\mathbf{E}_{t,y,s}$ or $\mathbf{E}_{t,y}$.

Definition 2.2 We say that p is the *buyer's utility-based price* at time t if it solves the identity

$$W(t,x,y) = V(t,x-p,y,s). \tag{2.5}$$

Remark 2.3

(a) When the utility function is of *exponential type*, as given above, the expression for p shall be given in terms of the value functions of two different risk sensitive optimal control problems. These kind of problems have been studied in an independent way by many authors; the interested reader is referred in particular to [2, 7, 11, 12, 16, 20, 25], and references therein.

(b) As a consequence of the arguments given below, it shall be concluded that p is well defined, and it is a function of (t, y, s).

In order to derive the expression for p, let us calculate the following expectation for a given admissible strategy $\alpha \in \mathcal{A}$,

$$\begin{aligned}
\mathbf{E}_{t,x,y,s}[e^{-\gamma(X_T+h(S_T))}] &= e^{-\gamma x}\mathbf{E}_{t,x,y,s}\exp\{-\gamma\int_t^T \alpha_u\mu(Y_u)du \\
&\quad -\gamma\int_t^T \alpha_u\sigma(Y_u)dW_u^1 - \gamma h(S_T)\} \\
&= e^{-\gamma x}\mathbf{E}_{t,x,y,s}^{\mathbf{P}^\alpha}\exp\{\int_t^T [\frac{1}{2}\gamma^2\alpha_u^2\sigma^2(Y_u) - \gamma\alpha_u\mu(Y_u)]du \\
&\quad -\gamma h(S_T)\}.
\end{aligned} \tag{2.6}$$

The second equality was obtained by changing in $\mathcal{F}_T$ the original probability measure $\mathbf{P}$ by the absolutely continuous measure $\mathbf{P}^\alpha$, with Radon-Nikodym derivative

$$\frac{d\mathbf{P}^\alpha}{d\mathbf{P}}|_{\mathcal{F}_T} = \exp\left\{\int_0^T [-\gamma\alpha_u\sigma(Y_u)dW_u^1 - \frac{1}{2}\gamma^2\alpha_u^2\sigma^2(Y_u)du]\right\}. \tag{2.7}$$

By Girsanov's theorem, under measure $\mathbf{P}^\alpha$ the process defined as

$$\tilde{W}_t^1 = W_t^1 + \int_0^t \gamma\alpha_u\sigma(Y_u)du, \quad \tilde{W}_t^2 = W_t^2$$

is a Brownian motion in $I\!R^2$ and, hence, the dynamics of S_t and Y_t can be written, respectively, as

$$dS_t = S_t[(\mu(Y_t) - \gamma\alpha_t\sigma^2(Y_t))dt + \sigma(Y_t)d\tilde{W}_t^1]$$
$$dY_t = [g(Y_t) - \gamma\rho\alpha_t\sigma(Y_t)\beta(Y_t)]dt + \beta(Y_t)[\rho d\tilde{W}_t^1 + \sqrt{1-\rho^2}d\tilde{W}_t^2].$$

Multiplying by -1 in both sides of (2.6) and maximizing with respect to the set of admissible strategies $\mathcal{A}$, we write the value function $V(t,x,y,s)$ in (2.4) as

$$\begin{aligned} V(t,x,y,s) &= -e^{-\gamma x} \inf_{\alpha\in\mathcal{A}} \mathbf{E}_{t,y,s}^{\mathbf{P}^\alpha} \exp\{\int_t^T (\frac{1}{2}\gamma^2\alpha_u^2\sigma^2(Y_u) - \gamma\alpha_u\mu(Y_u))du - \gamma h(S_T)\} \\ &= -e^{-\gamma x - \gamma\bar{\psi}(t,y,s)}, \end{aligned} \tag{2.8}$$

where

$$\bar{\psi}(t,y,s) = \sup_{\alpha\in\mathcal{A}}\{-\frac{1}{\gamma}\log \mathbf{E}_{t,y,s}^{\mathbf{P}^\alpha} \exp\{-\gamma\int_t^T l(Y_u,\alpha_u)du - \gamma h(S_T)\}\}, \tag{2.9}$$

with $l(y,\alpha) = \alpha\mu(y) - \frac{1}{2}\gamma\alpha^2\sigma^2(y)$. Observe that defining $Q(y) = \frac{\mu^2(y)}{2\gamma\sigma^2(y)}$ and $\tilde{l}(y,\alpha) = l(y,\alpha) - \|Q\|$, with $\|Q\| = \sup_{y\in I\!R} Q(y)$, the value function $\bar{\psi}$ can be written as

$$\bar{\psi}(t,y,s) = \|Q\|(T-t) + \psi(t,y,s),$$

with

$$\psi(t,y,s) = \sup_{\alpha\in\mathcal{A}}\{-\frac{1}{\gamma}\log \mathbf{E}_{t,y,s}^{\mathbf{P}^\alpha} \exp\{-\gamma\int_t^T \tilde{l}(Y_u,\alpha_u)du - \gamma h(S_T)\}\}. \tag{2.10}$$

Analogously, for the value function $W(t,x,y)$ we have the expression

$$\begin{aligned} W(t,x,y) &= -e^{-\gamma x} \inf_{\alpha\in\mathcal{A}} \mathbf{E}_{t,y}^{\mathbf{P}^\alpha} \exp\{\int_t^T (\frac{1}{2}\gamma^2\alpha_u^2\sigma^2(Y_u) - \gamma\mu(Y_u)\alpha_u)du\} \\ &= -e^{-\gamma x - \gamma\bar{\phi}(t,y)}, \end{aligned} \tag{2.11}$$

where

$$\begin{aligned} \bar{\phi}(t,y) &= \sup_{\alpha\in\mathcal{A}}\{-\frac{1}{\gamma}\log \mathbf{E}_{t,y}^{\mathbf{P}^\alpha} \exp\{-\gamma\int_t^T l(Y_u,\alpha_u)du\}\} \qquad (2.12)\\ &= \|Q\|(T-t) + \sup_{\alpha\in\mathcal{A}}\{-\frac{1}{\gamma}\log \mathbf{E}_{t,y}^{\mathbf{P}^\alpha} \exp\{-\gamma\int_t^T \tilde{l}(Y_u,\alpha_u)du\}\} \\ &= \|Q\|(T-t) + \phi(t,y). \end{aligned}$$

Remark 2.4 These representations of the value functions shall be important in the following sections because both $\psi(\cdot, y, s)$ and $\phi(\cdot, y)$ are increasing, since $\tilde{l} \leq 0$.

Therefore, going back to the definition of the utility-based price of the contingent claim $h(S_T)$ in (2.5), expressions (2.8) and (2.11) yield the existence of a unique p, dependent on (t, y, s), such that (2.5) holds, and is given by

$$p = p(t, y, s) = \psi(t, y, s) - \phi(t, y). \tag{2.13}$$

Remark 2.5 In this note we have taken the perspective of the buyer to define the price of the derivative $h(S_T)$, but a similar analysis can be made for the seller's utility-based price; see, for instance, [9, 27]. On the other hand, it is well known by now that this pricing rule has important properties, like monotonicity, convexity and translation invariance; see [10], where these and other important properties of this pricing methodology are presented.

In the next sections both risk sensitive control problems, defined in (2.10) and (2.12), will be solved using dynamic programming methods, proving that the value functions $\psi(t, y, s)$ and $\phi(t, y)$ are smooth solutions of the associated HJB equations and obtaining the corresponding gradient estimates.

3 Hamilton-Jacobi-Bellman Equation

We begin this section with the description of the HJB equations associated to the value functions (2.10) and (2.12). In the first case, the function ψ satisfies at least formally the parabolic semilinear PDE

$$\begin{aligned} 0 = {} & \psi_t + s\mu(y)\psi_s + g(y)\psi_y + \frac{1}{2}s^2\sigma^2(y)\psi_{ss} + \frac{1}{2}\beta^2(y)\psi_{yy} + \rho s\sigma(y)\beta(y)\psi_{ys} \\ & + \gamma(1-\rho)s\sigma(y)\beta(y)\psi_s\psi_y - \frac{\gamma}{2}(s\sigma(y)\psi_s + \beta(y)\psi_y)^2 \\ & + \max_{\alpha \in I\!R}\{-\gamma\alpha\sigma^2(y)s\psi_s - \gamma\rho\alpha\sigma(y)\beta(y)\psi_y + \tilde{l}(y, \alpha)\}, \end{aligned} \tag{3.1}$$

with boundary condition $\psi(T, y, s) = h(s)$. The equation for the function ϕ is analogous to the previous one, removing all the terms involving s, i.e.

$$\begin{aligned} 0 = {} & \phi_t + g(y)\phi_y + \frac{1}{2}\beta^2(y)\phi_{yy} - \frac{\gamma}{2}\beta^2(y)\phi_y^2 \\ & + \max_{\alpha \in I\!R}\{-\gamma\rho\alpha\sigma(y)\beta(y)\phi_y + \tilde{l}(y, \alpha)\}, \end{aligned} \tag{3.2}$$

with $\phi(T, y) = 0$.

The maximum in the r.h.s. of (3.1) and (3.2) is achieved, respectively, at

$$\alpha^{\psi} = -s\psi_s - \frac{\rho\beta(y)}{\sigma(y)}\psi_y + \frac{\mu(y)}{\gamma\sigma^2(y)} \quad \text{and} \quad \alpha^{\phi} = -\frac{\rho\beta(y)}{\sigma(y)}\phi_y + \frac{\mu(y)}{\gamma\sigma^2(y)}. \tag{3.3}$$

Substituting these values in their respective equations, and after some calculations, we obtain the equations

$$0 = \psi_t + \tilde{g}(y)\psi_y + \frac{1}{2}\beta^2(y)\psi_{yy} - \frac{1}{2}\tilde{\beta}^2(y)\psi_y^2 + \frac{1}{2}s^2\sigma^2(y)\psi_{ss} \tag{3.4}$$
$$+\rho s\sigma(y)\beta(y)\psi_{sy} + \tilde{Q}(y)$$

and

$$0 = \phi_t + \tilde{g}(y)\phi_y + \frac{1}{2}\beta^2(y)\phi_{yy} - \frac{1}{2}\tilde{\beta}^2(y)\phi_y^2 + \tilde{Q}(y), \tag{3.5}$$

where

$$\tilde{g}(y) = g(y) - \rho\frac{\mu(y)\beta(y)}{\sigma(y)}, \quad \tilde{\beta}(y) = \sqrt{\gamma(1-\rho^2)}\beta(y) \text{ and } \tilde{Q}(y) = Q(y) - \|Q\|.$$

Remark 3.1 These PDEs will be studied in a separate way. The relationship between the solution that we will find for these equations and the original value functions for the risk sensitive control problems will be established by proving a verification theorem in the next section.

Remark 3.2 When $\rho^2 = 1$ the market is complete, and the PDE equation obtained for the utility-based price $p(t, y, s)$ from (3.4)–(3.5) is the Black-Scholes equation. This PDE is linear and has been studied extensively in the literature. Hereafter we assume that $0 \leq \rho^2 < 1$, and the PDE solved by $p(t, y, s)$ is presented in Sect. 5.

Equation (3.5) can be solved using a log transformation argument, which has been successfully implemented in the theory of PDEs to linearize some equations. Within the context of utility-based pricing it has been used before in [27]. Consider the function

$$f(t, y) = \exp\{-\gamma(1-\rho^2)\phi(t, y)\}.$$

Then, direct calculations yield

$$f_t + \tilde{g}f_y + \frac{1}{2}\beta f_{yy} = \gamma(1-\rho^2)\tilde{Q}f.$$

The solution of this linear equation can be represented in the following way. Let

$$d\tilde{Y}_t = \tilde{g}(\tilde{Y}_t)dt + \beta(\tilde{Y}_t)dW_t^1,$$

where W_t^1 is the first component of the Brownian motion (W^1, W^2) introduced in Sect. 2. Hence,

$$f(t,y) = \mathbf{E}_{t,y}[\exp\{-\gamma(1-\rho^2)\int_t^T \tilde{Q}(\tilde{Y}_u)du\}],$$

where $\mathbf{E}_{t,y}$ is the expectation conditioned on $\tilde{Y}_t = y$.

Since $\tilde{Q} \leq 0, f$ is decreasing in t, and the next estimate holds

$$-\gamma(1-\rho^2)\|\tilde{Q}\| \leq \frac{f_t}{f} \leq 0.$$

Now, a proof of existence of solution to the PDE (3.4) is presented. First, the singularity on $s = 0$ will be removed from that equation to obtain a uniformly elliptic equation. This can be done by making the change of variable $z = \log s$. Defining $\tilde{\psi}(t,y,z) = \psi(t,y,e^z)$ and $\tilde{h}(z) = h(e^z)$, after some calculations the PDE (3.4) can be written in terms of $\tilde{\psi}$ as the following semi-linear equation

$$\tilde{\psi}_t + \mathcal{L}\tilde{\psi} = \mathcal{H}(y, \tilde{\psi}_y), \quad \text{with} \quad \tilde{\psi}(T,y,z) = \tilde{h}(z). \tag{3.6}$$

The differential operator $\mathcal{L}$ is defined for a smooth function $w(t,y,z)$ as

$$\mathcal{L}w = \frac{1}{2}\sigma^2(y)w_{zz} + \frac{1}{2}\beta^2(y)w_{yy} + \rho\sigma(y)\beta(y)w_{zy} - \frac{1}{2}\sigma^2(y)w_z + \tilde{g}(y)w_y,$$

and $\mathcal{H}(y,p) = \frac{1}{2}\tilde{\beta}^2(y)p^2 - \tilde{Q}(y)$; recall that $\tilde{Q}(y) = Q(y) - \|Q\|$.

Definition 3.3 Let $\underline{w}, \bar{w} : [0,T] \times I\!R \times I\!R \to I\!R$ be smooth functions. We say that

(i) $\underline{w}(t,y,z)$ is a subsolution of (3.6) if

$$\underline{w}_t + \mathcal{L}\underline{w} \geq \mathcal{H}(y, \underline{w}_y), \quad \text{with} \ \underline{w}\,(T,y,z) \leq \tilde{h}(z).$$

(ii) Analogously, $\bar{w}(t,y,z)$ is a supersolution if

$$\bar{w}_t + \mathcal{L}\bar{w} \leq \mathcal{H}(y, \bar{w}_y), \quad \text{with} \ \bar{w}(T,y,z) \geq \tilde{h}(z).$$

In fact, it can be verified easily that a subsolution $\underline{w}\,(t,y,z)$ is given by

$$\underline{w}\,(t,y,z) = \inf_{z \in I\!R} \tilde{h}(z) - \|Q\|(T-t),$$

satisfying the terminal condition. On the other hand, we can also construct a supersolution $\bar{w}(t,y,z)$ as the solution of the linear equation

$$\bar{w}_t + \mathcal{L}\bar{w} = -\tilde{Q}, \tag{3.7}$$

with $\bar{w}(T, y, z) = \tilde{h}(z)$. The function $\bar{w}(t, y, z)$ has an explicit form, in terms of the processes $\tilde{Y}_t$, Z_t satisfying the equations

$$d\tilde{Y}_t = \tilde{g}(\tilde{Y}_t)dt + \beta(\tilde{Y}_t)(\rho dW_t^1 + \sqrt{1-\rho^2}dW_t^2)$$
$$dZ_t = -\frac{1}{2}\sigma^2(\tilde{Y}_t)dt + \sigma(\tilde{Y}_t)dW_t^1.$$

Here (W^1, W^2) is the two dimensional Brownian motion introduced in Sect. 2. Equation (3.7) has a solution given by

$$\bar{w}(t, y, z) = \mathbf{E}_{t,y,z}[\int_t^T \tilde{Q}(u, \tilde{Y}_u, Z_u)du + \tilde{h}(Z_T)]. \tag{3.8}$$

Furthermore, the above sub and supersolutions $\underline{w}$ and $\bar{w}$ are ordered, i.e. $\underline{w} \leq \bar{w}$. Next theorem states that a smooth solution to (3.6) exists between these two functions. This is the main result of this paper.

Theorem 3.4 *Under Assumption A, there exists a unique classical solution* $\tilde{\psi} : [0, T] \times I\!R \times I\!R \to I\!R$ *to (3.6) with*

$$\underline{w} \leq \tilde{\psi} \leq \bar{w}.$$

Moreover, its gradient $(\tilde{\psi}_y, \tilde{\psi}_z)$ *is bounded.*

Remark 3.5

1. Notice that the last assertion on the derivatives of $\tilde{\psi}$ is very important in order for the Markov control policy $\alpha^{\psi}(t, y)$ defined by (3.3) to be admissible.
2. One of the important consequences of the previous theorem is that, in an indirect way, we provide the optimal solution α^{ψ} to the optimal investment problem (2.4) when a liability $h(S_T)$ is part of the investor's portfolio and his preferences are of exponential type. Viscosity solution methods have been implemented by Sircar and Zariphopoulou [27] to analyse a similar problem. However, the gradient estimates established in this note are not provided, which are crucial to solve the optimal investment problem mentioned above.

The proof of this theorem is provided in the rest of this section through a sequence of preliminary results listed as lemmas. The first step is to study the problem in a constraint domain, getting some useful estimates on the gradient.

Given $R > 0$, let $B_R := \{(y, z) \mid y^2 + z^2 < R^2\}$ be the open ball of radius R in $I\!R^2$, and denote by $\bar{B}_R$ and ∂B_R its closure and boundary, respectively.

Lemma 3.6 *There exists a unique classical solution* $w^R : [0, T] \times \bar{B}_R \to I\!R$ *to the following terminal boundary value problem*

$$\begin{aligned} &w_t + \mathcal{L}w = \mathcal{H}(y, w_y), && t \in (0, T), \ (y, z) \in B_R, \\ &w(T, y, z) = \tilde{h}(z), && (y, z) \in \bar{B}_R, \\ &w(t, y, z) = \bar{w}(t, y, z), && t \in [0, T], \ (y, z) \in \partial B_R, \end{aligned} \tag{3.9}$$

with

$$\underline{w} \le w^R \le \bar{w}.$$

This implies that there is a constant $M > 0$ (dependent on $T > 0$ but independent of R) such that

$$\sup_{[0,T]\times B_R} \{|w^R(t, y, z)|\} \le M.$$

Also, for any $0 < r \le R - 1$, there is $M_1(r)$ such that

$$\sup_{[0,T]\times B_r} \{|(w^R)_y|, |(w^R)_z|\} \le M_1(r), \ R \ge r + 1.$$

Remark 3.7 In addition to Assumption A, we assume that for an integer $n_0 \ge 0$ and $0 < \alpha < 1$, the derivatives of the coefficients of order up to n_0 are of α-Hölder continuous and h has continuous derivatives of order up to $n_0 + 2$ that are also α-Hölder continuous. Then, for every $r > 0$, in $[0, T] \times B_r$ the α- Hölder norm of the derivatives of w^R with $R \ge r + 1$ of order up to $n_0 + 2$ are uniformly bounded.

Sketch of the Proof of Lemma 3.6 The first part of this lemma is a consequence of a result in the theory of partial differential equation. See Theorem 6.1, Chapter V in [22] (1968 edition). The last statement is a consequence of Theorem 5.2, Chapter IV [22]. We give some explanations of these results within our context in the following.

To adopt the notations in [22], we denote $x = (y, z)$,

$$u(t, x) = w(T - t, y, z).$$

The equation for u is given by

$$u_t(t, x) = \mathcal{L}u(t, x) - \mathcal{H}(y, u_y), \ x = (y, z). \tag{3.10}$$

The boundary value of the equation is given by the function ψ defined by

$$\psi(0, x) = \tilde{h}(z), \ x = (y, z) \in B_R,$$

$$\psi(t, x) = \bar{w}(T - t, y, z), \ 0 \le t \le T, x = (y, z) \in \partial B_R.$$

Hence, ψ is a smooth function satisfying the compatibility condition of first order:

$$\psi_t(t, x) = \mathcal{L}\psi(t, x) - \mathcal{H}(y, \psi_y), \ t = 0. \tag{3.11}$$

See (6.3) in Chapter V, [22]. This important property ensures that the solution w^R obtained has good regularity properties, including at the boundary of $[0,T] \times B_R$. For example, the first order derivative in t and the derivatives in y, z up to the second order are Hölder continuous. The sup-norm of these quantities (including their Hölder norms of proper order) has bounds in compact sets independent of R that can also be derived. The bound depends on the upper bound of the sup-norm of the solution (in each bounded region), the elliptic constant of $\mathcal{L}$, and the Hölder continuity of the coefficients of the equation. The result is stated in Theorem 6.1 (Chapter V, [22]). The relation $\underline{w} \le w^R \le \bar{w}$ will provide the upper bound for the sup-norm of w^R (in compact sets). Hence we can prove the uniform boundness in compact sets for the derivatives of the solutions when $R \to \infty$. In particular, we can take a subsequence $R_n \to \infty$ such that w^{R_n} converges to a smooth solution of (3.6). This will be needed in the next lemma.

If we do not have (3.11) for the boundary value, it is also possible to prove the existence of solution of (3.10) in $(0,T) \times B_R$ under suitable conditions. The solution is of Hölder continuous in the closure and is smooth in the interior. Such results can be found also in Section 6 (Chapter V, [22]).

The relation $\underline{w} \le w^R \le \bar{w}$ follows by the maximum principle in differential equation. For example, $\underline{w} \le w^R$ can be shown as follows. Assume to the contrary that

$$\max_{[0,T]\times\bar{B}_R} \{\underline{w}(t,y,z) - w^R(t,y,z)\} > 0.$$

We consider for a small positive $\epsilon > 0$,

$$\underline{w}_\epsilon(t,y,z) = \underline{w}(t,y,z) - \epsilon(T-t).$$

Then, $\underline{w}_\epsilon$ is also a subsolution for $\epsilon > 0$. It satisfies

$$(\underline{w}_\epsilon)_t + \mathcal{L}\underline{w}_\epsilon > \mathcal{H}(y, (\underline{w}_\epsilon)_y),$$

a strict inequality for $\epsilon > 0$. For small $\epsilon > 0$, we also have

$$\max_{[0,T]\times\bar{B}_R} \{\underline{w}_\epsilon(t,y,z) - w^R(t,y,z)\} > 0.$$

The maximum is taken at $t_0 \in [0,T]$, $(y_0, z_0) \in \bar{B}_R$. Then we must have $(y_0, z_0) \in B_R$ and $0 \le t_0 < T$. At (t_0, y_0, z_0), we have the relations:

$$(\underline{w}_\epsilon)_t - w^R_t \le 0, \ (\underline{w}_\epsilon)_y - w^R_y = 0.$$

We also have

$$\frac{1}{2}\sigma^2((\underline{w}_\epsilon)_{zz} - w^R_{zz}) + \frac{1}{2}\beta^2((\underline{w}_\epsilon)_{yy} - w^R_{yy}) + \rho\sigma\beta((\underline{w}_\epsilon)_{yz} - w^R_{yz}) \le 0.$$

Using these, we have

$$\begin{aligned}&\mathcal{L}\underline{w}_\epsilon(t_0,y_0,z_0)-\mathcal{H}(y_0,(\underline{w}_\epsilon)_y(t_0,y_0,z_0))\\&\le \mathcal{L}w^R(t_0,y_0,z_0)-\mathcal{H}(y_0,w^R_y(t_0,y_0,z_0)).\end{aligned}$$

Then

$$\begin{aligned}0&<(\underline{w}_\epsilon)_t(t_0,y_0,z_0)+\mathcal{L}\underline{w}_\epsilon(t_0,y_0,z_0)-\mathcal{H}(y_0,(\underline{w}_\epsilon)_y(t_0,y_0,z_0))\\&\le w^R_t(t_0,y_0,z_0)+\mathcal{L}w^R(t_0,y_0,z_0)-\mathcal{H}(y_0,w^R_y(t_0,y_0,z_0))=0.\end{aligned}$$

This is a contradiction. □

In the next result the monotonicity with respect to R of the functions w^R is analyzed.

Lemma 3.8 *Let $R > 0$ be fixed. Then, we have that*

$$\underline{w}(t,y,z)\le w^{R+1}(t,y,z)\le w^R(t,y,z)\le \bar{w}(t,y,z),\quad \textit{for}\ \ t\in[0,T],\ (y,z)\in B_R.$$

Proof From the previous lemma we know that w^R solves the Dirichlet problem (3.9) in B_R. On the other hand, using again the same result $w^{R+1}(t,y,z)\le \bar{w}(t,y,z)$ for $t\in[0,T],\ (y,z)\in B_{R+1}$, which yields

$$w^{R+1}(t,y,z)\le \bar{w}(t,y,z)=w^R(t,y,z),\quad \text{for}\ \ t\in[0,T],\ (y,z)\in\partial B_R.$$

Using now the maximum principle, we obtain that $w^{R+1}(t,y,z)\le w^R(t,y,z)$ in $[0,T]\times B_R$. Therefore, fixing $R'>0$, for any $R>R'$ the sequence w^R is decreasing in R within $B_{R'}$. □

Since $\{w^R\}$ is decreasing in R and bounded from below by $\underline{w}$, the limit function $w(t,y,z)=\lim_{R\to\infty}w^R(t,y,z)$, is well defined, and also satisfies $w(t,y,z)\ge \underline{w}\ (t,y,z)$ in $[0,T]\times B_R$, which is a solution of the Eq. (3.6). Hence, w is a bounded function, that is, there is a constant C such that

$$|w(t,y,z)|\le C.$$

Now, some estimates of w^R_y and w^R_z are shown. We use Bernstein's argument from the theory of PDEs, which is a very useful (and relatively easy) approach to prove such estimate. See [22, p. 414].

Lemma 3.9 *For $0\le t\le T$ and $(y,z)\in B_R$, $w^R_y(t,y,z), w^R_z(t,y,z)$ are uniformly bounded on R. That is, there is $c_T>0$ such that for all $R>0$, we have*

$$|w^R_y(t,y,z)|\le c_T,\ |w^R_z(t,y,z)|\le c_T,\ 0\le t\le T, |y|^2+|z|^2<R^2.$$

Proof We omit R in w^R and write w for w^R throughout this proof. In the following, we consider $0 \le t \le T$, $(y,z) \in B_R$.

For $\gamma > 0$, define

$$G_0(t,y,z) = \tfrac{1}{2}(w_y^2(t,y,z) + w_z^2(t,y,z)),$$
$$G_1(t,y,z) = \exp(\gamma w(t,y,z))$$

We want to consider G_0G_1 and calculate

$$\frac{d}{dt}(G_0G_1) + L(G_0G_1),$$

denoting

$$\frac{dG_0}{dt} + LG_0 = H_0.$$

Then,

$$\begin{aligned} H_0 &= w_y(\tfrac{d}{dt}w_y + Lw_y) + w_z(\tfrac{d}{dt}w_z + Lw_z) \\ &\quad + \tfrac{1}{2}\sigma^2 w_{yz}^2 + \tfrac{1}{2}\beta^2 w_{yy}^2 + \rho\sigma\beta w_{yy}w_{yz} + \tfrac{1}{2}\sigma^2 w_{zz}^2 + \tfrac{1}{2}\beta^2 w_{yz}^2 + \rho\sigma\beta w_{zy}w_{zz}. \end{aligned}$$

In the above and also in the following calculation, we often omit (t,y,z). Using the equation for w and taking derivative with respect to y, we obtain

$$\frac{d}{dt}w_y + Lw_y = \tilde{\beta}^2 w_y w_{yy} + \tilde{\beta}\tilde{\beta}_y - \tilde{Q}_y - \sigma\sigma_y w_{zz} - \beta\beta_y w_{yy} - \rho(\sigma\beta)_y w_{yz} + \sigma\sigma_y w_z - \tilde{g}_y w_y.$$

Similarly,

$$\frac{d}{dt}w_y + Lw_y = \tilde{\beta}^2 w_y w_{yz}.$$

Then,

$$\begin{aligned} H_0 &= w_y(\tilde{\beta}^2 w_y w_{yy}) + w_z(\tilde{\beta}^2 w_y w_{yz}) \\ &\quad + w_y(\tilde{\beta}\tilde{\beta}_y - \tilde{Q}_y - \sigma\sigma_y w_{zz} - \beta\beta_y w_{yy} - \rho(\sigma\beta)_y w_{yz} + \sigma\sigma_y w_z - \tilde{g}_y w_y) \\ &\quad + \tfrac{1}{2}\sigma^2 w_{yz}^2 + \tfrac{1}{2}\beta w_{yy}^2 + \rho\sigma\beta w_{yy}w_{yz} + \tfrac{1}{2}\sigma^2 w_{zz}^2 + \tfrac{1}{2}\beta w_{yz}^2 + \rho\sigma\beta w_{zy}w_{zz}. \end{aligned}$$

Notice that

$$\begin{aligned}
&\tfrac{1}{2}\sigma^2 w_{yz}^2 + \tfrac{1}{2}\beta w_{yy}^2 + \rho\sigma\beta w_{yy}w_{yz} + \tfrac{1}{2}\sigma^2 w_{zz}^2 + \tfrac{1}{2}\beta w_{yz}^2 + \rho\sigma\beta w_{zy}w_{zz} \\
&= (1-|\rho|)(\tfrac{1}{2}\sigma^2 w_{yz}^2 + \tfrac{1}{2}\beta^2 w_{yy}^2) + |\rho|(\tfrac{1}{2}\sigma^2 w_{yz}^2 + \tfrac{\rho}{|\rho|}\sigma\beta w_{yz}w_{yy} + \tfrac{1}{2}\beta^2 w_{yy}^2) \\
&+(1-|\rho|)(\tfrac{1}{2}\sigma^2 w_{zz}^2 + \tfrac{1}{2}\beta^2 w_{yz}^2) + |\rho|(\tfrac{1}{2}\sigma^2 w_{zz}^2 + \tfrac{\rho}{|\rho|}\sigma\beta w_{zz}w_{yz} + \tfrac{1}{2}\beta^2 w_{yz}^2) \\
&\geq (1-|\rho|)(\tfrac{1}{2}\sigma^2 w_{yz}^2 + \tfrac{1}{2}\beta^2 w_{yy}^2) + \tfrac{1}{2}\sigma^2 w_{zz}^2 + \tfrac{1}{2}\beta^2 w_{yz}^2 \\
&\geq c_1(w_{yz}^2 + w_{yy}^2 + w_{zz}^2 + w_{yz}^2),
\end{aligned}$$

where $c_1 > 0$ is a constant depending on ρ and the constant c in Assumption A.3. Then,

$$\begin{aligned}
H_0 \geq\ & c_1(w_{yz}^2 + w_{yy}^2 + w_{zz}^2 + w_{yz}^2) - c_2(|w_y|^3 + |w_y|^2 + |w_y| + |w_y||w_z| \\
&+|w_y||w_{yy}| + |w_y||w_{yz}| + |w_y||w_{yy}|),
\end{aligned}$$

where

$$c_2 = \max\{\|\tilde{\beta}\tilde{\beta}_y\|, \|\tilde{g}_y\|, \|\tilde{Q}_y\|, \|\sigma\sigma_y\|, \|\beta\beta_y\|, |\rho|\|(\sigma\beta)_y\|\}.$$

Here $\|\tilde{\beta}\tilde{\beta}_y\|$ denotes the sup-norm of the function $\tilde{\beta}\tilde{\beta}_y$, and similar notation is used for other terms.

We next consider

$$\frac{d}{dt}G_1 + LG_1 = H_1.$$

After some calculations we have the expression,

$$\begin{aligned}
H_1 &= \gamma G_1(\tfrac{d}{dt}w + Lw) + \gamma^2 G_1(\tfrac{1}{2}\sigma^2 w_z^2 + \tfrac{1}{2}\beta w_y^2 + \rho\sigma\beta w_y w_z) \\
&= \gamma G_1(\tfrac{1}{2}\tilde{\beta}^2 w_y^2 - \tilde{Q}) + \gamma^2 G_1(\tfrac{1}{2}\sigma^2 w_z^2 + \tfrac{1}{2}\beta w_y^2 + \rho\sigma\beta w_y w_z).
\end{aligned}$$

Now, we consider

$$\frac{d}{dt}(G_0G_1) + L(G_0G_1) = H_2.$$

Here

$$\begin{aligned}
H_2 &= H_0G_1 + H_1G_0 + \tfrac{1}{2}\sigma^2(D_zG_0D_zG_1) + \tfrac{1}{2}\beta^2(D_yG_0D_yG_1) \\
&\quad+\rho\sigma\beta(DyG_0D_zG_1 + D_zG_0D_yG_1) \\
&= H_0G_1 + G_0G_1(\gamma(\tfrac{1}{2}\tilde{\beta}^2 w_y^2 - \tilde{Q}) + \gamma^2(\tfrac{1}{2}\sigma^2 w_z^2 + \tfrac{1}{2}\beta w_y^2 + \rho\sigma\beta w_y w_z)) \\
&\quad+\gamma G_1(\sigma^2(w_y w_z w_{yz} + w_z^2 w_{zz}) + \beta^2(w_y^2 w_{yy} + w_y w_z w_{yz})) \\
&\quad+\rho\sigma\beta(w_y w_z w_{yy} + w_z^2 w_{yz} + w_y^2 w_{yz} + w_y w_z w_{zz}).
\end{aligned}$$

Observe that $H_2 = G_1 H_3$, and

$$\begin{aligned} H_3 \geq\, & c_1(w_{yz}^2 + w_{yy}^2 + w_{zz}^2 + w_{yz}^2) - c_2(|w_y|^3 + |w_y|^2 + |w_y| + |w_y||w_z| \\ & +|w_y||w_{yy}| + |w_y||w_{yz}| + |w_y||w_{yy}|) + c_3\gamma^2(w_y^2 + w_z^2)^2 \\ & -c_4\gamma w_y^2(w_y^2 + w_z^2) - c_5\gamma(|w_y||w_z||w_{yz}| + w_z^2|w_{zz}| + w_y^2|w_{yy}| \\ & +|w_y||w_z||w_{yz}| + |w_y|w_z||w_{yy}| + w_z^2|w_{yz}| + w_y^2|w_{yz}| + |w_y||w_z||w_{yz}|). \end{aligned}$$

Here we used that $\tilde{Q} \leq 0$. Note that the constant $c_3 > 0$ depends on the constant c in Assumption A.3.

$$c_4 = \frac{1}{2}\|\tilde{\beta}\|^2, \; c_5 = \max\{\|\sigma\|^2, \|\beta\|^2, |\rho|\|\sigma\|\|\beta\|\}.$$

From this, it is easy to see, with γ large enough, that there are positive constants, c_6, c_7, c_8, such that

$$H_3 \geq c_6(w_{yz}^2 + w_{yy}^2 + w_{zz}^2 + w_{yz}^2) + c_7\gamma^2(w_y^2 + w_z^2)^2 - c_8.$$

Finally, take

$$G = G_0 G_1 + 2c_8 t.$$

Then

$$\frac{d}{dt}G + LG > 0, \; 0 \leq t \leq T, \; (y, z) \in B_R.$$

Now, we can apply the maximum principle and boundness of $\bar{w}_y, \bar{w}_z$ to show for $0 \leq t \leq T, \; (y, z) \in B_R$, that $G(t, y, z)$ is less or equal than

$$\max\{G(s, \xi, \zeta); s = T, (\xi, \zeta) \in B_R \text{ or } 0 \leq s \leq T \text{ and } |\xi|^2 + |\zeta|^2 = R^2\}.$$

Here $\bar{w}$ is the supersolution given in (3.8). This shows the boundness of w_y, w_z, completing the proof of the lemma. □

As a consequence of the previous two lemmas, the next result follows by taking $R \to \infty$. Recall that $w(t, y, z) = \lim_{R\to\infty} w^R(t, y, z)$.

Lemma 3.10 *w_y, w_z are bounded functions on $[0, T]\times R^2$. That is, there is a constant c_T, depending on T, such that*

$$|w_y(t, y, z)| \leq c_T, |w_z(t, y, z)| \leq c_T, \; 0 \leq t \leq T, (y, z) \in I\!R^2.$$

Finally, Theorem 3.4 follows from Lemmas 3.6 to 3.10.

4 Verification Theorem

In this section we present a verification theorem for the optimal control problem (2.10), which states that the classical solution founded in Theorem 3.4 corresponds to the value function defined in (2.10); a straightforward corollary of this result is the uniqueness of classical solutions to the PDE (3.4) between the class of functions defined in Theorem 3.4. Similar results can be stated for the value function $\phi(t, y)$, but we will present only the proof for the first one to avoid unnecessary repetitions.

Theorem 4.1 *Let* $\Psi \in C^{1,2}((0,T) \times I\!R) \cap C((0,T] \times I\!R)$ *be the smooth solution found in Theorem* 3.4. *Then, it is the value function* (2.10). *Moreover, the Markov control* α^{Ψ} *defined in* (3.3) *is optimal, i.e.*

$$\Psi(t,y,s) = -\frac{1}{\gamma}\log \mathbf{E}^{\mathbf{P}^{\alpha^{\Psi}}}_{t,y,s} \exp\{-\gamma \int_t^T \tilde{l}(Y_u, \alpha^{\Psi}_u)du - \gamma h(S_T)\}.$$

Proof Let α_t be an arbitrary strategy in $\mathcal{A}$. Then, in view of (2.2) the change of measure in (2.7) associated with α_t is well defined. Let Ψ be as in the statement of the theorem, and define $\tau_n := \inf\{u \geq t \mid |Y_u| > n \text{ or } S_u > n\}$. Then, by Ito's formula and (3.4),

$$\Psi(T\wedge\tau_n, Y_{T\wedge\tau_n}, S_{T\wedge\tau_n}) - \Psi(t,y,s) = \int_t^{T\wedge\tau_n} [\Psi_u + S_u\Psi_s(\mu(Y_u) - \gamma\alpha_u\sigma^2(Y_u))$$
$$+\Psi_y(g(Y_u) - \gamma\rho\alpha_u\sigma(Y_u)\beta(Y_u)) + \frac{1}{2}S_u^2\sigma^2(Y_u)\Psi_{ss} + \frac{1}{2}\beta^2(Y_u)\Psi_{yy}$$
$$+\rho S_u\sigma(Y_u)\beta(Y_u)\Psi_{ys}] + \int_t^{T\wedge\tau_n} S_u\sigma(Y_u)\Psi_s d\tilde{W}^1_u + \beta(Y_u)\Psi_y(\rho d\tilde{W}^1_u$$
$$+\sqrt{1-\rho^2}d\tilde{W}^2_u)$$
$$\leq -\int_t^{T\wedge\tau_n} \tilde{l}(Y_u, S_u)du + \int_t^{T\wedge\tau_n} [\frac{\gamma}{2}(S_u\sigma(Y_u)\Psi_s + \beta(Y_u)\Psi_y)^2$$
$$-\gamma(1-\rho)S_u\sigma(Y_u)\beta(Y_u)\Psi_s\Psi_y]du$$
$$+\int_t^{T\wedge\tau_n} S_u\sigma(Y_u)\Psi_s d\tilde{W}^1_u + \beta(Y_u)\Psi_y(\rho d\tilde{W}^1_u + \sqrt{1-\rho^2}d\tilde{W}^2_u).$$

Here (and in the next) we omit (u, Y_u, S_u) in Ψ_u, Ψ_y, Ψ_s, etc. Then,

$$\mathbf{E}^{\mathbf{P}^{\alpha}}_{t,y,s}\exp\{-\gamma\int_t^{T\wedge\tau_n} \tilde{l}(Y_u, S_u)du - \gamma\Psi(T\wedge\tau_n, Y_{T\wedge\tau_n}, S_{T\wedge\tau_n})\} \geq e^{-\gamma\Psi(t,y,s)}.$$

$$\mathbf{E}^{\mathbf{P}^{\alpha}}_{t,y,s}\exp\int_t^{T\wedge\tau_n} -\{\frac{\gamma^2}{2}(S_u\sigma(Y_u)\Psi_s + \beta(Y_u)\Psi_y)^2$$

$$-\gamma^2(1-\rho)S_u\sigma(Y_u)\beta(Y_u)\Psi_s\Psi_y\}du\cdot$$

$$\exp\{-\gamma\int_t^{T\wedge\tau_n} S_u\sigma(Y_u)\Psi_s d\tilde{W}_u^1+\beta(Y_u)\Psi_y(\rho d\tilde{W}_u^1+\sqrt{1-\rho^2}d\tilde{W}_u^2)\}$$

$$=e^{-\gamma\Psi(t,y,s)}\mathbf{E}_{t,y,s}^{\mathbf{P}^\alpha}\exp\{-\gamma\int_t^{T\wedge\tau_n} S_u\sigma(Y_u)\Psi_s d\tilde{W}_u^1+\beta(Y_u)\Psi_y(\rho d\tilde{W}_u^1$$

$$+\sqrt{1-\rho^2}d\tilde{W}_u^2)\}\cdot\exp\{-\frac{\gamma^2}{2}\int_t^{T\wedge\tau_n}(S_u^2\sigma^2\Psi_s^2+\beta\Psi_y^2+2\rho S_u\sigma\beta\Psi_s\Psi_y)du\}.$$

Then, using the terminal condition $\Psi(T,Y_T,S_T)=h(S_T)$, the boundness of Ψ and the fact that $\tilde{\ell}$ is negative, taking the limit when $n\to\infty$ we get that

$$-\frac{1}{\gamma}\log\mathbf{E}_{t,y,s}^{\mathbf{P}^\alpha}e^{-\gamma\int_t^T\tilde{l}(Y_u,S_u)du-\gamma h(S_T)}\le\Psi(t,y,s).$$

Since α_t was chosen arbitrarily, it follows that

$$\Psi(t,y,s)\ge\psi(t,y,s).$$

To verify the reverse inequality, we only need to check that the candidate for being optimal control, defining a Markov policy through (3.3), belongs to the set of admissible strategies $\mathcal{A}$. However, this is straightforward, in view of the estimation obtained in Theorem 3.4. □

5 Relations with BSDEs

As it was proved in the previous sections, the solution of the quasilinear PDE (3.4) corresponds to the value function of the optimal control problem in (2.10). This provides a stochastic representation of that solution. In this section another representation is given using forward backward stochastic differential equations (FBSDE's). This relation can be used to prove existence of solutions to the FBSDE's, when there exists a smooth solution to the PDE; see, for instance, [1, 8, 21] and references therein.

Using the results of Sect. 3, a straightforward application of Ito's formula makes the connection with the BSDE. Note that the analysis of the exponential hedging problem is also an interesting problem where a quadratic BSDE arises, see [26].

The class of BSDE's associated with (3.4) is of the form

$$-dy_t=[-\frac{1}{2}\gamma(z_{2,t})^2+\tilde{Q}(Y_t)]dt-z_t\cdot dW_t, \tag{5.1}$$

with terminal condition $y_T = h(S_T)$; here $y_t \in I\!R$ and $z_t = (z_{1,t}, z_{2,t}) \in I\!R^2$ are the solutions that are described below.

The processes Y_t and S_t appearing in (5.1) and the terminal condition satisfy the forward SDE

$$dS_t = S_t(Y_t)\sigma(Y_t)dW_t^1$$

and

$$dY_t = \tilde{g}(Y_t)dt + \beta(Y_t)[\rho dW_t^1 + \sqrt{1-\rho^2}dW_t^2],$$

with initial conditions at time t_0 given by $S_{t_0} = s$ and $Y_{t_0} = y$.

This type of FBSDE's have been studied extensively in recent years. The quadratic growth of the driver in the variable z is not covered by the standard existence and uniqueness theorems and different arguments are needed, see for instance [8] and [6]. In this particular case, a solution to (5.1) can be derived easily, thanks to Theorem 2.1 and using Ito's formula.

More precisely, the unique solution of (5.1) is given by

$$y_t = \psi(t, Y_t, S_t)$$

and $z_t = (z_{1,t}, z_{2,t})$ is defined as

$$z_t = (\rho\beta(Y_t)\psi_y(t, Y_t, S_t) + \sigma(Y_t)\psi_s(t, Y_t, S_t), \sqrt{1-\rho^2}\beta(Y_t)\psi_y(t, Y_t, S_t)).$$

Observe that in view of the gradient estimate obtained in Theorem 3.4, $(y_s, \; z_s)$ belongs to the set of square integrable adapted processes. Uniqueness follows from the comparison results in [8].

Using the definition of $p(t, y, s)$ given in (2.13) and the expressions of the PDEs solved by the functions ψ and ϕ (see (3.4) and (3.5)) we can write the PDE for the utility-based price $p(t, y, s)$,

$$\begin{aligned} 0 = p_t + (\tilde{g}(y) - \tilde{\beta}^2\phi_y(y))p_y - \frac{1}{2}\tilde{\beta}^2(y)p_y^2 + \frac{1}{2}s^2\sigma^2(y)p_{ss} + \frac{1}{2}\beta^2(y)p_{yy} \quad (5.2)\\ +\rho s\sigma(y)\beta(y)p_{sy}. \end{aligned}$$

This equation is very similar to (3.4) and the same analysis developed in Sect. 3 can be repeated, and hence, it is possible to read the solution of the associated BSDE with quadratic growth.

6 Risk Neutral Valuation

Recall that a risk neutral measure is an equivalent measure to **P**, such that the risky asset is a martingale. In this section we present an explicit form of a martingale measure associated with the utility-based price methodology defined in Sect. 2. Notice that from (2.13) we know that $p(t,y,s) = \psi(t,y,s) - \phi(t,y)$, with ψ and ϕ being classical solutions of (3.4) and (3.5), respectively. From these PDEs we can write the equation satisfied by $p(t,y,s)$,

$$p_t + [\tilde{g} - \frac{1}{2}\tilde{\beta}^2(\psi_y + \phi_y)]p_y + \frac{1}{2}\beta^2 p_{yy} + \frac{1}{2}\sigma^2 s^2 p_{ss} + \rho s\sigma\beta p_{ys} = 0, \tag{6.1}$$

with final condition $p(T,y,s) = h(s)$.

From the previous equation we can read the form of the risk neutral measure P^* associated with $p(t,y,s)$.

Proposition 6.1 *There exists a martingale measure P^* such that the next stochastic representation of $p(t,y,s)$ holds*

$$p(t,y,s) = \mathbf{E}^{P^*}_{t,y,s}[h(S_T)]. \tag{6.2}$$

Proof Define the process

$$M_t = \exp\left\{ -\int_0^t [\frac{\mu}{\sigma} dW_u^1 + \frac{\mu^2}{2\sigma^2} du] - \int_0^t [\frac{\gamma}{2}\bar{\rho}\beta(\psi_y + \phi_y) dW_u^2 \right.$$
$$\left. + \frac{\gamma^2}{4}\bar{\rho}^2\beta^2(\psi_y + \phi_y)^2 du] \right\},$$

with $\bar{\rho} = \sqrt{1-\rho^2}$. In view of the results obtained in the previous sections on the gradient estimates, it follows that this process is a martingale and induces a probability measure.

From Girsanov's theorem, $\tilde{W}_t^1 := W_t^1 + \frac{\mu}{\sigma}$ and $\tilde{W}_t^2 := W_t^2 + \frac{\gamma}{2}\bar{\rho}\beta(\psi_y + \phi_y)$ is a Brownian motion in $I\!R^2$ under measure P^*, and the dynamics of S_t and Y_t can be written as

$$dS_t = S_t\sigma(Y_t)d\tilde{W}_t^1$$

and

$$dY_t = \left[g(Y_t) - \rho\frac{\mu(Y_t)\beta(Y_t)}{\sigma(Y_t)} - \frac{1}{2}\gamma\bar{\rho}^2\beta^2(Y_t)(\psi_y(t,Y_t,Z_t) - \phi(t,Y_t)) \right] dt$$
$$+\beta(Y_t)[\rho d\tilde{W}_t^1 + \bar{\rho} d\tilde{W}_t^2].$$

From the Feynman-Kac stochastic representation we obtain (6.2). □

Observe that the nonlinear property of the utility-based pricing method (see [10]) is expressed in (6.2), since the measure P^* depends on the option $h(S_T)$. The measure associated with utility-based valuation have been studied by different authors [15], in particular the form of the minimal relative entropy measure is defined in [27], Section 2.1.

Acknowledgements The authors thank the anonymous referee for his/her insightful comments that help to improve the presentation of the paper. This work was partially developed during the visit of the first author to Academia Sinica, which support and hospitality is greatly appreciated.

References

1. G. Barles, E. Lesigne, SDE, BSDE and PDE, in *Backward Stochastic Differential Equations*, ed. by N. El Karoui, L. Mazliak. Pitman Research Notes, vol. 364 (Longman, Harlow, 1997), pp. 47–80
2. G. Barles, H.M. Soner, Option pricing with transaction costs and a nonlinear Black-Scholes equation. Financ. Stoch. **2**, 369–397 (1998)
3. D. Becherer, Rational hedging and valuation of integrated risks under constant absolute risk aversion. Insur. Math. Econ. **33**, 1–28 (2003)
4. D. Becherer, Bounded solutions to backwards SDEs with jumps for utility maximization and indifference pricing. Ann. Appl. Probab. **16**, 2027–2054 (2006)
5. T.R. Bielecki, S.R. Pliska, Risk sensitive dynamic asset management. Appl. Math. Optim. **39**, 337–360 (1999)
6. P. Briand, Y. Hu, BSDE with quadratic growth and unbounded terminal value. Probab. Theory Relat. Fields **141**, 543–567 (2008)
7. R. Cavazos-Cadena, D. Hernández-Hernández, Characterization of the optimal risk sensitive average cost in finite controlled Markov chains. Ann. Appl. Probab. **15**, 175–212 (2005)
8. N. El Karoui, Backward stochastic differential equations a general introduction, in *Backward Stochastic Differential Equations*, ed. by N. El Karoui, L. Mazliak. Pitman Research Notes, vol. 364 (Longman, Harlow, 1997), pp. 7–26
9. N. El Karoui, M.C. Quenez, Dynamic programming and pricing of contingent claims in a incomplete market. SIAM J. Control Optim. **33**, 29–69 (1993)
10. R. Rouge, N. El Karoui, Pricing by utility maximization. Math. Financ. **10**, 259–276 (2000)
11. W.H. Fleming, D. Hernández-Hernández, The tradeoff between consumption and investment in incomplete financial markets. Appl. Math. Optim. **52**, 219–235 (2005)
12. W.H. Fleming, S.J. Sheu, Risk sensitive and an optimal investment model II. Ann. Appl. Probab. **12**, 730–767 (2002)
13. W.H. Fleming, H.M. Soner, *Controlled Markov Processes and Viscosity Solutions*, 2nd edn. (Springer, New York, 2006)
14. J.-P. Fouque, G. Papanicolaou, K.R. Sircar, *Derivatives in Financial Markets with Stochastic Volatility* (Cambridge University Press, Cambridge, 2000)
15. M. Frittelli, Introduction to a theory of value coherent with the no-arbitrage principle. Financ. Stoch. **4**, 275–297 (2000)
16. H. Hata, S.-J. Sheu, Hamilton-Jacobi-Bellman equation for an optimal consumption problem: I. Existence of solution. SIAM J. Control Optim. **50**, 2373–2400 (2012)
17. V. Henderson, Valuation of claims on nontraded assets using utility maximization. Math. Financ. **12**, 351–373 (2002)
18. V. Henderson, D. Hobson, Utility indifference pricing: an overview, in *Volume on Indifference Pricing*, ed. by R. Carmona (Princeton University Press, Princeton, 2004)

19. S.D. Hodges, A. Neuberger, Optimal replication of contingent claims under transaction costs. Rev. Future Mark. **8**, 222–239 (1989)
20. H. Kaise, S.-J. Sheu, On the structure of solutions of ergodic type Bellman type equation related to risk sensitive control. Ann. Probab. **34**, 284–320 (2006)
21. M. Kobylanski, Backward stochastic differential equations with quadratic growth. Ann. Probab. **28**, 558–602 (2000)
22. O.A. Ladyzenskaya, V.A. Solonikov, N.N. Uralseva, *Linear and Quasilinear Equations of Parabolic Type*. AMS Translations of Mathematical Monographs (American Mathematical Society, Providence, 1968)
23. M. Mania, M. Schweizer, Dynamic exponential utility indifference valuation. Ann. Appl. Probab. **15**, 2113–2143 (2005)
24. M. Musiela, T. Zariphopoulou, A valuation algorithm for indifference prices in incomplete markets. Financ. Stoch. **8**, 339–414 (2004)
25. H. Nagai, Bellman equations of risk sensitive control. SIAM J. Control Optim. **34**, 74–101 (1996)
26. J. Sekine, On exponential hedging and related quadratic backward stochastic differential equations. Appl. Math. Optim. **54**, 131–158 (2006)
27. R. Sircar, T. Zariphopoulou, Bounds and asymptotic approximations when volatility is random. SIAM J. Control Optim. **43**, 1328–1353 (2005)

The Backbone Decomposition for Superprocesses with Non-local Branching

Antonio Murillo-Salas and José Luis Pérez

Abstract We provide a path-wise "backbone" decomposition for supercritical superprocesses with non-local branching. Our result complements a related result obtained for supercritical superprocesses without non-local branching in Berestycki et al. (Stoch Proc Appl 121:1315–1331, 2011). Our approach relies heavily on the use of so-called Dynkin-Kuznetsov $\mathbb{N}$-measures.

Keywords Superprocesses • Backbone decomposition • Non-local branching

Mathematics Subject Classification (2010). 60J80; 60J68.

1 Introduction

In this note we consider any superprocess $X = \{X_t : t \geq 0\}$ on $\mathbb{R}^d$ which is well defined for initial configurations $\mu \in \mathcal{M}_C(\mathbb{R}^d)$, the space of finite and compactly supported measures, having associated a conservative diffusion semigroup $\mathcal{P} = \{\mathcal{P}_t : t \geq 0\}$ on $\mathbb{R}^d$ and a branching mechanism ψ of the form

$$\psi(x,f,z) = \psi^L(x,z) + \psi^{NL}(x,f), \quad x \in \mathbb{R}^d,\ z \geq 0, f \in B^+(\mathbb{R}^d), \tag{1.1}$$

where $B^+(\mathbb{R}^d)$ denotes the set of positive measurable functions on $\mathbb{R}^d$, i.e., we consider superprocesses with non-local branching (see [3]). The first term corresponds

A. Murillo-Salas
Departamento de Matemáticas, Universidad de Guanajuato, Jalisco s/n, Mineral de Valenciana, Guanajuato, Gto. C.P. 36240, México
e-mail: amurillos@ugto.mx

J.L. Pérez (✉)
Department of Probability and Statistics, IIMAS-UNAM, 01000 Mexico, D.F., Mexico
e-mail: garmendia@sigma.iimas.unam.mx

R.H. Mena et al. (eds.), *XI Symposium on Probability and Stochastic Processes*, Progress in Probability 69, DOI 10.1007/978-3-319-13984-5_10

to the branching mechanism related to the local branching of the superprocess X, and according to [3] it takes the following form

$$\psi^L(x,z) = \alpha(x)z + \beta(x)z^2 + \int_0^\infty (e^{-zu} - 1 + zu)\Pi^L(x,du), \quad x \in \mathbb{R}^d,\ z \geq 0,$$

for bounded measurable functions $\alpha : \mathbb{R}^d \to \mathbb{R}$, $\beta : \mathbb{R}^d \to \mathbb{R}_+$, and $(u \wedge u^2)\Pi^L$ is a bounded kernel from $\mathbb{R}^d$ to $(0,\infty)$ (i.e. the application $x \to \int_{\mathbb{R}^d}(u \wedge u^2)\Pi^L(x,du)$ is bounded on $\mathbb{R}^d$). On the other hand, the second term in the right hand side of (1.1) is related to non-local branching which takes the form (cf. [3])

$$\psi^{NL}(x,f) = (f(x) - \zeta(x,f)), \qquad x \in \mathbb{R}^d, f \in B^+(\mathbb{R}^d),$$

with

$$\zeta(x,f) = \int_{M_0(\mathbb{R}^d)} \left(\gamma(x,\pi)\pi(f) + \int_0^\infty (1 - e^{-u\pi(f)})\Pi^{NL}(x,\pi,du)\right) G(x,d\pi),$$

where $\gamma \in B^+(\mathbb{R}^d \times M_0(\mathbb{R}^d))$ ($M_0(\mathbb{R}^d)$ denotes the set of probability measures on $\mathbb{R}^d$), $u\Pi^{NL}(x,\pi,du)$ is a bounded kernel from $\mathbb{R}^d \times M_0(\mathbb{R}^d)$ to $(0,\infty)$ and $G(x,d\pi)$ is a probability kernel from $\mathbb{R}^d$ to $M_0(\mathbb{R}^d)$ with

$$\gamma(x,\pi) + \int_0^\infty u\Pi^{NL}(x,\pi,du) \leq 1.$$

In fact, X is a Markovian $\mathcal{M}_C(\mathbb{R}^d)$-valued process whose one-dimensional distributions are characterised by the following result

Lemma 1.1 (Lemma 3.3 in [3]) *For all $f \in bp(\mathbb{R}^d)$, the space of non-negative, bounded measurable functions on $\mathbb{R}^d$,*

$$-\log \mathbb{E}_\mu(e^{-\langle f, X_t\rangle}) = \int_{\mathbb{R}^d} u_f(x,t)\mu(dx), \qquad \mu \in \mathcal{M}_C(\mathbb{R}^d),\ t \geq 0.$$

where $u_f(x,t)$ is the unique non-negative solution to the integral equation

$$u_f(x,t) = \mathcal{P}_t[f](x) - \int_0^t \mathcal{P}_s[\psi^L(\cdot, u_f(\cdot, t-s)) + \psi^{NL}(\cdot, u_f(\cdot, t-s))](x). \tag{1.2}$$

We call $(X, \mathbb{P}_\mu)$ a $(\mathcal{P}, \psi^L, \psi^{NL})$-superprocess started at $\mu \in \mathcal{M}_C(\mathbb{R}^d)$.

The goal of this note is to give a path-wise backbone decomposition for a $(\mathcal{P}, \psi^L, \psi^{NL})$-superprocess, similar to the work [1] where the non-local branching is not considered. Loosely speaking, the backbone decomposition is a way to reconstruct a supercritical superprocess from a branching particle system (*called the backbone*) together with some sources ($(\mathcal{P}, \psi^L, \psi^{NL})$-superprocesses conditioned to die) of Poissonian immigration along the paths of the particles in the backbone. Such a decomposition has been done in [8] for a quadratic superprocess from the analitic point of view. Since then there has been a lot of interest in finding a path-wise backbone decomposition for several different models of superprocesses due to a variety of applications that have been found (e.g. [11, 13]).

Very recently, in [12], the authors provide the backbone decomposition for a quite general spatially dependent supercritical superprocess without non-local branching. See [12] Section 2 for a summary of some backbone decompositions found in the literature. Here, we are interested in the effects that the non-local branching has on the backbone decomposition, hence throughout this paper we drop out the assumption of having a spatially dependent branching mechanism. Namely, we consider

$$\psi^L(z) = \alpha z + \beta z^2 + \int_0^\infty (e^{-zu} - 1 + zu)\Pi^L(du), \ z \geq 0,$$

with $\alpha \in \mathbb{R}$, $\beta \geq 0$, and Π^L a measure concentrated in $(0, \infty)$ such that $\int_0^\infty (u \wedge u^2)\Pi^L(du) < \infty$. For the non-local branching we assume that the probability kernel $G(x, d\pi) \equiv$ unit mass at some $\pi(x, \cdot) \in M_0(\mathbb{R}^d)$. For a measurable function f we set $\pi(x,f) \equiv \int_{\mathbb{R}^d} f(y)\pi(x, dy)$. In this case, the non-local branching mechanism is given by

$$\psi^{NL}(x,f) = f(x) - \zeta(\pi(x,f)), \qquad x \in \mathbb{R}^d, f \in B^+(\mathbb{R}^d),$$

where

$$\zeta(\lambda) = \gamma\lambda + \int_0^\infty (1 - e^{-\lambda u})\Pi^{NL}(du), \qquad \lambda \geq 0,$$

where $\gamma \geq 0$ and $\int_0^\infty u\Pi^{NL}(du) < \infty$ is such that

$$\gamma + \int_0^\infty u\Pi^{NL}(du) \leq 1.$$

Putting all together the above assumptions, we get that the mild equation (1.2) satisfied by the semigroup u_f can be written as

$$u_f(x, t) = \mathcal{P}_t[f](x) - \int_0^t \mathcal{P}_s[\phi^L(u_f(\cdot, t-s)) + \phi^{NL}(\cdot, u_f(\cdot, t-s))](x), \qquad (1.3)$$

where $\phi^L(z) = \psi^L(z) + z$ for $z \geq 0$, and $\phi^{NL}(x,f) = \psi^{NL}(x,f) - f(x)$ for $x \in \mathbb{R}^d$, $f \in B^+(\mathbb{R}^d)$.

The note is organised as follows. Section 2 contains the backbone decomposition given in [1], when non-local branching is not taken into account. In Sect. 3 we obtain the superprocess X conditioned to die and characterise the prolific individuals which are responsible for the infinite growth of the total mass. Finally, Sect. 4 provides the backbone decomposition.

2 The Backbone Decomposition Without Non-local Branching

The so-called backbone decompositions have been known in the earlier and more analytical setting of semigroup decompositions through the works [6–8] as well as in the pathwise setting in the work of [14, 15]. The purpose of this section is to introduce the pathwise backbone decomposition for a supercritical superprocess without non-local branching given in [1], we hope this will make the rest of the paper easier to follow.

To describe the backbone decomposition in detail, consider the process $\{\Lambda_t^X : t \geq 0\}$ which has the following pathwise construction. First, sample from a branching particle diffusion with branching generator

$$F(r) = q\left(\sum_{n\geq 0} p_n r^n - r\right) = \frac{1}{\lambda^*}\psi(\lambda^*(1-r)),\ r \in [0,1], \tag{2.1}$$

and particle motion which is that of a Markov process with semigroup $\mathcal{P}$. Note that in the above generator, we have that q is the rate at which individuals reproduce and $\{p_n : n \geq 0\}$ is the offspring distribution. With the particular branching generator given by (2.1), $q = \psi'(\lambda^*)$, $p_0 = p_1 = 0$, and for $n \geq 2$, $p_n := p_n[0,\infty)$ where for $y \geq 0$, we defined the measure $p_n(\cdot)$ on $\{2,3,4,\ldots\} \times [0,\infty)$ by

$$p_n(\mathrm{d}y) = \frac{1}{\lambda^*\psi'(\lambda^*)}\left\{\beta(\lambda^*)^2\delta_0(\mathrm{d}y)\mathbf{1}_{\{n=2\}} + (\lambda^*)^n\frac{y^n}{n!}e^{-\lambda^* y}\Pi(\mathrm{d}y)\right\}.$$

If we denote the aforesaid branching particle diffusion by $Z^X = \{Z_t^X : t \geq 0\}$ then we shall also insist that the configuration of particles in space at time zero, Z_0, is given by an independent Poisson random measure with intensity $\lambda^*\mu$. Next, *dress* the branches of the spatial tree that describes the trajectory of Z^X in such a way that a particle at the space-time position $(\xi, t) \in [0,\infty)^2$ has an independent $\mathcal{X}$-valued trajectory grafted on to it with rate

$$2\beta\mathrm{d}\mathbb{N}_\xi^* + \int_0^\infty ye^{-\lambda^* y}\Pi(\mathrm{d}y)\mathrm{d}\mathbb{P}_{\xi\delta_y}^*.$$

The measures $\{\mathbb{N}_x, x \in \mathbb{R}^d\}$ are the so-called Dynkin-Kuznetsov measures (see [5]), which satisfy

$$\mathbb{N}_x\left(1 - e^{-\langle f, X_t\rangle}\right) = -\log\mathbb{E}_x\left(e^{-\langle f, X_t\rangle}\right), \tag{2.2}$$

for all $f \in bp(\mathbb{R}^d)$ and $t \geq 0$. The measures $\{\mathbb{N}_x, x \in \mathbb{R}^d\}$ play the role of the Lévy-measure (in the space of measure-valued cadlag paths $\mathcal{X}$) for the infinite divisible measure $\mathbb{P}_{\delta_x}$. The measure $\mathbb{N}_x^*$ denotes the Dynkin-Kuznetsov measure associated to the superprocess conditioned to die. Moreover, on the event that an individual in Z^X dies and branches into $n \geq 2$ offspring at spatial position $\xi \in [0, \infty)$, with probability $p_n(\mathrm{d}y)\mathbb{P}^*_{y\delta_\xi}$, an additional independent $\mathcal{X}$-valued trajectory is grafted on to the space-time branching point. The quantity Λ_t^X is now understood to be the total dressed mass present at time t together with the mass present at time t of an independent copy of $(X, \mathbb{P}^*_\mu)$ issued at time zero. We denote the law of (Λ^X, Z^X) by $\mathbf{P}_\mu$.

The backbone decomposition is now summarised by the following theorem lifted from Berestycki et al. [1].

Theorem 2.1 *For any $\mu \in \mathcal{M}_F(\mathbb{R}^d)$, the process $(\Lambda^X, \mathbf{P}_\mu)$ is Markovian and has the same law as $(X, \mathbb{P}_\mu)$. Moreover, for each $t \geq 0$, the law of Z_t^X given Λ_t^X is that of a Poisson random measure with intensity measure $\lambda^* \Lambda_t^X$.*

3 The Conditioned Superprocess and Prolific Individuals

3.1 The Conditioned Superprocess

We note that the total mass process, $\|X\| := \{\|X_t\| \equiv \langle 1, X_t \rangle, t \geq 0\}$, is a continuous state branching process with branching mechanism $\bar{\psi}$ given by

$$\bar{\psi}(\lambda) := (\alpha + 1)\lambda + \beta\lambda^2 + \int_0^\infty (e^{-\lambda u} - 1 + \lambda u)\Pi^L(du) - \gamma\lambda - \int_0^\infty (1 - e^{-\lambda u})\Pi^{NL}(du). \tag{3.1}$$

In order to avoid explosion of the total mass in finite time we assume that

$$\int_{0+} 1/|\bar{\psi}(\xi)| d\xi = \infty,$$

(see [9]). We will assume that the branching mechanism (3.1) is supercritical in the sense that $0 < -\bar{\psi}'(0+) < \infty$, thus the mean-total mass grows exponentially at rate $-\bar{\psi}'(0+)$. Under the above assumptions, and recalling the fact that $\bar{\psi}$ is strictly convex [9], there exists a unique $\lambda^* > 0$ such that $\bar{\psi}(\lambda^*) = 0$. Moreover, for all $\mu \in \mathcal{M}_C(\mathbb{R}^d)$,

$$\mathbb{P}_\mu(\lim_{t \uparrow \infty} \|X_t\| = 0) = e^{-\|\mu\|\lambda^*}.$$

We also assume the condition

$$\int^{\infty} \frac{1}{\overline{\psi}(\xi)} d\xi < \infty,$$

which ensures that the event $\{\lim_{t\uparrow\infty} \|X_t\| = 0\}$ agrees with the event of extinction $\{\zeta < \infty\}$, with $\zeta = \inf\{t > 0 : \|X_t\|\}$ (e.g. see [9] and [2]).

We can express the probability of survival in terms of the so-called Dynkin-Kuznetsov measures $\{\mathbb{N}_x, x \in \mathbb{R}^d\}$ as follows. Set $\mathcal{E} := \{\lim_{t\uparrow\infty} \|X_t\| = 0\}$ then we have

$$\mathbb{P}_\mu(\mathcal{E}) = e^{-\mathbb{N}_\mu(\mathbf{1}_S)} = e^{-\int_{\mathbb{R}^d} \mathbb{N}_x(\mathbf{1}_S)\mu(dx)} = e^{-\lambda^* \|\mu\|},$$

where S denotes the event of survival. Using the probability of extinction for the superprocess X we can now prove the following

Lemma 3.1 *For each $\mu \in \mathcal{M}_C(\mathbb{R}^d)$, define the law of X with initial configuration μ conditioned on becoming extinct by $\mathbb{P}^*_\mu$. Specifically, for all events A, measurable in the natural sigma algebra of X,*

$$\mathbb{P}^*_\mu(A) = \mathbb{P}_\mu(A| \lim_{t\uparrow\infty} \|X_t\| = 0).$$

Then, for all bounded f

$$-\log \mathbb{E}^*_\mu(e^{-\langle f, X_t\rangle}) = \int_D u^*_f(x,t)\mu(dx),$$

with

$$u^*_f(x,t) = u_{f+\lambda^*}(x,t) - \lambda^*,$$

where $u^*_f(x,t)$ is the unique solution of the integral equation

$$u^*_f(x,t) = \mathcal{P}_t(f)(x) - \int_0^t \mathcal{P}_s[\phi^{L,*}(u^*_f(\cdot, t-s)) + \phi^{NL,*}(\cdot, u^*_f(\cdot, t-s))](x), \qquad (3.2)$$

where $\phi^{L,*}(\lambda) = \phi^L(\lambda + \lambda^*)$ for $\lambda \geq -\lambda^*$ and $\phi^{NL,*}(x,f) = \phi^{NL}(x, f + \lambda^*)$ for any positive measurable function f such that $f + \lambda^* \in B^+(\mathbb{R}^d)$. That is to say $(X, \mathbb{P}^*_\mu)$ is a $(\mathcal{P}, \phi^{L,*}, \phi^{NL,*})$-superprocess.

Proof Set $\mathcal{E} = \{\lim_{t\uparrow\infty} \|X_t\| = 0\}$, then

$$\begin{aligned}\mathbb{E}^*_\mu(e^{-\langle f, X_t\rangle}) &= \mathbb{E}_\mu(e^{-\langle f, X_t\rangle}|\mathcal{E}) \\ &= e^{\lambda^* \|\mu\|}\mathbb{E}_\mu(e^{-\langle f, X_t\rangle} 1_{\mathcal{E}})\end{aligned}$$

$$= e^{\lambda^* \|\mu\|}\mathbb{E}_\mu(e^{-\langle f, X_t\rangle}\mathbb{E}_{X_t}(1_{\mathcal{E}}))$$
$$= e^{\lambda^* \|\mu\|}\mathbb{E}_\mu(e^{-\langle f+\lambda^*, X_t\rangle})$$
$$= e^{-\langle u_{f+\lambda^*}(\cdot,t)-\lambda^*, \mu\rangle}.$$

Now, using (1.3) it is easy to check that $u_f^*(x,t) = u_{f+\lambda^*}(x,t) - \lambda^*$ is a solution to

$$u_f^*(x,t) = \mathcal{P}_t(f)(x) - \int_0^t \mathcal{P}_s[\phi^{L,*}(u_f^*(\cdot, t-s)) + \phi^{NL,*}(\cdot, u_f^*(\cdot, t-s))](x),$$

where $\phi^{L,*}(\lambda) = \phi^L(\lambda + \lambda^*)$ and $\phi^{NL,*}(\cdot, f) = \phi^{NL}(\cdot, f + \lambda^*)$. □

3.2 Prolific Individuals

We will now identify the branching mechanism of the backbone for the superprocess X, i.e., we will give the generator of the continuous-time Galton-Watson process related to the genealogies responsible for the infinite growth of the process, in the form

$$F(x,s) = q\left(\sum_{n\geq 0} p_n^L + \sum_{n\geq 0} p_n^{NL}\right)(s^n - s),$$

where $q > 0$ is the common rate of splitting and $\{p_n^L : n \geq 0\}$ is the offspring distribution related to local branching, i.e. p_n^L is the probability of having n offspring at the position in which the parent dies. Respectively, p_n^{NL} is the probability of having n offspring displaced from the position x of the death of the parent according to a random variable Θ such that $\Theta + x$ has distribution $\pi(x, \cdot)$.

Moreover the branching rate is given by $q \equiv (\phi^L)'(\lambda^*)$ (we leave it to the reader to verify that $q > 0$), $p_0^L = p_1^L = 0$, and for any $n \geq 2$,

$$p_n^L = \frac{1}{\lambda^* q}\left\{\beta(\lambda^*)^2 \mathbf{1}_{\{n=2\}} + \int_{(0,\infty)} \frac{(y\lambda^*)^n}{n!} e^{-\lambda^* y}\Pi^L(dy)\right\}. \tag{3.3}$$

For the non-local offspring distribution we have that $p_0^{NL} = 0$ and for $n \geq 1$,

$$p_n^{NL} = \frac{1}{\lambda^* q}\left\{\lambda^* \gamma \mathbf{1}_{\{n=1\}} + \int_{(0,\infty)} \frac{(y\lambda^*)^n}{n!} e^{-\lambda^* y}\Pi^{NL}(dy)\right\}. \tag{3.4}$$

We leave to the reader to verify that effectively $\sum_{n\geq 1}(p_n^L + p_n^{NL}) = 1$. On the other hand, to describe the law related to the discontinuous immigration along the backbone, once again we will deal with the local and non-local immigration separately. For the local immigration we have that

$$\eta_n^L(dy) = \frac{1}{p_n^L \lambda^* q}\left\{\beta(\lambda^*)^2 \delta_0(dy)\mathbf{1}_{\{n=2\}} + \frac{(y\lambda^*)^n}{n!} e^{-\lambda^* y}\Pi^L(dy)\right\}, \ n \geq 2, \tag{3.5}$$

whereas for the non-local type of immigration we have

$$\eta_n^{NL}(dy) = \frac{1}{p_n^{NL}\lambda^* q}\left\{\lambda^* \gamma \delta_0(dy)\mathbf{1}_{\{n=1\}} + \frac{(y\lambda^*)^n}{n!} e^{-\lambda^* y}\Pi^{NL}(dy)\right\}, \ n \geq 1. \tag{3.6}$$

4 Backbone Decomposition

4.1 A Branching Particle System with Four Types of Immigration

Let $\mathcal{M}_a(\mathbb{R}^d)$ be the space of finite atomic measures on $\mathbb{R}^d$. Now suppose that $\xi = \{\xi_t : t \geq 0\}$ is the stochastic process whose semi-group is given by $\mathcal{P}$. We shall use the expectation operators $\{E_x : x \in \mathbb{R}^d\}$ defined by $E_x(f(\xi_t)) = \mathcal{P}_t[f](x)$. Let $Z = \{Z_t, t \geq 0\}$ be a $\mathcal{M}_a(\mathbb{R}^d)$-valued process in which individuals, from the moment of birth, live for an independent and exponentially distributed time with parameter q during which they execute a $\mathcal{P}$-diffusion issued from their position of birth and at death they give birth at the same position to an independent number of offspring locally with probabilities $\{p_n^L : n \geq 2\}$, and non-locally with probabilities $\{p_n^{NL}(\cdot) : n \geq 1\}$. Hence, Z is a non-local branching particle system such that

$$-\log E_x(e^{-\langle f, Z_t\rangle}) = v_f(x,t),$$

where the semigroup v_f satisfies the following integral equation

$$\begin{aligned} e^{-v_f(x,t)} &= e^{-qt}\mathcal{P}_t\left[e^{-f}\right](x) + \int_0^t ds\, qe^{-qs} \\ &\times \mathcal{P}_s\left[\sum_{n=0}^{\infty} p_n^L e^{-nv_f(\cdot,t-s)} + \int_{\mathcal{M}_F(\mathbb{R}^d)} \sum_{n=0}^{\infty} e^{-\langle v_f(\cdot,t-s),\nu\rangle} p_n^{NL}(l\pi)^{*n}(d\nu)\right](x), \end{aligned}$$

where $l\pi(d\nu)$ denotes the image of π under the map $y \to \delta_y$ from $\mathbb{R}^d$ to $\mathcal{M}_F(\mathbb{R}^d)$ (the space of finite measures on $\mathbb{R}^d$) and $(l\pi)^{*n}$ denotes the n-fold convolution of $l\pi$. Thus, a parent particle at the position $x \in \mathbb{R}^d$ when branches it gives birth to a random number of offspring in the following fashion: it produces n new individuals, which are initially located at x, with probability p_n^L; and produces m new individuals, which choose their locations in $\mathbb{R}^d$ independently of each other according to the (non-random) distribution $\pi(x, \cdot)$, with probability p_m^{NL}. We shall refer to Z as the backbone with initial configuration denoted by $\nu \in \mathcal{M}_a(\mathbb{R}^d)$. We will use the Ulam-Harris notation, i.e., that the individuals in Z are uniquely identifiable amongst $\mathcal{T}$, the set labels of individuals realised in Z. For each individual $u \in \mathcal{T}$ we shall write τ_u and σ_u for its birth and death times respectively, $\{z_u(r) : r \in [\tau_u, \sigma_u]\}$ for its spatial trajectory and N_u for the number of offspring it has at time σ_u.

With these elements at hands we are able to express the backbone decomposition of the superprocess X. We are interested in immigrating $(\mathcal{P}, \phi^{L,*}, \psi^{NL,*})$-superprocesses along the backbone Z in a way that the rate of immigration is related to the subordinator (i.e. a Lévy process with a.s. increasing paths), whose Laplace exponent is given by

$$\begin{aligned}\Phi(\lambda) &= (\phi^L)'(\lambda + \lambda^*) - (\phi^L)'(\lambda^*) \\ &= 2\beta\lambda + \int_{(0,\infty)} (1 - e^{-\lambda y}) y e^{-\lambda^* y} \Pi^L(dy), \end{aligned} \tag{4.1}$$

together with some additional immigration at the splitting times of Z.

Definition 4.1 For $\nu \in \mathcal{M}_a(\mathbb{R}^d)$ and $\mu \in \mathcal{M}_C(\mathbb{R}^d)$ let Z be a $(\mathcal{P}, F)$-branching diffusion with initial configuration ν and $\bar{X}$ an independent copy of X under $\mathbb{P}^*_\mu$. Then we define the measure-valued stochastic process $\Delta = \{\Delta_t : t \geq 0\}$ on $\mathbb{R}^d$ by

$$\Delta = \bar{X} + I^{\mathbb{N}^*} + I^{\mathbb{P}^*} + I^{\eta,L} + I^{\eta,NL},$$

where the processes $I^{\mathbb{N}^*}$, $I^{\mathbb{P}^*}$, $I^{\eta,L}$, and $I^{\eta,NL}$ are independent of $\bar{X}$ and, conditionally on Z, are independent of each other. More precisely, these processes are described as follows:

1 **Continuous immigration:** The process $I^{\mathbb{N}^*}$ is measure-valued on $\mathbb{R}^d$ such that

$$I_t^{\mathbb{N}^*} = \sum_{u \in \mathcal{T}} \sum_{\tau_u < r \leq t \wedge \sigma_u} X_{t-r}^{(1,u,r)}$$

where, given Z, independently for each $u \in \mathcal{T}$ such that $\tau_u < t$, the processes $X_\cdot^{(1,u,r)}$ are countable in number and correspond to χ-valued, Poissonian immigration along the space-time trajectory $\{(z_u(r), r) : r \in (\tau_u, t \wedge \sigma_u]\}$ with rate $2\beta dr \times d\mathbb{N}^*_{z_u(r)}$.

2 **Discontinuous immigration:** The process $I^{\mathbb{P}^*}$ is measure-valued on $\mathbb{R}^d$ such that

$$I_t^{\mathbb{P}^*} = \sum_{u \in \mathcal{T}} \sum_{\tau_u < r \leq t \wedge \sigma_u} X_{t-r}^{(2,u,r)}$$

where, given Z, independently for each $u \in \mathcal{T}$ such that $\tau_u < t$, the processes $X_\cdot^{(2,u,r)}$ are countable in number and correspond to χ-valued, Poissonian immigration along the space-time trajectory $\{(z_u(r), r) : r \in (\tau_u, t \wedge \sigma_u]\}$ with rate

$$dr \times \int_{y\in(0,\infty)} ye^{-\lambda^* y}\Pi^L(dy) \times d\mathbb{P}^*_{y\delta_{z_u(r)}}.$$

3 **Local Branch point biased immigration:** The process $I^{\eta,L}$ is measure-valued on $\mathbb{R}^d$ such that

$$I_t^{\eta,L} = \sum_{u\in\mathcal{T}} 1_{\{\sigma_u \le t\}} X^{(3,u)}_{t-\sigma_u}$$

where, given Z, independently for each $u \in \mathcal{T}$ such that $\sigma_u < t$, the processes $X^{(3,u)}_\cdot$ is an independent copy of X issued at time σ_u with law $\mathbb{P}_{Y_u\delta_{z_u(\sigma_u)}}$ where Y_u is an independent random variable with distribution $\eta^L_{N_u}(dy)$.

4 **Non-local Branch point biased immigration:** The process $I^{\eta,NL}$ is measure-valued on $\mathbb{R}^d$ such that

$$I_t^{\eta,NL} = \sum_{u\in\mathcal{T}} 1_{\{\sigma_u \le t\}} X^{(3,u)}_{t-\sigma_u}$$

where, given Z, independently for each $u \in \mathcal{T}$ such that $\sigma_u < t$, the processes $X^{(3,u)}_\cdot$ is an independent copy of X issued at time σ_u with law $\mathbb{P}_{Y_u\pi(z_u(\sigma_u),\cdot)}$ where Y_u is an independent random variable with distribution $\eta^{NL}_{N_u}(dy)$.

Moreover, we denote the law of Δ by $\mathbf{P}_{\mu\times\nu}$.

We will now state our first theorem

Theorem 4.2 *For every $\mu \in \mathcal{M}_C(\mathbb{R}^d)$, $\nu \in \mathcal{M}_a(\mathbb{R}^d)$ and $f, h \in bp(\mathbb{R}^d)$ we have that*

$$\mathbf{E}_{\mu\times\nu}(e^{-\langle f,\Delta_t\rangle - \langle f,Z_t\rangle}) = e^{-\langle u_f^*(\cdot,t),\mu\rangle - \langle v_{f,h}(\cdot,t),\nu\rangle}, \tag{4.2}$$

where $\exp\{-v_{f,h}(x,t)\}$ *is the unique* $[0, 1]$*-valued solution to the integral equation*

$$\begin{aligned} e^{-v_{f,h}(x,t)} &= \mathcal{P}_t\left[e^{-h}\right](x) + \frac{1}{\lambda^*}\int_0^t \mathcal{P}_s[\phi^{L,*}(-\lambda^* e^{-v_{f,h}(\cdot,t-s)} + u_f^*(\cdot,t-s)) \\ &\quad - \phi^{L,*}(u_f^*(\cdot,t-s)) + \phi^{NL,*}(-\lambda^* e^{-v_{f,h}(\cdot,t-s)} + u_f^*(\cdot,t-s)) \\ &\quad - \phi^{NL,*}(u_f^*(\cdot,t-s))](x), \end{aligned} \tag{4.3}$$

for all $x \in \mathbb{R}^d$ and $t \ge 0$.

In order to prove the Theorem 4.2 we will need to prove first some preliminary results, uniqueness to Eq. (4.3) will be proven in a more general setting in Lemma 6.

Lemma 4.3 *For all $f \in bp(\mathbb{R}^d)$, $\mu \in \mathcal{M}_C(\mathbb{R}^d)$, $\nu \in \mathcal{M}_a(\mathbb{R}^d)$ and $t \geq 0$, we have*

$$\mathbf{E}_{\mu\times\nu}(e^{-\langle f, I_t^{\mathbb{N}^*} + I_t^{\mathbb{P}^*}\rangle}|\{Z_s : s \leq t\}) = \exp\left\{-\int_0^t \langle \Phi(u_f^*(\cdot, t-s)), Z_s\rangle ds\right\},$$

where Φ is given by (4.1).

Proof We write

$$\langle f, I_t^{\mathbb{N}^*} + I_t^{\mathbb{P}^*}\rangle = \sum_{u\in\mathcal{T}} \sum_{\tau_u < r \leq t\wedge\sigma_u} \langle f, X_{t-r}^{(1,u,r)}\rangle + \sum_{u\in\mathcal{T}} \sum_{\tau_u < r \leq t\wedge\sigma_u} \langle f, X_{t-r}^{(2,u,r)}\rangle.$$

Hence conditioning on Z, appealing to the independence of the immigration processes together with Campbell's formula (see e.g. Theorem 2.7 in [10])

$$\mathbf{E}_{\mu\times\nu}(e^{-\langle f, I_t^{\mathbb{N}^*}\rangle}|\{Z_s : s \leq t\}) = \exp\left\{-\sum_{u\in\mathcal{T}} 2\int_{\tau_u}^{t\wedge\sigma_u} \beta\cdot\mathbb{N}^*_{z_u(r)}(1 - e^{-\langle f, X_{t-r}\rangle})dr\right\}.$$

Now using that $\mathbb{N}^*_{z_u(r)}(1 - e^{-\langle f, X_{t-r}\rangle}) = u_f^*(z_u(r), t-r)$ (see (2.2)), we have

$$\mathbf{E}_{\mu\times\nu}(e^{-\langle f, I_t^{\mathbb{N}^*}\rangle}|\{Z_s : s \leq t\}) = \exp\left\{-\sum_{u\in\mathcal{T}} 2\beta\int_{\tau_u}^{t\wedge\sigma_u} u_f^*(z_u(r), t-r)dr\right\}. \quad (4.4)$$

On the other hand

$$\begin{aligned}
&\mathbf{E}_{\mu\times\nu}(e^{-\langle f, I_t^{\mathbb{P}^*}\rangle}|\{Z_s : s \leq t\}) \\
&\quad = \exp\left\{-\sum_{u\in\mathcal{T}}\int_{\tau_u}^{t\wedge\sigma_u} dr\int_0^\infty ye^{-\lambda^*}\Pi^L(dy)\mathbb{P}^*_{y\delta_{z_u(r)}}(1 - e^{-\langle f, X_{t-r}\rangle})\right\} \\
&\quad = \exp\left\{-\sum_{u\in\mathcal{T}}\int_{\tau_u}^{t\wedge\sigma_u} dr\int_0^\infty ye^{-\lambda^*}\Pi^L(dy)(1 - e^{-yu_f^*(z_u(r), t-r)})\right\}. \quad (4.5)
\end{aligned}$$

Then, using (4.1), (4.4) and (4.5) we get that

$$\begin{aligned}
\mathbf{E}_{\mu\times\nu}(e^{-\langle f, I_t^{\mathbb{N}^*} + I_t^{\mathbb{P}^*}\rangle}|\{Z_s : s \leq t\}) = \exp\Bigg\{ &-\sum_{u\in\mathcal{T}}\int_{\tau_u}^{t\wedge\sigma_u} dr\Bigg(2\beta u_f^*(z_u(r), t-r) \\
&+ \int_0^\infty ye^{-\lambda^*}\Pi^L(dy)(1 - e^{-yu_f^*(z_u(r), t-r)})\Bigg)\Bigg\}
\end{aligned}$$

$$= \exp\left\{-\sum_{u\in\mathcal{T}}\int_{\tau_u}^{t\wedge\sigma_u} dr(\Phi(u_f^*(z_u(r),t-r)))\right\}$$

$$= \exp\left\{-\int_0^t \langle\Phi(u_f^*(\cdot,t-r)),Z_r\rangle dr\right\}.$$

□

In the next lemma we shall use the notation

$$\pi(\cdot,f(\circ,t)) \equiv \int_{\mathbb{R}^d} f(y,t)\pi(\cdot,dy),$$

for a measurable function f.

Lemma 4.4 *Suppose that* $f,h\in bp(\mathbb{R}^d)$ *and* $g_s(x)$ *is jointly measurable in* (s,x) *and bounded on finite time horizonts of* s. *Then for all* $x\in\mathbb{R}^d$ *and* $t\geq 0$,

$$\mathbf{E}_{\mu\times\nu}\left(\exp\left\{-\int_0^t\langle g_{t-s},Z_s\rangle ds-\langle f,I_t^{\eta,L}\rangle-\langle f,I_t^{\eta,NL}\rangle-\langle h,Z_t\rangle\right\}\right)=e^{-\langle w(\cdot,t),\nu\rangle},$$

where $\exp\{-w(x,t)\}$ *is the unique* $[0,1]$*-valued solution to the integral equation*

$$e^{-w(x,t)}=\mathcal{P}_t[e^{-h}](x)+\frac{1}{\lambda^*}\int_0^t\mathcal{P}_s[H_{t-s}(\cdot,-\lambda^*e^{-w(\cdot,t-s)})-\lambda^*e^{-w(\cdot,t-s)}g_{t-s}(\cdot)](x)ds. \tag{4.6}$$

where

$$H_{t-s}(\cdot,-\lambda^*e^{-w(\cdot,t-s)})=-\lambda^*qe^{-w(\cdot,t-s)}+\beta(\lambda^*)^2e^{-2w(\cdot,t-s)}+\gamma\lambda^*\pi(\cdot,e^{-w(\circ,t-s)})$$
$$+\int_{(0,\infty)}(e^{\lambda^*ye^{-w(\cdot,t-s)}}-1-\lambda^*ye^{-w(\cdot,t-s)})e^{-(\lambda^*+u_f^*(\cdot,t-s))y}\Pi^L(y)$$
$$+\int_{(0,\infty)}(e^{\lambda^*\pi(\cdot,e^{-w(\circ,t-s)})y}-1)e^{-\pi(\cdot,\lambda^*+u_f^*(\circ,t-s))}\Pi^{NL}(dy).$$

Proof Following the proof of Theorem 2.2 in [8] it is enough to prove the result for g being time-independent. Recall that $\xi=\{\xi_t:t\geq 0\}$ is the stochastic process whose semi-group is given by $\mathcal{P}$. Let us define a new semigroup

$$\mathcal{P}_t^g[f](x)=E_x\left(e^{-\int_0^t g(\xi_s)}f(\xi_s)\right),$$

for $f, g \in bp(\mathbb{R}^d)$. Standard Feynman-Kac manipulations (cf. Lemma 2.3 in [8]) give us that

$$\mathcal{P}_t^g[f](x) = \mathcal{P}_t[f](x) - \int_0^t ds \mathcal{P}_s[g(\cdot)\mathcal{P}_{t-s}^g[f](\cdot)](x). \tag{4.7}$$

Conditioning on the first branching time, and recalling that the branching occurs at rate $q = (\phi^L)'(\lambda^*)$ we get that

$$e^{-w(x,t)} = e^{-qt}\mathcal{P}_t^g[e^{-h}](x) + \int_0^t dsqe^{-qs}\mathcal{P}_s^g\Bigg[\sum_{n\geq 2} p_n^L e^{-nw(\cdot,t-s)} \int_{(0,\infty)} \eta_n^L(dy)e^{-yu_f^*(\cdot,t-s)}$$
$$+ \int_{\mathcal{M}_C(\mathbb{R}^d)} \sum_{n\geq 1} e^{-\langle w(\cdot,t-s),\nu\rangle} p_n^{NL}(l\pi)^{*n}(d\nu) \int_0^\infty \eta_n^{NL}(dy)e^{-y\pi(\cdot,u_f^*(\circ,t-s))}\Bigg](x). \tag{4.8}$$

Using (3.4), (3.6) and performing similar computations to the ones in [3] (cf. Sect. 3) we have that

$$\sum_{n\geq 1} p_n^{NL}(l\pi)^{*n}(d\nu) \int_0^\infty \eta_n^{NL}(dy)e^{-y\pi(\cdot,u_f^*(\circ,t-s))}$$
$$= \sum_{n\geq 1} \int_0^\infty \pi(\cdot, e^{-w(\circ,t-s)})^n \frac{1}{\lambda^* q}\left\{\gamma\lambda^*\delta_0(dy)1_{\{n=1\}} + \frac{(y\lambda^*)^n}{n!}e^{-\lambda^* y}\Pi^{NL}(dy)\right\} e^{-y\pi(\cdot,u_f^*(\circ,t-s))}$$
$$= \frac{1}{\lambda^* q}\left\{\int_0^\infty (e^{\lambda^* y\pi(\cdot,e^{-w(\circ,t-s)})} - 1)e^{-\pi(\cdot,\lambda^*+u_f^*(\circ,t-s))}\Pi^{NL}(dy) + \gamma\lambda^*\pi(\cdot, e^{-w(\circ,t-s)})\right\}. \tag{4.9}$$

Now for the local branching term we obtain, by proceeding as in the proof of Lemma 4 in [1] and using (3.3) and (3.5), the following

$$\sum_{n\geq 2} p_n^L e^{-nw(\cdot,t-s)} \int_{(0,\infty)} \eta_n^L(dy)e^{-yu_f^*(\cdot,t-s)}$$
$$= \frac{1}{\lambda^* q}\left\{\beta(\lambda^*)^2 e^{-2w(\cdot,t-s)} + \int_0^\infty (e^{\lambda^* ye^{-w(\cdot,t-s)}} - 1 - \lambda^* ye^{-w(\cdot,t-s)})e^{-y(\lambda^*+u_f^*(\cdot,t-s))}\Pi^L(dy)\right\}. \tag{4.10}$$

Using (4.9) and (4.10) in (4.8) we have that

$$e^{-w(x,t)}$$
$$= e^{-qt}\mathcal{P}_t^g[e^{-h}](x) + \int_0^t ds\cdot e^{-qs}\mathcal{P}_s^g\left[\frac{1}{\lambda^*}(H_{t-s}(\cdot, -\lambda^* e^{-w(\cdot,t-s)}) + \lambda^* qe^{-w(\cdot,t-s)})\right](x)$$
$$= \mathcal{P}_t^g[e^{-h}](x) + \int_0^t ds\mathcal{P}_s^g\left[\frac{1}{\lambda^*}H_{t-s}(\cdot, -\lambda^* e^{-w(\cdot,t-s)})\right](x). \tag{4.11}$$

where the second inequality follows from a standard technique found for example in Lemma 4.1.1 of [4]. Now making the same computations as in [1] we obtain that

$$\int_0^t ds\mathcal{P}_s\left[g(\cdot)\mathcal{P}^g_{t-s}[e^{-h}](\cdot)\right](x)$$
$$+\frac{1}{\lambda^*}\int_0^t ds\int_0^s dr\mathcal{P}_r\left[g(\cdot)\mathcal{P}^g_{s-r}[H_{t-s}(\cdot,-\lambda^* e^{-\omega(\cdot,t-s)})]\right](x)$$
$$=\int_0^t ds\mathcal{P}_s\left[g(\cdot)\mathcal{P}^g_{t-s}[e^{-h}](\cdot)\right](x)$$
$$+\frac{1}{\lambda^*}\int_0^t dr\mathcal{P}_r\left[g(\cdot)\int_r^t ds\mathcal{P}^g_{s-r}[H_{t-s}(\cdot,-\lambda^* e^{-\omega(\cdot,t-s)})](\cdot)\right](x)$$
$$=\int_0^t ds\mathcal{P}_s\left[g(\cdot)\mathcal{P}^g_{t-s}[e^{-h}](\cdot)\right](x)$$

$$+\frac{1}{\lambda^*}\int_0^t dr\mathcal{P}_r\left[g(\cdot)\int_0^{t-r} d\theta\mathcal{P}^g_\theta[H_{t-\theta-r}(\cdot,-\lambda^* e^{-\omega(\cdot,t-s)})](\cdot)\right](x)$$
$$=\int_0^t dr\mathcal{P}_r\Bigg[g(\cdot)\Bigg\{\mathcal{P}^g_{t-r}[e^{-h}](\cdot)$$
$$+\frac{1}{\lambda^*}\int_0^{t-r} d\theta\mathcal{P}^g_\theta[H_{t-r-\theta}(\cdot,-\lambda^* e^{-\omega(\cdot,t-s)})](\cdot)\Bigg\}\Bigg](x)$$
$$=\int_0^t ds\mathcal{P}_s\left[g(\cdot)e^{-\omega(\cdot,t-s)}\right](x). \tag{4.12}$$

Next, we use (4.7) and (4.12) in (4.11) to obtain that

$$e^{-w(x,t)}=\mathcal{P}_t[e^{-h}](x)-\int_0^t ds\mathcal{P}_s[g(\cdot)\mathcal{P}^g_{t-s}[e^{-h}](\cdot)](x)$$
$$+\frac{1}{\lambda^*}\int_0^t ds\Bigg\{\mathcal{P}_s[H_{t-s}(\cdot,-\lambda^* e^{-w(\cdot,t-s)})](x)$$
$$-\int_0^s dr\mathcal{P}_r[g(\cdot)\mathcal{P}^g_{s-r}[H_{t-s}(\cdot,-\lambda^* e^{-w(\cdot,t-s)})]](x)\Bigg\}$$
$$=\mathcal{P}_t[e^{-h}](x)+\frac{1}{\lambda^*}\int_0^t ds\mathcal{P}_s\left[H_{t-s}(\cdot,-\lambda^* e^{-w(\cdot,t-s)})-\lambda^* g(\cdot)e^{-w(\cdot,t-s)}\right](x).$$

The proof is complete as soon as we can establish uniqueness to (4.6). The proof is guided by the same arguments as in the proof of Lemma 4 in [1], i.e., it suffices to check that for each fixed $T>0$, there exists $K>0$ such that

$$\sup_{s\le T}\sup_{y\in\mathbb{R}^d}|H_s(y,-u(y))-H_s(y,-v(y))|\le K\sup_{y\in\mathbb{R}^d}|u(y)-v(y)|,$$

where u and v are any two measurable mappings from $\mathbb{R}^d$ to $[0, \lambda^*]$, then Lemma 2.1 in [8] gives the result. To this end we define for $\lambda \geq -\lambda^*$ and $u \geq 0$,

$$\chi_u^1(\lambda) = \lambda q + \beta(\lambda)^2 + \int_{(0,\infty)} (e^{-\lambda y} - 1 + \lambda y)e^{-(\lambda^*+u)y}\Pi^L(y),$$

and for any positive measurable function such that $f + \lambda^* \in B(\mathbb{R}^d)$, and $v \geq 0$

$$\chi_u^2(\lambda) = \gamma\lambda + \int_{(0,\infty)} (e^{-\lambda z} - 1)e^{-(\lambda^*+u)z}\Pi^{NL}(dz).$$

Therefore by definition we have that

$$H_s(y, -v(y)) = \chi^1_{u_f^*(y,t-s)}(-v(y)) + \chi^2_{\pi(y,u_f^*(\circ,t-s))}(-\pi(y, v(\circ))),$$

for any measurable mapping v from $\mathbb{R}^d$ to $[0, \lambda^*]$.

With the help of Lemma 5 in [1] and the fact that $\pi(x, \cdot)$ is a probability measure for every $x \in \mathbb{R}^d$, we can see that for fixed $T > 0$,

$$\begin{aligned}
&\sup_{s\leq T}\sup_{y\in\mathbb{R}^d}|H_s(y, -u(y)) - H_s(y, -v(y))| \\
&\quad\leq \sup_{s\leq T}\sup_{y\in\mathbb{R}^d} |\chi^1_{u_f^*(y,t-s)}(-u(y)) - \chi^1_{u_f^*(y,t-s)}(-v(y))| \\
&\quad+ \sup_{s\leq T}\sup_{y\in\mathbb{R}^d} |\chi^2_{\pi(y,u_f^*(\circ,t-s))}(-\pi(y, -u(\circ))) - \chi^2_{\pi(y,u_f^*(\circ,t-s))}(-\pi(y, -v(\circ)))| \\
&\quad\leq \sup_{0\leq u^*\leq \bar{u}_T}\sup_{y\in\mathbb{R}^d} |\chi^1_{u^*}(-u(y)) - \chi^1_{u^*}(-v(y))| \\
&\quad+ \sup_{0\leq u^*\leq \bar{u}_T}\sup_{y\in\mathbb{R}^d} |\chi^2_{u^*}(-\pi(y, -u(\circ))) - \chi^2_{u^*}(-\pi(y, -v(\circ)))| \\
&\quad\leq K \sup_{y\in\mathbb{R}^d} |u(-y) - v(-y)|,
\end{aligned}$$

where u and v are any two measurable mappings from $\mathbb{R}^d$ to $[0, \lambda^*]$,

$$K = \sup_{0\leq u^*\leq \bar{u}_T}\sup_{y\in\mathbb{R}^d} (|(\chi^1_{u^*})'(-\lambda)| + |(\chi^2_{u^*})'(-\lambda)|) < \infty, \tag{4.13}$$

(observe that using Lemma 5 in [1] we have that (4.13) is true if and only if $\bar{\psi}'(0+) > -\infty$) and

$$\bar{u}_T = \sup_{s\leq T}\sup_{y\in\mathbb{R}^d} u_f^*(y, s) < \infty.$$

Following the same steps as in the proof of Lemma 4 in [1] the finiteness of $\bar{u}_T$ can be deduced from the fact that if we assume, without loss of generality, that f is bounded by $\theta \geq 0$, then

$$u_f^*(y, s) \leq U_\theta^*(s), \qquad \text{for all } y \in \mathbb{R}^d \text{ and } s \geq 0,$$

where $U_\theta^*(s)$ is the unique solution to

$$U_\theta^*(s) + \int_0^s \bar{\psi}^*(U_\theta^*(u))du = \theta,$$

with

$$\bar{\psi}^*(\lambda) = \bar{\psi}(\lambda + \lambda^*), \qquad \text{for all } \lambda \geq -\lambda^*.$$

This implies that $\bar{u}_T \leq \sup_{s \leq T} U_\theta^*(s) < \infty$, and thus the proof is complete. □

Proof of Theorem 4 It just suffices to prove, thanks to Lemma 3.1, that

$$\mathbf{E}_{\mu \times \nu}(e^{\langle f, I_t \rangle - \langle h, Z_t \rangle}) = e^{-\langle v_{f,h}(\cdot,t), \nu \rangle}$$

where $I := I^{\mathbb{N}^*} + I^{\mathbb{P}^*} + I^{\eta,L} + I^{\eta,NL}$, and $v_{f,h}$ solves (4.3). Putting Lemma 4.3 and Lemma 4.4 together it suffices to show that when $g_{t-s}(\cdot) = \Phi(u_f^*(\cdot, t-s))$ (where Φ is given in (4.1)) we have that $\exp\{-w(x,t)\}$ is the solution to (4.3). So making the computations as in [1] it is easy to see that

$$\begin{aligned}
&H_{t-s}(\cdot, -\lambda^* e^{-w(\cdot,t-s)}) - \lambda^* \Phi(u_f^*(\cdot, t-s))e^{-w(\cdot,t-s)} \\
&\quad = -\lambda^* q e^{-w(\cdot,t-s)} + \gamma\lambda^* \pi(\cdot, e^{-w(\circ,t-s)}) + \beta(\lambda^*)^2 e^{-2w(\cdot,t-s)} + \lambda^* q e^{-w(\cdot,t-s)} \\
&\quad -\lambda^* e^{-w(\cdot,t-s)} \left((\alpha+1) + 2\beta(u_f^*(\cdot, t-s) + \lambda^*) - \int_0^\infty (x e^{-x(u_f^*(\cdot,t-s)+\lambda^*)} - x)\Pi^L(dx) \right) \\
&\quad + \int_0^\infty (e^{\lambda^* y e^{-w(\cdot,t-s)}} - 1 - \lambda^* y e^{-w(\cdot,t-s)}) e^{-(\lambda^* + u_f^*(\cdot,t-s))y} \Pi^L(dy) \\
&\quad + \int_0^\infty (e^{\lambda^* y \pi(\cdot, e^{-w(\circ,t-s)})} - 1) e^{-y\pi(\cdot, \lambda^* + u_f^*(\circ,t-s))} \Pi^{NL}(dy) \\
&\quad = \phi^{L,*}(-\lambda^* e^{-w(\cdot,t-s)} + u_f^*(\cdot, t-s)) + \phi^{NL,*}(\cdot, -\lambda^* e^{-w(\cdot,t-s)} + u_f^*(\cdot, t-s)) \\
&\quad - (\phi^{L,*}(u_f^*(\cdot, t-s)) + \phi^{NL,*}(\cdot, u_f^*(\cdot, t-s))).
\end{aligned}$$

4.2 Backbone Decomposition

Finally with all those elements we are able to prove the following theorem which is the main result of this work. We will deal with the case when we randomize the law $\mathbf{P}_{\mu\times\nu}$ for $\mu \in \mathcal{M}_C(\mathbb{R}^d)$ by replacing the deterministic measure ν with a Poisson random measure having intensity measure $\lambda^*\mu$. We denote the resulting law by $\mathbf{P}_\mu$.

Theorem 4.5 *For any $\mu \in \mathcal{M}_C(\mathbb{R}^d)$, the process $(\Delta, \mathbf{P}_\mu)$ is Markovian and has the same law as $(X, \mathbb{P}_\mu)$.*

Proof The proof is guided by the calculations found in the proof of Theorem 2 of [1]. We start by addressing the claim that $(\Delta, \mathbf{P}_\mu)$ is a Markov process. Given the Markov property of the pair (Δ, Z), it suffices to show that given Δ_t the atomic measure Z_t is equal in law to a Poisson random measure with intensity $\lambda^*\Delta_t(dx)$. Thanks to Campbell's formula for Poisson random measures, this is equivalent to showing that for all $h \in bp(\mathbb{R}^d)$,

$$\mathbf{E}_\mu(e^{-\langle h, Z_t\rangle}|\Delta_t) = \exp\{-\langle\lambda^*(1-e^{-h}), \Delta_t\rangle\},$$

which in turn is equivalent to showing that for all $f, h \in bp(\mathbb{R}^d)$,

$$\mathbf{E}_\mu(e^{-\langle h, Z_t\rangle - \langle f, \Delta_t\rangle}) = \mathbf{E}_\mu(e^{-\langle\lambda^*(1-e^{-h})+f, \Delta_t\rangle}). \tag{4.14}$$

Note from (4.2) however that when we randomize ν so that it has the law of a Poisson random measure with intensity $\lambda^*\mu(dx)$, we find the identity

$$\mathbf{E}_\mu(e^{-\langle h, Z_t\rangle - \langle f, \Delta_t\rangle}) = e^{-\langle u_f^*(\cdot,t)+\lambda^*(1-e^{-v_{f,h}(\cdot,t)}), \mu\rangle}.$$

Moreover, if we replace f by $\lambda^*(1-e^{-h})+f$ and h by 0 in (4.2) and again randomize ν so that it has the law of a Poisson random measure with intensity $\lambda^*\mu(dx)$ then we get

$$\mathbf{E}_\mu(e^{\langle\lambda^*(1-e^{-h})+f, \Delta_t\rangle}) = \exp\left\{\langle u^*_{\lambda^*(1-e^{-h})+f}(\cdot,t) + \lambda^*(1-\exp\{-v_{\lambda^*(1-e^{-h})+f,0}(\cdot,t)\}), \mu\rangle\right\}.$$

These last two observations indicate that (4.14) is equivalent to showing that for all $f, h \in bp(\mathbb{R}^d)$, $x \in \mathbb{R}^d$ and $t \geq 0$,

$$u_f^*(x,t) + \lambda^*(1-e^{-v_{f,h}(x,t)}) = u^*_{\lambda^*(1-e^{-h})+f}(x,t) + \lambda^*(1-e^{-v_{\lambda^*(1-e^{-h})+f,0}(x,t)}). \tag{4.15}$$

Note that both left and right hand side of the equality above are necessarily non-negative given that they are Laplace exponents of the left and right hand sides of (4.14). Making use of (1.3), (3.2), and (4.3), it is computationally straightforward to show that both left and right hand side of (4.15) solve (1.3) with initial condition $f+\lambda^*(1-e^{-h})$. Since (1.3) has a unique solution with this initial condition, namely

$u_{f+\lambda^*(1-e^{-h})}(x,t)$, we conclude that (4.15) holds true. The proof of the claimed Markov property is thus complete.

Having now established the Markov property, the proof is complete as soon as we can show that $(\Delta, \mathbf{P}_\mu)$ has the same semi-group as $(X, \mathbb{P}_\mu)$. However, from the previous part of the proof we have already established that when $f, h \in bp(\mathbb{R}^d)$,

$$\mathbf{E}_\mu(e^{-\langle h,Z_t\rangle - \langle f,\Delta_t\rangle}) = e^{-\langle u_{\lambda^*(1-e^{-h})+f},\mu\rangle} = \mathbb{E}_\mu(e^{-\langle f+\lambda^*(1-e^{-h}),X_t\rangle}).$$

In particular, choosing $h = 0$ we find

$$\mathbf{E}_\mu(e^{-\langle f,\Delta_t\rangle}) = \mathbb{E}_\mu(e^{-\langle f,X_t\rangle}),$$

which is equivalent to the equality of the semigroups of $(\Delta, \mathbf{P}_\mu)$ and $(X, \mathbb{P}_\mu)$. □

Acknowledgements The authors want to thank the comments of the anonymous referee which improved the presentation of the paper.

References

1. J. Berestycki, A.E. Kyprianou, A. Murillo-Salas, The prolific backbone for supercritical superdiffusions. Stoch. Proc. Appl. **121**, 1315–1331 (2011)
2. N.H. Bingham, Continuous branching processes and spectral positivity. Stoch. Process. Appl. **4**, 217–242 (1976)
3. D.A. Dawson, L.G. Gorostiza, Z. Li, Non-local branching superprocesses and some related models. Acta Appl. Math. **74**, 93–112 (2002)
4. E.B. Dynkin, *Diffusions, Superprocesses and Partial Differential Equations*. Colloquium Publications, vol. 50 (American Mathematical Society, Providence, 2002)
5. E.B. Dynkin, S.E. Kuznetsov. $\mathbb{N}$-measures for branching exit Markov systems and their applications to differential equations. Probab. Theory Relat. Fields. **130**, 135–150 (2004)
6. J. Engländer, R.G. Pinsky, On the construction and support properties of measure-valued diffusions on $D \subseteq R^d$ with spatially dependent branching. Ann. Probab. **27**, 684–730 (1999)
7. S.N. Evans, Two representations of a superprocess. Proc. R. Soc. Edinb. **123A**, 959–971 (1993)
8. S.N. Evans, N. O'Connell, Weighted occupation time for branching particle systems and a representation for the supercritical superprocess. Can. Math. Bull. **37**, 187–196 (1994)
9. D. Grey, Asymptotic behaviour of continuous time continuous state-space branching processes. J. Appl. Probab. **11**, 669–677 (1974)
10. A. Kyprianou, *Fluctuations of Lévy Processes with Applications. Introductory Lectures*, 2nd edn. (Springer, Berlin, 2014)
11. A.E. Kyprianou, A. Murillo-Salas, J.L. Pérez, An application of the backbone decomposition to supercritical super-Brownian motion with a barrier. J. Appl. Probab. **49**, 671–684 (2012)
12. A.E. Kyprianou, J-L. Perez, Y-X. Ren, The backbone decomposition for spatially dependent supercritical superprocesses. Sém. Probab. XLVI **2123**, 33–59 (2014)
13. P. Milos, Spatial CLT for the supercritical Ornstein-Uhlenbeck superprocess. arXiv:1203.6661 (2014+)
14. T. Salisbury, J. Verzani, On the conditioned exit measures of super Brownian motion. Probab. Theory Relat. Fields. **115**, 237–285 (1999)
15. T. Salisbury, J. Verzani Non-degenerate conditionings of the exit measure of super Brownian motion. Stoch. Proc. Appl. **87**, 25–52 (2000)

On Lévy Semistationary Processes with a Gamma Kernel

Jan Pedersen and Orimar Sauri

Abstract This paper studies some probabilistic properties of a Lévy semistationary process when the kernel is given by $\varphi_{\alpha,\lambda}(s) = e^{-\lambda s}s^{\alpha}$ for $\alpha > -1$ and $\lambda > 0$. We study the stationary distribution induced by this process. In particular, we show that this distribution is self-decomposable for $-1 < \alpha < 0$ and under certain conditions it can be characterized by the so-called cancellation property.

Keywords Lévy semistationary processes • Self-decomposable distributions • Stationary processes • Cancellation property • Multiplicative convolutions • Type G distributions

1 Introduction

Lévy semistationary processes ($\mathcal{LSS}$) were introduced as a class of models for energy spot prices by Barndorff-Nielsen et al. [2] taking as a starting point the works of Barndorff-Nielsen and Schmiegel [6–8], where the authors present a tempospatial model for the velocity field of a fluid in a turbulent environment. Recall that a stochastic process $(Y_t)_{t\in\mathbb{R}}$ on a filtered probability space $(\Omega, \mathcal{F}, (\mathcal{F}_t)_{t\in\mathbb{R}}, \mathbb{P})$, is said to be an $\mathcal{LSS}$ if it is described by the following dynamics

$$Y_t = \theta + \int_{-\infty}^{t} g(t-s)\,\sigma_s dL_s + \int_{-\infty}^{t} q(t-s)\,a_s ds, \quad t \in \mathbb{R},$$

where $\theta \in \mathbb{R}$, L is a Lévy process with triplet (γ, b, ν), g and q are deterministic functions such that $g(x) = q(x) = 0$ for $x \leq 0$, and σ and a are adapted càdlàg processes. When L is a Brownian motion Y is called a *Brownian semistationary process* ($\mathcal{BSS}$). For further references to theory and applications of Lévy semistationary

J. Pedersen
Department of Mathematical Sciences, Aarhus University, Aarhus, Denmark
e-mail: jan@imf.au.dk

O. Sauri (✉)
Department of Economics and CREATES, Aarhus University, Aarhus, Denmark
e-mail: osauri@econ.au.dk

R.H. Mena et al. (eds.), *XI Symposium on Probability and Stochastic Processes*, Progress in Probability 69, DOI 10.1007/978-3-319-13984-5_11

processes, see for instance Veraart and Veraart [23] and Benth et al. [11]. See also Brockwell et al. [12].

In this paper, we study some probabilistic properties of the $\mathcal{LSS}$ given by

$$Y_t := \int_{-\infty}^{t} \varphi_{\alpha,\lambda}(t-s)\,\sigma_s dL_s, \quad t \in \mathbb{R}, \tag{1.1}$$

where

$$\varphi_{\alpha,\lambda}(x) := x^{\alpha} e^{-\lambda x}, \quad x > 0, \tag{1.2}$$

$\alpha > -1$, $\lambda > 0$ and σ a strongly stationary càdlàg predictable process.

Corcuera et al. [13] consider the Gaussian case of Y in (1.1) as a model for the longitudinal component of the wind velocity in the atmospheric boundary layer. The authors give necessary conditions for the existence and the semimartingale property of Y while they describe some asymptotics for a general $\mathcal{BSS}$ process. See also Barndorff-Nielsen et al. [1]. In the present paper, for a very general σ, we provide necessary conditions, in terms of the triplet of L_1, for the existence of Y which are also sufficient when $\sigma \equiv 1$.

In absence of stochastic volatility Y is just a continuous time moving average process and its stationary distribution corresponds to the law of the random variable $\int_0^{\infty} \varphi_{\alpha,\lambda}(s)\,dL_s$ which is infinitely divisible. In particular, when $\alpha = 0$, Y becomes a stationary Ornstein-Uhlenbeck process driven by L. In this situation, the mapping which transforms $\mathcal{L}(L_1)$ into $\mathcal{L}(Y_0)$ (here $\mathcal{L}(X)$ denotes the law of the random variable X) creates a bijection between its own domain and the space of self-decomposable distributions (for further details see Sato [20]). Note that in this case, the parameter λ does not affect $\mathcal{L}(Y_0)$. On the other hand, when $\alpha \neq 0$ it turns out, not surprisingly, that the properties of $\mathcal{L}(Y_0)$ depends on α. In fact, we show that for $\alpha > -1$, $\mathcal{L}(Y_0)$ has absolutely continuous Lévy measure. Moreover, for $0 > \alpha > -1$ it is actually self-decomposable.

As in the traditional case $\alpha = 0$ mentioned above, we study injectivity of the mapping $\Phi_{\alpha} : \mathcal{L}(L_1) \mapsto \mathcal{L}\left(\int_0^{\infty} \varphi_{\alpha,\lambda}(s)\,dL_s\right)$ in the case $-1 < \alpha < 0$. Our approach uses *multiplicative convolutions* (introduced by Jacobsen et al. [15]), and we show in fact that injectivity of Φ_{α} is closely related to the so-called cancellation property of a concrete multiplicative convolution. It turns out that Φ_{α} is injective on a quite large subset of its domain when $-1/2 < \alpha < 0$ and on a somewhat smaller subset when $-1 < \alpha \leq -1/2$.

Let us finally remark that at even greater level of generality, given a measurable function $f : \mathbb{R}^+ \to \mathbb{R}$, it is of interest to study injectivity of the mapping $\Phi_f : \mathcal{L}(L_1) \mapsto \mathcal{L}\left(\int_0^{\infty} f(s) dL_s\right)$. For instance, in Sato [22], Barndorff-Nielsen et al. [4] and Pedersen and Sato [18] examples have been given where the functional Φ_f is in fact one-to-one for a rich class of functions. However, the inverse of Φ_f is not known except for the cases $f(s) = e^{-s}$ and $f(s) = e^{-s}\mathbf{1}_{[0,T]}(s)$, for some $T > 0$. In our case, the inverse of Φ_{α} will not be given in closed form.

The paper is organized as follows: Sect. 2 introduces the basic notation and definitions. We give a brief review of stochastic integration on the real line for Lévy processes. To finish the section we provide necessary and sufficient conditions on the triplet of L for the existence of Y. In absence of stochastic volatility, in Sect. 3 we describe the characteristic triplet of Y_0. In particular, we prove that its Lévy measure is absolutely continuous with respect to the Lebesgue measure. Using this, we conclude that for $0 > \alpha > -1$, the law of Y_0 is self-decomposable. Later on, we solve partially the cancellation property for a particular multiplicative convolution in order to describe the injectivity of Φ_α. The last section discusses the law of Y_0 in presence of stochastic volatility. In particular we show that in general Y is not infinitely divisible anymore. In view of this, we provide necessary conditions for which Y remains infinitely divisible (type G) as well as a description of its characteristic triplet.

2 Preliminaries and Basic Results

Throughout this paper $(\Omega, \mathcal{F}, (\mathcal{F}_t)_{t\in\mathbb{R}}, \mathbb{P})$ denotes a filtered probability space satisfying the usual conditions of right-continuity and completeness. A two-sided Lévy process $(L_t)_{t\in\mathbb{R}}$ on $(\Omega, \mathcal{F}, \mathbb{P})$ is a stochastic process with independent and stationary increments with càdlàg paths. We say that $(L_t)_{t\in\mathbb{R}}$ is an $(\mathcal{F}_t)$-Lévy process if for all $t > s$, $L_t - L_s$ is $\mathcal{F}_t$-measurable and independent of $\mathcal{F}_s$.

Denote by $ID(\mathbb{R})$ the space of infinitely divisible distributions on $\mathbb{R}$. Any Lévy process is infinitely divisible in the sense of finite-dimensional distributions, and L_1 has a Lévy-Khintchine representation given by

$$\log \hat{\mu}(z) = i\gamma z - \frac{1}{2}b^2 z^2 + \int_{\mathbb{R}} \left[e^{ixz} - 1 - iz\tau(x)\right] \nu(dx), \quad z \in \mathbb{R},$$

where $\hat{\mu}$ is the characteristic function of the law of L_1, $\gamma \in \mathbb{R}$, $b \geq 0$ and ν is a Lévy measure, i.e. $\nu(\{0\}) = 0$ and

$$\int_{\mathbb{R}} 1 \wedge |x|^2 \nu(dx) < \infty.$$

Here, we assume that the truncation function τ is given by $\tau(x) = \frac{x}{1+|x|^2}$, $x \in \mathbb{R}$. By $SD(\mathbb{R})$, we mean the subset of $ID(\mathbb{R})$ of self-decomposable distributions on $\mathbb{R}$. More precisely, $\mu \in ID(\mathbb{R})$ belongs to $SD(\mathbb{R})$ if and only if its Lévy measure ν can be written as

$$\nu(dr) = \frac{k(r)}{|r|} dr, \quad r \in \mathbb{R}\setminus\{0\},$$

where k is decreasing on $(0, \infty)$ and increasing on $(-\infty, 0)$.

Below we discuss how the stochastic integral in (1.1) is to be understood.

2.1 Stochastic Integration on the Real Line

In the following, we present a short review of Basse-O'Connor et al. [10] concerning to the existence of stochastic integrals of the form $\int_{\mathbb{R}} \psi_s dL_s$, where $(\psi_t)_{t\in\mathbb{R}}$ is a predictable process and $(L_t)_{t\in\mathbb{R}}$ is an $(\mathcal{F}_t)$-Lévy process with triplet (γ, b, ν).

Let $\mathcal{L}^0(\Omega, \mathcal{F}, \mathbb{P})$ be the space of real-valued random variables. For any $Z \in \mathcal{L}^0(\Omega, \mathcal{F}, \mathbb{P})$, put $\|Z\|_0 := \mathbb{E}(1 \wedge |Z|)$. Denote by $\mathcal{P}$ the predictable σ-field on $\mathbb{R}\times\Omega$, i.e.

$$\mathcal{P} := \sigma\{(u,t] \times A : -\infty < u \leq t < \infty, A \in \mathcal{F}_u\}.$$

Consider ϑ, the space of simple predictable processes, i.e. $\psi \in \vartheta$ if and only if ψ can be written as

$$\psi_t(\omega) = \sum_{i=1}^{k} a_i \mathbf{1}_{(u_i,t_i]}(t)\,\mathbf{1}_{A_i}(\omega), \quad t \in \mathbb{R}, \omega \in \Omega,$$

where $a_i \in \mathbb{R}$, $-\infty < u_i \leq t_i < \infty$ and $A_i \in \mathcal{F}_{u_i}$, for $i = 1, \ldots, k$. Given $\psi \in \vartheta$, define the linear operator $m : \vartheta \to \mathcal{L}^0(\Omega, \mathcal{F}, \mathbb{P})$ by

$$m(\psi) := \sum_{i=1}^{k} a_i \mathbf{1}_{A_i} (L_{t_i} - L_{u_i}). \tag{2.1}$$

In stochastic integration theory, commonly one is looking for a linear extension of operators of the form (2.1) to a suitable space, let's say L_m, such that $m(\psi)$ can be approximated by simple integrals of elements of ϑ. More precisely, if m can be extended to L_m and ϑ is dense in this set, we say that ψ is L-integrable or $\psi \in L_m$ and we define its stochastic integral with respect to L as

$$\int_{\mathbb{R}} \psi_s dL_s := \mathbb{P}\text{-}\lim_{n\to\infty} m(\psi_n), \tag{2.2}$$

provided that $\psi_n \in \vartheta$, $|\psi_n| \leq \phi$, $\phi \in L_m$ and $\psi_n \to \psi$ pointwise.

It can be shown that the simple integral can be extended to the set

$$L_m = \left\{\psi : (\mathbb{R}\times\Omega, \mathcal{P}) \to (\mathbb{R}, \mathcal{B}(\mathbb{R})) \mid \lim_{r\to 0} \|r\psi\|_m = 0\right\}, \tag{2.3}$$

where

$$\|\psi\|_m := \sup_{\zeta\in\vartheta, |\zeta|\leq\psi} \|m(\zeta)\|_0.$$

Basse-O'Connor et al. [10] showed that $\psi \in L_m$ if and only if the following three conditions hold almost surely:

$$1.\ b^2 \int_{\mathbb{R}} \psi_s^2 ds < \infty; \quad 2.\ \int_{\mathbb{R}} \int_{\mathbb{R}} 1 \wedge |x\psi_s|^2 \nu(dx)\, ds < \infty; \tag{2.4}$$

$$3.\ \int_{\mathbb{R}} \left| \gamma \psi_s + \int_{\mathbb{R}} [\tau(x\psi_s) - \tau(x)\psi_s] \nu(dx) \right| ds < \infty.$$

It is important to remark that when ψ is deterministic, conditions 1.-3. correspond to the conditions found by Rajput and Rosiński [19, Theorem 2.7]. Furthermore, it should be noted that Rajput and Rosiński [19] not only integrate w.r.t. Lévy processes but also w.r.t. independently scattered random measures. In particular, when ψ is deterministic such integral coincides with (2.2).

2.2 Existence of Y and Its Stationary Structure

Following Basse-O'Connor et al. [10] and Basse-O'Connor [9], in this subsection we investigate necessary and sufficient conditions for the existence of $(Y_t)_{t\in\mathbb{R}}$ in Eq. (1.1) based on the triplet of L.

In view of (2.4), Y is a well defined stochastic process if and only if for each $t \in \mathbb{R}$ almost surely $b^2 \int_0^\infty [\varphi_{\alpha,\lambda}(r)\sigma_{t-r}]^2 dr < \infty$ and

$$\int_{\mathbb{R}} \int_0^\infty 1 \wedge |x\varphi_{\alpha,\lambda}(r)\sigma_{t-r}|^2 dr\nu(dx) < \infty, \tag{2.5}$$

and

$$\int_0^\infty \left| \gamma\varphi_{\alpha,\lambda}(r)\sigma_{t-r} + \int_{\mathbb{R}} [\tau(x\varphi_{\alpha,\lambda}(r)\sigma_{t-r}) - \varphi_{\alpha,\lambda}(r)\sigma_{t-r}\tau(x)] \nu(dx) \right| dr < \infty. \tag{2.6}$$

Recently, Basse-O'Connor [9] has given necessary and sufficient conditions for (2.5), (2.6) and the semimartingale property of Y when $\sigma \equiv 1$. On the other hand, Sato [21] characterized the integrability of a certain family of mappings $s \longmapsto f(s)$, where $f(s) \asymp \varphi_{\alpha,\lambda}(s)$ when $s \to \infty$, i.e. there exist $a_1, a_2 > 0$ such that $a_2\varphi_{\alpha,\lambda}(s) \le f(s) \le a_1\varphi_{\alpha,\lambda}(s)$ for s large. We exploit their results in order to obtain necessary and sufficient conditions for the existence of Y when σ is a strongly stationary process. Recall that a process $(X_t)_{t\in\mathbb{R}}$ is said to be strongly stationary if for any $h \in \mathbb{R}$, $(X_{t+h})_{t\in\mathbb{R}} \overset{d}{=} (X_t)_{t\in\mathbb{R}}$.

Theorem 2.1 *Let $(\sigma_t)_{t\in\mathbb{R}}$ be a non-zero predictable càdlàg process and $(L_t)_{t\in\mathbb{R}}$ an $(\mathcal{F}_t)$-Lévy process with triplet (γ, b, ν). Assume that σ is strongly stationary, square integrable and for $-1/2 < \alpha < 0$, $\mathbb{E}\left(|\sigma_0|^{-\frac{1}{\alpha}}\right) < \infty$. Then, we have that $s \longmapsto$*

$\varphi_{\alpha,\lambda}(t-s)\,\sigma_s \mathbf{1}_{\{s\leq t\}}$ *is L-integrable for any* $t \in \mathbb{R}$, *if the following two conditions are satisfied:*

1. $\int_{|x|>1} \log(|x|)\,\nu(dx) < \infty$,
2. *One of the following conditions are satisfied:*
   (a) $\alpha > -1/2$;
   (b) $\alpha = -1/2$, $b = 0$ *and* $\int_{|x|\leq 1} |x|^2 |\log(|x|)|\,\nu(dx) < \infty$;
   (c) $\alpha \in (-1, -1/2)$, $b = 0$ *and* $\int_{|x|\leq 1} |x|^{-1/\alpha}\,\nu(dx) < \infty$.

Conversely, assume that $\sigma \equiv 1$. *If* $s \longmapsto \varphi_{\alpha,\lambda}(t-s)\,\mathbf{1}_{\{s\leq t\}}$ *is L-integrable for any* $t \in \mathbb{R}$, *then* 1.–2. *hold.*

For the proof of the previous theorem we will need the next lemma, the proof of which can be found in Lemma 2.8 of Rajput and Rosiński [19].

Lemma 2.2 *Define*

$$V(u) := \left|\gamma u + \int_{\mathbb{R}} [\tau(xu) - u\tau(x)]\,\nu(dx)\right|, \quad u \in \mathbb{R}. \tag{2.7}$$

Then for any $d > 0$

$$\sup_{-d\leq c\leq d} V(cu) \leq dV(u) + K_d \int_{\mathbb{R}} 1 \wedge |xu|^2\,\nu(dx), \quad u \in \mathbb{R},$$

where K_d *is a positive constant which only depends on* τ *and* d.

Proof of Theorem 2.1 For $r, c > 0, \alpha > -1$, define

$$\phi_{c,\alpha}(r) := \begin{cases} r^{\alpha}\mathbf{1}_{\{0<r\leq 1\}} + e^{-cr}\mathbf{1}_{\{r>1\}} & \text{for } -1 < \alpha < 0; \\ e^{-cr} & \text{for } \alpha \geq 0. \end{cases}.$$

Then, there exists $c_1 > 0$ such that for all $r > 0$

$$\varphi_{\alpha,\lambda}(r) \leq c_1 \phi_{\lambda/2,\alpha}(r). \tag{2.8}$$

Due to this and (2.3), it is enough to verify that the mapping $r \mapsto \phi_{c,\alpha}(r)\,\sigma_{t-r}\mathbf{1}_{\{r>0\}}$ is L-integrable when $c = \lambda/2$. Firstly note that the stationarity of σ implies that

$$\mathbb{E}\left(\int_0^{\infty} [\phi_{c,\alpha}(r)\,\sigma_{t-r}]^2\,dr\right) = \mathbb{E}\left(\sigma_0^2\right)\int_0^{\infty} \phi_{c,\alpha}^2(r)\,dr, \quad t \in \mathbb{R},$$

which is finite if and only if $\alpha > -1/2$, so if $b > 0$, Y is well defined if $\alpha > -1/2$ and (2.5)–(2.6) hold. Therefore, we only need to verify that in this case equations (2.5) and (2.6) are implied by 1. and 2.

By stationarity and the inequality

$$1 \wedge |ab| \leq \begin{cases} 1 \wedge |b| & \text{if } |a| \leq 1; \\ |a| (1 \wedge |b|) & \text{if } |a| > 1, \end{cases}$$

we get for all $t, x \in \mathbb{R}$ and $r > 0$

$$\mathbb{E}\left[1 \wedge |x\phi_{c,\alpha}(r)\, \sigma_{t-r}|^2\right] \leq \left(1 \wedge |x\phi_{c,\alpha}(r)|^2\right) \mathbb{E}\left[\left(1 + \sigma_0^2\right)\right]. \tag{2.9}$$

Moreover, from the proof of Theorem 2 in Basse-O'Connor [9]

$$\int_{\mathbb{R}} \int_0^\infty 1 \wedge |x\phi_{c,\alpha}(r)|^2 \, dr \nu(dx) < \infty,$$

if and only if conditions 1. and 2. are satisfied. Therefore, by (2.9) Eq. (2.5) holds if 1.–2. are assumed to be true.

Now we proceed to verify (2.6). By the strong stationarity of σ, (2.6) can be obtained if

$$\mathbb{E}\left\{\int_0^\infty V\left[\sigma_0 \phi_{c,\alpha}(r)\right] dr\right\} < \infty. \tag{2.10}$$

It follows from Lemma 2.2 that there exists a $K_1 > 0$ only depending on τ, such that for any $r > 0$

$$V\left[\sigma_0 \phi_{c,\alpha}(r)\right] \mathbf{1}_{\{|\sigma_0| \leq 1\}} \leq V\left[\phi_{c,\alpha}(r)\right] \mathbf{1}_{\{|\sigma_0| \leq 1\}} + K_1 \int_{\mathbb{R}} 1 \wedge |x\phi_{c,\alpha}(r)|^2 \nu(dx).$$

Thus

$$\begin{aligned} V\left[\sigma_0 \phi_{c,\alpha}(r)\right] &\leq V\left[\sigma_0 \phi_{c,\alpha}(r)\right] \mathbf{1}_{\{|\sigma_0| > 1\}} + V\left[\phi_{c,\alpha}(r)\right] \mathbf{1}_{\{|\sigma_0| \leq 1\}} \\ &\quad + K_1 \int_{\mathbb{R}} 1 \wedge |x\phi_{c,\alpha}(r)|^2 \nu(dx) \\ &\leq \int_{\mathbb{R}} H_{c,\alpha}(\sigma_0, x, r) \nu(dx) \mathbf{1}_{\{|\sigma_0| > 1\}} \\ &\quad + \int_{\mathbb{R}} H_{c,\alpha}(1, x, r) \nu(dx) \\ &\quad + |\gamma \phi_{c,\alpha}(r)| + K_1 \int_{\mathbb{R}} 1 \wedge |x\phi_{c,\alpha}(r)|^2 \nu(dx), \end{aligned} \tag{2.11}$$

where, using that $\tau(x) = \frac{x}{1+|x|^2}$,

$$H_{c,\alpha}(v,x,r) := |\tau(xv\phi_{c,\alpha}(r)) - \phi_{c,\alpha}(r)\, v\tau(x)|$$
$$= \frac{|x|^2}{1+|x|^2}\frac{|vx\phi_{c,\alpha}(r)|}{1+|vx\phi_{c,\alpha}(r)|^2}\left|1-|v\phi_{c,\alpha}(r)|^2\right|, \quad r>0, v,x\in\mathbb{R}.$$

Suppose that $-1<\alpha<0$. In view that in this case $\left|1-|v\phi_{c,\alpha}(r)|^2\right| \leq (1+v^2)$ for $r>1$, we have

$$\int_1^\infty H_{c,\alpha}(v,x,r)\,dr \leq \frac{1}{c}\left(1+v^2\right)\frac{|x|^2}{1+|x|^2}\int_c^\infty \frac{|vxe^{-r}|}{1+|vxe^{-r}|^2}dr \tag{2.12}$$
$$= \frac{1}{c}\left(1+v^2\right)\frac{|x|^2}{1+|x|^2}\int_0^{|vxe^{-c}|}\frac{1}{1+s^2}ds$$
$$\leq \frac{\pi}{2c}\left(1+v^2\right)\frac{|x|^2}{1+|x|^2}, \quad \text{for } v,x\in\mathbb{R},$$

and

$$\int_0^1 H_{c,\alpha}(\sigma_0,x,r)\,\mathbf{1}_{\{|\sigma_0|>1\}}dr = |\alpha|^{-1}|\sigma_0|^{-\frac{1}{\alpha}}\frac{|x|^2}{1+|x|^2}|x|^{-\frac{1}{\alpha}}$$
$$\times\int_{|\sigma_0 x|}^\infty \frac{s^{\frac{1}{\alpha}}}{1+s^2}\left|\left(s|x|^{-1}\right)^2-1\right|\mathbf{1}_{\{|\sigma_0|>1\}}ds$$
$$\leq |\alpha|^{-1}|\sigma_0|^{-\frac{1}{\alpha}}\mathbf{1}_{\{|\sigma_0|>1\}}\frac{|x|^2}{1+|x|^2}|x|^{-\frac{1}{\alpha}}$$
$$\times\int_{|x|}^\infty \frac{s^{\frac{1}{\alpha}}}{1+s^2}\left|\left(s|x|^{-1}\right)^2-1\right|ds$$
$$= |\sigma_0|^{-\frac{1}{\alpha}}\mathbf{1}_{\{|\sigma_0|>1\}}\int_0^1 H_{c,\alpha}(1,x,r)\,dr, \quad \text{for } x\in\mathbb{R}.$$

Plugging the previous, after integrating, estimates into (2.11) give us

$$\mathbb{E}\left\{\int_0^\infty V[\sigma_0\phi_{c,\alpha}(r)]\,dr\right\} \leq |\alpha|^{-1}\mathbb{E}\left(1+|\sigma_0|^{-\frac{1}{\alpha}}\right)\int_{\mathbb{R}}\frac{|x|^2}{1+|x|^2}h(|x|)\,\nu(dx)$$
$$+\frac{\pi}{c}\frac{1}{2}\mathbb{E}\left(3+\sigma_0^2\right)\int_{\mathbb{R}}\frac{|x|^2}{1+|x|^2}\nu(dx)$$
$$+|\gamma|\int_0^\infty |\phi_{c,\alpha}(r)|\,dr$$
$$+K_1\int_0^\infty\int_{\mathbb{R}} 1\wedge|x\phi_{c,\alpha}(r)|^2\,\nu(dx)\,dr,$$

where

$$h(|x|) := |x|^{-\frac{1}{\alpha}} \int_{|x|}^{\infty} \frac{s^{\frac{1}{\alpha}}}{1+s^2}\left[\left(s|x|^{-1}\right)^2 - 1\right] ds, \quad x \in \mathbb{R}.$$

Obviously $|\gamma| \int_0^\infty |\phi_{c,\alpha}(r)|\, dr < \infty$ and $\int_{\mathbb{R}} \frac{|x|^2}{1+|x|^2} \nu(dx) < \infty$. Furthermore, $\int_0^\infty \int_{\mathbb{R}} 1 \wedge |x\phi_{c,\alpha}(r)|^2 \nu(dx)$ is finite if and only if 1. and 2. are assumed to be true (see the first part of the proof). By hypothesis, $\mathbb{E}\left(|\sigma_0|^{-\frac{1}{\alpha}}\right) < \infty$. Hence, (2.10) holds if the mapping $x \longmapsto \frac{|x|^2}{1+|x|^2} h(|x|)$ is ν-integrable whenever 1. and 2. are satisfied. Due to

$$h(|x|) \leq |x|^{-\frac{1}{\alpha}-2} \int_{|x|}^{\infty} \frac{s^{\frac{1}{\alpha}+2}}{1+s^2} ds, \quad x \in \mathbb{R}\backslash\{0\}, \tag{2.13}$$

it suffices to show that $x \longmapsto \frac{|x|^2}{1+|x|^2} h(|x|)$ is ν-integrable in a neighborhood of zero. Assume that $-1/2 < \alpha < 0$, then applying the L'Hôpital's rule to Eq. (2.13), we get that $h(|x|) \to 0$ as $x \to 0$, i.e. $x \longmapsto \frac{|x|^2}{1+|x|^2} h(|x|)$ is ν-integrable in a neighborhood of zero. Now, consider $-1 < \alpha \leq -1/2$. We claim that $|x|^2 h(|x|) \sim a_\alpha |x|^{-\frac{1}{\alpha}}$ as $x \to 0$ with $a_\alpha > 0$. Indeed, put $a_\alpha := \int_0^\infty \frac{s^{\frac{1}{\alpha}+2}}{1+s^2} ds < \infty$, by the Dominated Convergence Theorem, one easily sees that

$$\lim_{x \to 0} |x|^{2+\frac{1}{\alpha}} h(|x|) = \lim_{x \to 0} \int_{|x|}^{\infty} \frac{s^{\frac{1}{\alpha}}}{1+s^2} \left|s^2 - |x|^2\right| ds = a_\alpha,$$

as it was claimed. This implies that $\frac{|x|^2}{1+|x|^2} h(|x|) \sim a_\alpha |x|^{-\frac{1}{\alpha}}$ as $x \to 0$, which means that $x \longmapsto \frac{|x|^2}{1+|x|^2} h(|x|)$ is ν-integrable in a neighborhood of zero when $\alpha = -1/2$ or 2.(c) holds.

Finally, if $\alpha \geq 0$, then $\left|1 - |v\phi_{c,\alpha}(r)|^2\right| \leq \left(1 + v^2\right)$ for any $r, v > 0$. Thus, following the same reasoning as in (2.12), we observe that in this case

$$\int_0^\infty H_{c,\alpha}(\sigma_0, x, r)\, dr \leq \frac{\pi}{2c}\left(1 + \sigma_0^2\right) \frac{|x|^2}{1+|x|^2}, \quad \text{for any } x \in \mathbb{R},$$

which implies, by taking expectation and integrating with respect to ν, Eq. (2.10).

The converse is due to Basse-O'Connor [9]. □

From now on, we will assume that if $-1/2 < \alpha < 0$, then $\mathbb{E}\left(|\sigma_0|^{-\frac{1}{\alpha}}\right) < \infty$.

Example Suppose that the triplet of L is (γ, b, ν) with $\nu(dx) = c\delta_{\{1\}}(dx)$ for some $c > 0$, where δ denotes the Dirac delta measure. Then Y is well defined if $\alpha > -1/2$ or $-1/2 \geq \alpha > -1$ and $b = 0$.

Example Let $S^{\beta}(\mathbb{R}) \subset ID(\mathbb{R})$ be the set of β-stable distributions for $0 < \beta \leq 2$, i.e. $\mu \in S^{\beta}(\mathbb{R})$ for $0 < \beta < 2$ if and only if its triplet is $(\gamma, 0, \nu)$ with

$$\nu(dx) = c_1 |x|^{-\beta-1} \mathbf{1}_{\{x>0\}} dx + c_2 |x|^{-\beta-1} \mathbf{1}_{\{x<0\}} dx, \quad c_1, c_2 > 0,$$

and for $\beta = 2$, μ is Gaussian. Assume that L is a β-stable stable processes, that is, the law of L_1 belongs to $S^{\beta}(\mathbb{R})$. Then Y is well defined if $b = 0$ and $\beta < -1/\alpha$ or $\alpha > -1/2$.

Remark 2.3 Barndorff-Nielsen et al. [2] studied Y when $\alpha > -1/2$ and L is symmetric with second moment as well as independent of σ. In this case Y is weakly stationary with autocovariance function

$$\tilde{\gamma}(h) = \left(b^2 + \int_{\mathbb{R}} x^2 \nu(dx)\right) \mathbb{E}\left(\sigma_0^2\right) \frac{\Gamma(\alpha+1)}{\sqrt{\pi}} \left(\frac{h}{2\lambda}\right)^{\alpha+\frac{1}{2}} K_{\alpha+\frac{1}{2}}(\lambda h), \quad h \geq 0, \tag{2.14}$$

where K_p is the modified Bessel function of the third kind with index p. Thus, a sufficient condition for Y to be stationary relies in the independence between σ and L. However, such condition is not necessary, e.g. if for all $h \in \mathbb{R}$ we have

$$\left(\sigma_{t+h}, L_t^h\right)_{t\in\mathbb{R}} \stackrel{d}{=} (\sigma_t, L_t)_{t\in\mathbb{R}}, \tag{2.15}$$

where $L_t^h := L_{t+h} - L_h$, $t \in \mathbb{R}$, then Y is strongly stationary, but in this case its correlation structure is not known in explicit form.

3 The Class of Marginal Distributions of *Y*

In this section we study the class of marginal distributions induced by Y in absence of stochastic volatility. Recall that L_1 has characteristic triplet (γ, b, ν).

We start by noting that when $\sigma \equiv 1$ and 1.–2. of Theorem 2.1 hold, Y is a strongly stationary process and its marginal distribution is infinitely divisible and corresponds to the law of $\int_0^\infty \varphi_{\alpha,\lambda}(s)\, dL_s$, see Theorem 2.7 of Rajput and Rosiński [19]. Furthermore, we have

$$\int_0^\infty \varphi_{\alpha,\lambda}(s)\, dL_s = \int_0^\infty \varphi_{\alpha,1}(s)\, d\tilde{L}_s,$$

where $\tilde{L}_t := \frac{1}{\lambda^\alpha} L_{t/\lambda}$. This means that without loss of generality we may assume that $\lambda = 1$. The behavior of $\varphi_{\alpha,\lambda}$ changes drastically for $\alpha < 0$ and $\alpha \geq 0$. These kind of singularities are reflected in the properties of the distribution of Y_0 so we treat them separately in the next subsections. Observe that the case $\alpha = 0$ corresponds to the marginal law of an Ornstein-Uhlenbeck process which has been widely studied, so we omit this case. In the following φ_α denotes $\varphi_{\alpha,1}$.

3.1 The Case $\alpha > 0$

For $\alpha > 0$, φ_α is bounded but not invertible. However, the restrictions of φ_α to $(0, \alpha)$ and (α, ∞) are invertible with continuous derivatives. Denote by $\varphi_\alpha^{(1)} : (0, \varphi_\alpha(\alpha)) \to (0, \alpha)$ and $\varphi_\alpha^{(2)} : (0, \varphi_\alpha(\alpha)) \to (\alpha, \infty)$ these inverse functions. The following theorem describes the characteristic triplet of the law of $\int_0^\infty \varphi_\alpha(s)\, dL_s$.

Theorem 3.1 *Assume that* $\alpha > 0$ *and* $\int_{|x|>1} \log(|x|)\, \nu(dx) < \infty$. *Then*

1. *The triplet of* $\int_0^\infty \varphi_\alpha(s)\, dL_s$, *denoted by* $(\gamma_\alpha, b_\alpha, \nu_\alpha)$, *is given by*

$$
\begin{aligned}
\gamma_\alpha &= \gamma \Gamma(\alpha+1) + \int_0^\infty \int_{\mathbb{R}} [\tau(x\varphi_\alpha(r)) - \varphi_\alpha(r)\tau(x)]\, \nu(dx)\, dr; \\
b_\alpha &= b\Gamma(2\alpha+1); \\
\nu_\alpha(B) &= \int_0^\infty \int_{\mathbb{R}} \mathbf{1}_B(\varphi_\alpha(r)x)\, \nu(dx)\, dr, \quad B \in \mathcal{B}(\mathbb{R}).
\end{aligned} \tag{3.1}
$$

2. *The Lévy measure* ν_α *is absolutely continuous with respect to the Lebesgue measure with density given by*

$$
f_{\alpha,\nu}(r) = \begin{cases} \frac{1}{r} \int_{r/\varphi_\alpha(\alpha)}^\infty h_\alpha\left(\frac{r}{x}\right) \nu(dx), & r > 0; \\ \frac{1}{|r|} \int_{-\infty}^{r/\varphi_\alpha(\alpha)} h_\alpha\left(\left|\frac{r}{x}\right|\right) \nu(dx), & r < 0, \end{cases} \tag{3.2}
$$

   with

$$
h_\alpha(u) = u\left[\left(\varphi_\alpha^{(1)}\right)'(u) - \left(\varphi_\alpha^{(2)}\right)'(u)\right], \quad 0 < u < \varphi_\alpha(\alpha), \tag{3.3}
$$

   where $\varphi_\alpha^{(1)}$ *and* $\varphi_\alpha^{(2)}$ *are as before.*

Proof of Theorem 3.1 The first point follows from Rajput and Rosiński [19, Theorem 2.7] and Sato [21, Proposition 5.5]. To prove 2., let $g : \mathbb{R}^+ \to \mathbb{R}^+$ be a bounded measurable function. From (3.1), it follows that

$$
\begin{aligned}
\int_0^\infty g(y)\, \nu_\alpha(dy) &= \int_0^\infty \int_0^\infty g(\varphi_\alpha(s)x)\, ds \nu(dx) \\
&= \int_0^\infty \left[\int_0^\alpha g(\varphi_\alpha(s)x)\, ds + \int_\alpha^\infty g(\varphi_\alpha(s)x)\, ds\right] \nu(dx) \\
&= \int_0^\infty \int_0^{\varphi_\alpha(\alpha)x} g(r) \frac{r}{x} \left(\varphi_\alpha^{(1)}\right)'\left(\frac{r}{x}\right) \frac{dr}{r} \nu(dx)
\end{aligned}
$$

$$-\int_0^\infty \int_0^{\varphi_\alpha(\alpha)x} g(r)\frac{r}{x}\left(\varphi_\alpha^{(2)}\right)'\left(\frac{r}{x}\right)\frac{dr}{r}\nu(dx)$$

$$=\int_0^\infty g(r)\left\{\int_{\frac{r}{\varphi_\alpha(\alpha)}}^\infty \frac{r}{x}\left[\left(\varphi_\alpha^{(1)}\right)'\left(\frac{r}{x}\right)-\left(\varphi_\alpha^{(2)}\right)'\left(\frac{r}{x}\right)\right]\nu(dx)\right\}\frac{dr}{r},$$

where we made the change of variables $r=\varphi_\alpha(s)x$. Thus,

$$\int_0^\infty g(y)\,\nu_\alpha(dy)=\int_0^\infty g(r)f_{\alpha,\nu}(r)\,dr,$$

which shows that the restriction of ν_α to $\mathbb{R}^+\setminus\{0\}$ is absolutely continuous with density $f_{\alpha,\nu}$. Similarly, it follows that the restriction of ν_α to $\mathbb{R}^-\setminus\{0\}$ is absolutely continuous with density $f_{\alpha,\nu}$. □

In the following we describe the function h_α defined above.

Lemma 3.2 *Let h_α be as* (3.3). *Then, h_α is continuous and increasing on* $(0,\varphi_\alpha(\alpha))$ *with $h_\alpha(u)\to 1$ as $u\downarrow 0$ and*

$$h_\alpha(u)\sim\left(\frac{2\alpha\varphi_\alpha(\alpha)}{\varphi_\alpha(\alpha)-u}\right)^{1/2},\quad \text{as } u\uparrow\varphi_\alpha(\alpha). \tag{3.4}$$

Proof We start by noting that $\varphi_\alpha^{(1)}$ and $\varphi_\alpha^{(2)}$ are continuous and tend to 0 and ∞ as $u\to 0$, respectively. Now, since $\varphi_\alpha'(s)=\varphi_\alpha(s)\left(\frac{\alpha-s}{s}\right)$ we get that φ_α' never vanishes except for $s=\alpha$ and

$$\begin{aligned}h_\alpha(u)&=u\left[\frac{1}{\varphi_\alpha'\left(\varphi_\alpha^{(1)}(u)\right)}-\frac{1}{\varphi_\alpha'\left(\varphi_\alpha^{(2)}(u)\right)}\right]\\&=\frac{\varphi_\alpha^{(1)}(u)}{\alpha-\varphi_\alpha^{(1)}(u)}+\frac{\varphi_\alpha^{(2)}(u)}{\varphi_\alpha^{(2)}(u)-\alpha},\quad\text{for } u<\varphi_\alpha(\alpha).\end{aligned} \tag{3.5}$$

This means that h_α is continuous, differentiable and increasing on $(0,\varphi_\alpha(\alpha))$ with $h_\alpha(u)\to 1$ when $u\downarrow 0$ and $h_\alpha(u)\to\infty$ as $u\uparrow\varphi_\alpha(\alpha)$. Moreover, since $\varphi_\alpha^{(i)}(u)\to\alpha$ as $u\uparrow\varphi_\alpha(\alpha)$ for $i=1,2$, Eq. (3.5) implies that (3.4) holds if

$$\lim_{u\uparrow\phi(\alpha)}\frac{\left(\varphi_\alpha(\alpha)-u\right)^{1/2}}{\left|\alpha-\varphi_\alpha^{(i)}(u)\right|}=\sqrt{\varphi_\alpha(\alpha)/2\alpha},\quad\text{for } i=1,2.$$

This follows from

$$\lim_{s\to\alpha}\frac{\left(\varphi_\alpha(\alpha)-\varphi_\alpha(s)\right)}{(\alpha-s)^2}=\varphi_\alpha(\alpha)/2\alpha,$$

where we have used the L'Hôpital's rule, for $s = \varphi_{\alpha}^{(1)}(u), \varphi_{\alpha}^{(2)}(u)$. □

Example Suppose that $\nu = \delta_1$, the Dirac's delta measure concentrated in 1. We get in this case

$$f_{\alpha,\nu}(r) = \begin{cases} \frac{1}{r} h_{\alpha}(r) & r < \varphi_{\alpha}(\alpha); \\ +\infty & r = \varphi_{\alpha}(\alpha); \\ 0 & \text{otherwise.} \end{cases}$$

Note that $r \mapsto r f_{\alpha,\nu}(r)$ is neither increasing nor decreasing, has a singularity and is not continuous at $r = \varphi_{\alpha}(\alpha)$. However, it is still finite almost everywhere.

Remark 3.3 As the previous example shows, $f_{\alpha,\nu}$ is not continuous in general. Also, it could be infinite on a set of measure zero and with compact support. However, if $\nu \ll$ Leb with continuous Radon-Nikodym derivative, where Leb denotes the Lebesgue measure, then $f_{\alpha,\nu}$ is finite everywhere. Indeed, suppose that $\nu(dx) = g(x)\,dx$, with g continuous. Then by Lemma 3.2, for all $r > 0$ we have that there exist $c_1, c_2 > 0$ and $x_0 > r/\varphi_{\alpha}(\alpha)$, such that

$$c_1 \left(\frac{x}{x\varphi_{\alpha}(\alpha) - r} \right)^{1/2} \leq h_{\alpha}\left(\frac{r}{x}\right) \leq c_2 \left(\frac{x}{x\varphi_{\alpha}(\alpha) - r} \right)^{1/2}, \quad r/\varphi_{\alpha}(\alpha) < x \leq x_0.$$

Moreover, the following integral is finite

$$\int_{r/\varphi_{\alpha}(\alpha)}^{x_0} \left(\frac{x}{x\varphi_{\alpha}(\alpha) - r} \right)^{1/2} \nu(dx) = \frac{r}{\varphi_{\alpha}^{3/2}(\alpha)} \int_1^{\varphi_{\alpha}(\alpha)\frac{x_0}{r}} \left(\frac{y}{y-1} \right)^{1/2} g\left(\frac{ry}{\varphi_{\alpha}(\alpha)} \right) dy.$$

This, jointly with the fact that $h_{\alpha}(u) \to 1$ as $u \to 0$ (Lemma 3.2), give that $h_{\alpha}\left(\frac{r}{\cdot}\right)$ is ν-integrable for all $r > 0$, in other words $f_{\alpha,\nu}$ is finite.

3.2 The Case $-1 < \alpha < 0$

In this section we describe $(\gamma_{\alpha}, b_{\alpha}, \nu_{\alpha})$ for $-1 < \alpha < 0$. In particular, we show that the corresponding distribution is self-decomposable. Also, we study the Lévy density obtained in this case. Below we present the main result of this subsection.

Theorem 3.4 *Assume that $-1 < \alpha < 0$ and conditions 1. and 2. of Theorem 2.1 hold. Then $(\gamma_{\alpha}, b_{\alpha}, \nu_{\alpha})$, the characteristic triplet of $\int_0^{\infty} \varphi_{\alpha}(s)\,dL_s$, is given as in (3.1). Moreover*

1. *The Lévy measure of the distribution of* $\int_0^\infty \varphi_\alpha(s)\,dL_s$ *is absolutely continuous with respect to the Lebesgue measure with continuous density which is given by*

$$f_{\alpha,\nu}(r) = \begin{cases} \frac{1}{r}\int_0^\infty h_\alpha\left(\frac{r}{x}\right)\nu(dx), & r > 0; \\ \frac{1}{|r|}\int_{-\infty}^0 h_\alpha\left(\left|\frac{r}{x}\right|\right)\nu(dx), & r < 0, \end{cases} \tag{3.6}$$

with

$$h_\alpha(u) = \frac{u}{\left|\varphi'_\alpha\left(\varphi_\alpha^{-1}(u)\right)\right|}, \quad u > 0, \tag{3.7}$$

where φ'_α *and* φ_α^{-1} *are the derivative and the inverse of* φ_α*, respectively.*
2. *Let* μ_α *be the law of* $\int_0^\infty \varphi_\alpha(s)\,dL_s$*. Then* $\mu_\alpha \in SD(\mathbb{R})$*.*

The proof of this theorem is based on the next lemma.

Lemma 3.5 *Assume that* $-1 < \alpha < 0$*. Let* h_α *be as in (3.7) and put* $h_\alpha(0) = 1$*. Then* h_α *is continuous and decreasing on* $[0,\infty)$*. Moreover, the mapping* $u \mapsto h_\alpha(u)\,u^{-\frac{1}{\alpha}}$ *is increasing and bounded by* $|\alpha|^{-1}$ *for all* $u \geq 0$*, and*

$$h_\alpha(u) \sim |\alpha|^{-1} u^{\frac{1}{\alpha}}, \quad \text{as } u \to \infty. \tag{3.8}$$

Proof Firstly, we observe that in this case φ_α^{-1} is continuous on $(0,\infty)$ with $\lim_{z\to\infty}\varphi_\alpha^{-1}(z) = 0$ and $\lim_{z\downarrow 0}\varphi_\alpha^{-1}(z) = +\infty$. Further, since $\varphi'_\alpha(s) = \frac{\alpha-s}{s}\varphi_\alpha(s)$, we get that φ'_α is continuous on $(0,\infty)$ and

$$\lim_{s\to\infty}\left|\frac{\varphi_\alpha(s)}{\varphi'_\alpha(s)}\right| = 1, \quad \lim_{s\downarrow 0}\left|\alpha\frac{\varphi_\alpha(s)}{\varphi'_\alpha(s)}\varphi_\alpha(s)^{-\frac{1}{\alpha}}\right| = 1. \tag{3.9}$$

On the other hand, we see that for $s = \varphi_\alpha^{-1}(u)$

$$h_\alpha(u) = \frac{\varphi_\alpha(s)}{|\varphi'_\alpha(s)|}, \quad \text{and } h_\alpha(u)\,u^{-\frac{1}{\alpha}} = \frac{\varphi_\alpha(s)}{\varphi'_\alpha(s)}\varphi_\alpha(s)^{-\frac{1}{\alpha}} = e^{\frac{s}{\alpha}}(s-\alpha)^{-1},$$

i.e. h_α is continuous and decreasing on $(0,\infty)$ and since the mapping $s \mapsto e^{\frac{s}{\alpha}}(s-\alpha)^{-1}$ is decreasing, we get that $u \mapsto h_\alpha(u)\,u^{-\frac{1}{\alpha}}$ is increasing and bounded by $|\alpha|^{-1}$ for all $u \geq 0$. Moreover, plugging the previous equation into (3.9) gives us that $\lim_{u\downarrow 0} h_\alpha(u) = 1$, and

$$\lim_{u\to\infty}|\alpha|\,h_\alpha(u)\,u^{-\frac{1}{\alpha}} = \lim_{s\downarrow 0}\left|\alpha\frac{\varphi_\alpha(s)}{\varphi'_\alpha(s)}\varphi_\alpha(s)^{-\frac{1}{\alpha}}\right| = 1,$$

which concludes the proof. □

Proof of Theorem 3.4 As in the proof of Theorem 3.1, we describe the restriction of ν_α to $\mathbb{R}^+ \setminus \{0\}$ and $\mathbb{R}^- \setminus \{0\}$ separately.

Applying (3.1) and doing the change of variables $r = \varphi_\alpha(s)\, x$ give us that for any $g : \mathbb{R}^+ \to \mathbb{R}^+$ measurable function

$$\begin{aligned}
\int_0^\infty g(y)\, \nu_\alpha(dy) &= \int_0^\infty \int_0^\infty g(\varphi_\alpha(s)\, x)\, ds\nu(dx) \\
&= -\int_0^\infty \int_0^\infty \frac{g(r)}{x} \left[\frac{\partial}{\partial u} \varphi_\alpha^{-1}(u) \bigg|_{u=r/x} \right] dr\nu(dx) \\
&= \int_0^\infty \int_0^\infty g(r) \frac{r}{x} \left| \frac{\partial}{\partial u} \varphi_\alpha^{-1}(u) \bigg|_{u=r/x} \right| \frac{dr}{r} \nu(dx) \\
&= \int_0^\infty g(r) \int_0^\infty h_\alpha(r/x)\, \nu(dx)\, \frac{dr}{r}.
\end{aligned}$$

In order to get (3.6), we need to check that for all $r > 0$ we have $\int_0^\infty h_\alpha(r/x)\, \nu(dx) < \infty$. Lemma 3.5 guarantees that for any $r > 0$ the mappings $x \mapsto \left(\frac{r}{x}\right)^{-\frac{1}{\alpha}} h_\alpha(r/x)$ and $x \mapsto h_\alpha(r/x)$ are continuous and uniformly bounded by $|\alpha|^{-1}$ on $(0, \infty)$. Thus for any $r, x > 0$

$$h_\alpha(r/x) \le |\alpha|^{-1} \left[\left(\frac{x}{r}\right)^{-\frac{1}{\alpha}} \mathbf{1}_{\{|x| \le 1\}} + \mathbf{1}_{\{|x|>1\}} \right] \le |\alpha|^{-1} \left(r^{\frac{1}{\alpha}} + 1\right) \left(1 \wedge |x|^{-\frac{1}{\alpha}}\right).$$

Consequently, for any $r > 0$

$$\left(r^{\frac{1}{\alpha}} + 1\right)^{-1} \int_0^\infty h_\alpha(r/x)\, \nu(dx) \le \left|\alpha^{-1}\right| \int_{\mathbb{R}_+ \setminus \{0\}} 1 \wedge |x|^{-\frac{1}{\alpha}}\, \nu(dx) < \infty,$$

due to conditions 1. and 2. of Theorem 2.1. The continuity of $f_{\alpha,\nu}$ is obtained by an application of the Dominated Convergence Theorem. Following the same reasoning, we conclude that ν_α restricted to $\mathbb{R}_- \setminus \{0\}$ is absolutely continuous with respect to the Lebesgue measure for any $0 > \alpha > -1$ with density given by (3.6).

To complete the proof we only need to verify that for $-1 < \alpha < 0$ μ_α, the marginal distribution of Y, is self-decomposable. Thanks to part 1., the density of ν_α can be written as

$$f_{\alpha,\nu}(r) = \frac{1}{|r|} k_{\alpha,\nu}(r),$$

where

$$k_{\alpha,\nu}(r) = \int_0^\infty h_\alpha\left(\frac{r}{x}\right) \nu(dx)\, \mathbf{1}_{\{r>0\}} + \int_{-\infty}^0 h_\alpha\left(\left|\frac{r}{x}\right|\right) \nu(dx)\, \mathbf{1}_{\{r<0\}}.$$

Hence, it is enough to show that $k_{\alpha,\nu}$ is decreasing on $(0,\infty)$ and increasing on $(-\infty,0)$. This property follows immediately from Lemma 3.5. □

Example Consider ν as in the Example 2.2 and $-1<\alpha<0$. In this case, for any $r>0, f_{\alpha,\nu}(r)=\frac{c}{r}h_\alpha(r)$ and $\nu_\alpha((r,\infty))=c\varphi_\alpha^{-1}(r)$.

3.3 The Cancellation Problem of Multiplicative Convolutions and Injectivity of Φ_α

In this subsection we study the function mapping the law of L_1 to the law of $\int_0^\infty \varphi_\alpha(s)\,dL_s$ for $-1<\alpha<0$. We describe its domain and range as well as its injectivity. We relate this functional to a multiplicative convolution of measures and study its injectivity using the so-called cancellation property for multiplicative convolutions introduced by Jacobsen et al. [15].

Let us start by introducing the multiplicative operator for measures. Given η and μ two σ-finite measures on $\mathcal{B}(\mathbb{R}^+)$, as in Jacobsen et al. [15] we define and denote the *multiplicative convolution* between ν and η as

$$\nu\circledast\eta(B):=\int_0^\infty\int_0^\infty \mathbf{1}_B(yx)\,\nu(dx)\,\eta(dy),\quad B\in\mathcal{B}(\mathbb{R}^+). \tag{3.10}$$

This operator is a particular case of an Υ-transformation, see Barndorff-Nielsen et al. [5] for more details.

The name of multiplicative convolution comes from the fact that if X and Z are independent random variables with distribution μ_Z and μ_X, respectively, then $XZ\sim\mu_X\circledast\mu_Z$. However, if there is a probability measure μ such that $\mu_X\circledast\mu_Z=\mu\circledast\mu_Z$, it is not true in general that $\mu_X=\tilde{\mu}$. Generally, this holds under additional assumptions, such as that Z is positive and if the Fourier transform (Laplace transform) of $\log(Z)$ never vanishes. As in Jacobsen et al. [15], when a measure can be characterized in this way we say that it has the cancellation property. Being more precise, given ν and η σ-finite measures as before, we say that η has the *cancellation property (c.p.)* with respect to ν if

$$\nu\circledast\eta=\tilde{\nu}\circledast\eta,$$

for some σ-finite measure $\tilde{\nu}$, implies necessarily that $\nu=\tilde{\nu}$. Note that, for fixed η, the c.p. is equivalent to the injectivity of the operator $\cdot\circledast\eta$.

Now, we introduce a functional based on the marginal distribution of the process $(Y_t)_{t\in\mathbb{R}}$: Let μ be the law of L_1 and μ_α the distribution of $\int_0^\infty\varphi_\alpha(s)\,dL_s$. For $-1<\alpha<0$, define the mapping $\Phi_\alpha:\mathcal{D}(\Phi_\alpha)\to ID(\mathbb{R})$ by $\Phi_\alpha(\mu)=\mu_\alpha$, i.e. Φ_α transforms (γ,b,ν), the triplet of L_1, into $(\gamma_\alpha,b_\alpha,f_{\alpha,\nu})$ as in Theorem 3.4. Note that we have written the Lévy density instead of ν_α itself. Here $\mathcal{D}(\Phi_\alpha)$ is the domain of Φ_α, in other words $\mu\in\mathcal{D}(\Phi_\alpha)$ if and only if $\int_0^\infty\varphi_\alpha(s)\,dL_s$ exists. Theorem 2.1

shows that $\mu \in \mathcal{D}(\Phi_\alpha)$ if and only if (γ, b, ν), the characteristic triplet of μ, satisfies that $\int_{|x|>1} \log(|x|)\, \nu(dx) < \infty$ and

$$\begin{cases} b = 0 \text{ and } \int_{|x|\leq 1} |x|^2 |\log(|x|)|\, \nu(dx) < \infty, & \text{for } \alpha = -1/2; \\ b = 0 \text{ and } \int_{|x|\leq 1} |x|^{-1/\alpha}\, \nu(dx) < \infty, & \text{for } -1 < \alpha < -1/2. \end{cases} \tag{3.11}$$

Moreover, the range of Φ_α will be denoted as $\mathcal{R}(\Phi_\alpha)$.

Remark 3.4 Note that $\mathcal{D}(\Phi_\alpha)$ and $\mathcal{R}(\Phi_\alpha)$ are closed under convolutions. Indeed, since the Lévy measure of the convolution of two infinitely divisible distributions is the sum of their Lévy measures, we have that if $\mu^1, \mu^2 \in \mathcal{D}(\Phi_\alpha)$, then $\mu^1 * \mu^2 \in \mathcal{D}(\Phi_\alpha)$.

On the other hand, suppose that $\mu^1_\alpha, \mu^2_\alpha \in \mathcal{R}(\Phi_\alpha)$. Then there are $\mu^1, \mu^2 \in \mathcal{D}(\Phi_\alpha)$ such that

$$\Phi_\alpha\left(\mu^1\right) = \mu^1_\alpha \text{ and } \Phi_\alpha\left(\mu^2\right) = \mu^2_\alpha.$$

Moreover, by the above, it follows that $\mu^1 * \mu^2 \in \mathcal{D}(\Phi_\alpha)$. Now, by Proposition 2.6 of Rajput and Rosiński [19], we have that for any $\mu \in \mathcal{D}(\Phi_\alpha)$

$$\mathcal{C}_{\Phi_\alpha(\mu)}(z) = \int_0^\infty \mathcal{C}_\mu(\varphi_\alpha(s) z)\, ds, \quad z \in \mathbb{R},$$

where $\mathcal{C}_\mu(z) := \log \int_{\mathbb{R}} e^{izx} \mu(dx)$. Therefore, for all $z \in \mathbb{R}$

$$\mathcal{C}_{\Phi_\alpha(\mu^1 * \mu^2)}(z) = \int_0^\infty \mathcal{C}_{\mu^1 * \mu^2}(\varphi_\alpha(s) z)\, ds = \mathcal{C}_{\mu^1_\alpha * \mu^2_\alpha}(z),$$

which means that $\Phi_\alpha\left(\mu^1 * \mu^2\right) = \mu^1_\alpha * \mu^2_\alpha$. Hence $\mu^1_\alpha * \mu^2_\alpha \in \mathcal{R}(\Phi_\alpha)$.

We are interested in the injectivity of Φ_α. Let $\nu\left|_{\mathbb{R}^+ \setminus \{0\}}\right.$ and $\nu\left|_{\mathbb{R}^- \setminus \{0\}}\right.$ be the restrictions of ν to $\mathbb{R}^+ \setminus \{0\}$ and $\mathbb{R}^- \setminus \{0\}$, respectively. Defining $\nu^+ := \nu\left|_{\mathbb{R}^+ \setminus \{0\}}\right.$ and $\nu^-(dx) = \nu\left|_{\mathbb{R}^- \setminus \{0\}}\right.(-dx)$, one observes that from (3.6) the Lévy density of μ_α can be written as

$$f_{\alpha,\nu}(r) = \begin{cases} \frac{1}{r}\nu^+ \circledast \eta_\alpha(r, \infty), & r > 0; \\ \frac{1}{|r|}\nu^- \circledast \eta_\alpha(|r|, \infty), & r < 0, \end{cases} \tag{3.12}$$

where η_α is the measure on $\left(\mathbb{R}^+, \mathcal{B}\left(\mathbb{R}^+\right)\right)$ characterized by

$$\eta_\alpha((r, \infty)) := h_\alpha(r), \quad r \geq 0. \tag{3.13}$$

Since μ_α is entirely described by $(\gamma_\alpha, b_\alpha, f_{\alpha,\nu})$ (see part 1. of Theorem 3.4), if η_α has the c.p. w.r.t. ν^+ and ν^-, then $\Phi_\alpha(\mu) = \Phi_\alpha(\tilde{\mu})$ implies necessarily that $\mu = \tilde{\mu}$. Hence, in the following we study the c.p. for η_α.

We need the following sets:

$$\mathcal{H}_\alpha := \left\{\mu \in \mathcal{D}(\Phi_\alpha) : \eta_\alpha \text{ has the c.p. w.r.t. } \nu^+ \text{ and } \nu^-\right\}, \quad \text{for } -1 < \alpha < 0,$$

$$\mathcal{A}_1 := \left\{\mu \in ID(\mathbb{R}) : \nu(\mathbb{R}) < \infty \text{ and } \int_{\mathbb{R}} |x|^p \nu(dx) < \infty \text{ for some } p > 0\right\},$$

$$\mathcal{A}_2 := \left\{\mu \in ID(\mathbb{R}) : \int_{\mathbb{R}} |x|^a \mu(dx) < \infty \text{ for some } 2 < a\right\},$$

$$\mathcal{A}_\alpha := \left\{\mu \in ID(\mathbb{R}) : \int_{\mathbb{R}} |x|^p \vee |x|^q \nu(dx) < \infty \text{ for some } 0 < p < -\frac{1}{\alpha} \text{ and } p < q\right\}.$$

Above, ν denotes the Lévy measure of μ.

Theorem 3.7 *Let η_α be as in* (3.13). *Then*

1. *For any $-1 < \alpha < 0$, $\mathcal{A}_1 \subset \mathcal{H}_\alpha$. Moreover, $\mathcal{A}_2 \subset \mathcal{H}_\alpha$ for $-1/2 < \alpha < 0$ and $\mathcal{A}_\alpha \subset \mathcal{H}_\alpha$ for $-1 < \alpha \leq -1/2$.*
2. *The functional Φ_α is one-to-one on $\mathcal{A}_1 \cup \mathcal{A}_2$, for $-1/2 < \alpha < 0$ and on $\mathcal{A}_1 \cup \mathcal{A}_\alpha$, for $-1 < \alpha \leq -1/2$. In addition, we have that $S^\beta(\mathbb{R}) \subsetneq \mathcal{R}(\Phi_\alpha) \subsetneq SD(\mathbb{R})$ for any $\beta < -1/\alpha$, where $0 < \beta \leq 2$ and $-1 < \alpha < 0$, with*

$$\Phi_\alpha\left(S^\beta(\mathbb{R})\right) = S^\beta(\mathbb{R}). \tag{3.14}$$

Proof Firstly, let us observe that $\mathcal{A}_1 \subset \mathcal{D}(\Phi_\alpha)$ for any $-1 < \alpha < 0$ and $\mathcal{A}_2$ and $\mathcal{A}_\alpha$ belong to $\mathcal{D}(\Phi_\alpha)$ for $-1/2 < \alpha < 0$ and $-1 < \alpha \leq -1/2$, respectively. Furthermore, from Lemma 3.5, given $\alpha \in (-1,0)$, η_α as in (3.13) is a continuous probability measure with moments strictly smaller than $-1/\alpha$. Moreover, for any ν_1,ν_2 σ-finite measures on $(0,\infty)$, the relation $\nu_1 \circledast \eta_\alpha(r,\infty) = \nu_2 \circledast \eta_\alpha(r,\infty)$ for all $r > 0$ holds if and only if for any $g : \mathbb{R}^+ \to \mathbb{R}^+$ measurable function

$$\int_0^\infty \int_0^\infty g(xy)\,\eta_\alpha(dy)\,\nu_1(dx) = \int_0^\infty \int_0^\infty g(xy)\,\eta_\alpha(dy)\,\nu_2(dx). \tag{3.15}$$

For the rest of the proof, for any $\mu \in ID(\mathbb{R})$, ν will denote the Lévy measure of μ. Let $\mu \in \mathcal{A}_1$. Then, $\nu^+\left(\mathbb{R}^+\right) < \infty$ and there is $p > 0$, such that $\int_{\mathbb{R}^+} |x|^p \nu^+(dx) < \infty$. Suppose that there exists $\tilde{\nu}$ a σ-finite measure on $(0,\infty)$, such that

$$\tilde{\nu} \circledast \eta_\alpha(r,\infty) = \nu^+ \circledast \eta_\alpha(r,\infty), \quad r > 0. \tag{3.16}$$

As noticed above, for any $0 < q < -1/\alpha$, $0 < \int_0^\infty |y|^q \eta_\alpha(dy) < \infty$. Furthermore, applying (3.15) to $g(x) = x^\theta$ and $g \equiv 1$, we get that $\eta_\alpha\left(\mathbb{R}^+\right)\nu^+\left(\mathbb{R}^+\right) = \eta_\alpha\left(\mathbb{R}^+\right)\tilde{\nu}\left(\mathbb{R}^+\right)$ and for any $0 < \theta < p \wedge (-1/\alpha)$,

$$\left[\int_0^\infty |y|^\theta \eta_\alpha(dy)\right]\left[\int_0^\infty |x|^\theta \nu^+(dx)\right] = \left[\int_0^\infty |y|^\theta \eta_\alpha(dy)\right]\left[\int_0^\infty |x|^\theta \tilde{\nu}(dx)\right],$$

which implies that $\tilde{\nu}\left(\mathbb{R}^{+}\right)=\nu^{+}\left(\mathbb{R}^{+}\right)<\infty$ and for all $0<\theta<p\wedge(-1/\alpha)$

$$\int_{\mathbb{R}^{+}}|x|^{\theta}\,\nu^{+}(dx)=\int_{\mathbb{R}^{+}}|x|^{\theta}\,\tilde{\nu}(dx)<\infty. \tag{3.17}$$

Thus, $\tilde{\nu}$ and ν^{+} have the same mass, so without loss of generality we may assume that both are probability measures. Therefore, by uniqueness of the moment transform (see Section 4.25 in Hoffmann-Jørgensen [14]) $\nu^{+}\equiv\tilde{\nu}$, which means that the cancellation property w.r.t. ν^{+} holds. Similarly, it follows that η_{α} has the c.p. w.r.t. ν^{-}.

Now, assume that $-1/2<\alpha<0$ and consider $\mu\in\mathcal{A}_{2}$. Then, there exists $a>2$, such that $\int_{\mathbb{R}}|x|^{a}\,\mu(dx)<\infty$, which means $\int_{\mathbb{R}^{+}}|x|^{a}\,\nu^{+}(dx)<\infty$. Let $\tilde{\nu}$ be an arbitrary σ-finite measure on $(0,\infty)$ such that (3.16) holds. In this case, Eq. (3.17) remains true for $2\leq\theta<a\wedge(-1/\alpha)$. Hence, the measures

$$\nu_{2}^{+}(dx):=|x|^{2}\,\nu^{+}(dx);\quad\tilde{\nu}_{2}(dx):=|x|^{2}\,\tilde{\nu}(dx),$$

are finite with the same mass, so again we can assume that they are probability measures. Therefore

$$\int_{0}^{\infty}|x|^{\theta}\,\tilde{\nu}_{2}(dx)=\int_{0}^{\infty}|x|^{\theta}\,\nu_{2}^{+}(dx)<\infty,\quad\forall\,0\leq\theta<a\wedge(-1/\alpha)-2.$$

This implies necessarily that $\nu^{+}\equiv\tilde{\nu}$, the cancellation property. Again, it follows similarly that η_{α} has the c.p. w.r.t. ν^{-}.

Suppose now that $-1<\alpha\leq-1/2$. Analogously to before, we see that if $\mu\in\mathcal{A}_{\alpha}$, then there exists $0<p<-\frac{1}{\alpha}$ and $q>p$ such that for

$$\int_{0}^{\infty}|x|^{\theta}\,\tilde{\nu}_{p}(dx)=\int_{0}^{\infty}|x|^{\theta}\,\nu_{p}^{+}(dx)<\infty,\quad\forall\,0\leq\theta<q\wedge\left(-\frac{1}{\alpha}\right)-p,$$

where $\tilde{\nu}_{p}$ and ν_{p}^{+} are the finite measures given by

$$\nu_{p}^{+}(dx):=|x|^{p}\,\nu^{+}(dx);\quad\tilde{\nu}_{p}(dx):=|x|^{p}\,\tilde{\nu}(dx),$$

implying the cancellation property. Once again, ν^{-} is treated analogously.

As we mentioned before, the injectivity of Φ_{α} is implied by the cancellation property for ν^{+} and ν^{-}. By the reason that the Lévy density of μ_{α} is continuous, we get that $\mathcal{R}(\Phi_{\alpha})\subsetneq SD(\mathbb{R})$, so to finish the proof we only need to show that (3.14) holds. From Example 2.2 and (3.11), $S^{\beta}(\mathbb{R})\subset\mathcal{D}(\Phi_{\alpha})$ for $\beta<-1/\alpha$ or $\alpha>-1/2$. Take $\mu\in S^{\beta}(\mathbb{R})$, if $\beta=2$, the result is trivial so assume that $0<\beta<2$ and $\beta<-1/\alpha$ or $\alpha>-1/2$. Then $\nu^{+}(dx)=c^{+}|x|^{-\beta-1}dx$ with $c^{+}>0$. Since $f_{\alpha,\nu}$

can be written as

$$f_{\alpha,\nu}(r) = \frac{1}{r}\int_0^\infty \nu\left(ry^{-1},\infty\right)\eta_\alpha(dy)$$
$$= \frac{c^+}{\beta r}\left(\int_0^\infty y^\beta \eta_\alpha(dy)\right) r^{-\beta}, \quad r > 0.$$

we have that in this case the Lévy density of μ_α is

$$f_{\alpha,\nu}(r) = K_{\beta,\alpha} r^{-\beta-1}, \quad \text{for } r > 0,$$

with $K_{\beta,\alpha} = \frac{c^+}{\beta}\left(\int_0^\infty y^\beta \eta_\alpha(dy)\right)$. The case for ν^- is analogous, which means that $\Phi_\alpha\left(S^\beta(\mathbb{R})\right) \subset S^\beta(\mathbb{R})$. Conversely, given $\tilde{\mu} \in S^\beta(\mathbb{R})$, from the previous equation, there exists $\mu \in S^\beta(\mathbb{R})$ such that $\Phi_\alpha(\mu) = \tilde{\mu}$. □

4 Type G Distributions

In this section we will partially describe the stationary law of $(Y_t)_{t\in\mathbb{R}}$ when L is a Brownian motion and σ is non constant. We link such distribution to a class of Gaussian mixing (type G) laws, which depends on the behavior of the volatility process σ.

Let $\alpha \in (-1, 0)$. Assume that (2.15) holds with σ square integrable, then Y is strongly stationary and $Y_t \overset{d}{=} \int_0^\infty \varphi_{\alpha,\lambda}(s)\sigma_s dL_s$ for all $t \in \mathbb{R}$. However, in general such distribution is not infinitely divisible anymore, even when L and σ are independent. Indeed, consider σ to be a stationary and bounded, e.g. $\sigma_t = e^{-\xi_t}$ with ξ being an OU process driven by a subordinator, independent of L, and $L_t = t$. Then, by (2.8) and the boundedness of σ, there is a $c > 0$ such that almost surely

$$\int_0^\infty \varphi_{\alpha,\lambda}(s)\sigma_s dL_s \le c\int_0^\infty \left[s^\alpha \mathbf{1}_{\{0<s\le 1\}} + e^{-\lambda s/2}\mathbf{1}_{\{s>1\}}\right] ds.$$

Consequently, $\int_0^\infty \varphi_{\alpha,\lambda}(s)\sigma_s dL_s$ is bounded by a non-random constant, which means that it cannot be infinitely divisible.

Even in a simpler setting the infinite divisibility may fail: For $c > 1$, let $((\log c)N_t, L_t)_{t\ge 0}$ be a bivariate Lévy process where N and L are Poisson processes with parameters $a > 0$ and $b > 0$, respectively. Let $\overline{\nu}$ be the Lévy measure of such process and define $p := \frac{u}{u+v+w}$, $q := \frac{v}{u+v+w}$ and $r := \frac{w}{u+v+w}$, with u, v and w being the masses of $\overline{\nu}$ on the atoms $(1,0)$, $(0,1)$ and $(1,1)$, respectively. Put $\sigma_0 = 0$ and

$$\sigma_t = e^{-(\log c)N_{t-} + \lambda t - \log(t^\alpha)}, \quad t > 0.$$

Observe that in this case σ is not stationary and

$$\int_0^\infty \varphi_{\alpha,\lambda}(s)\,\sigma_s dL_s = \int_0^\infty c^{-N_{s-}} dL_s.$$

According to Proposition 5.1 in Lindner and Sato [16], for $0 < r < p$, the law of $\int_0^\infty \varphi_{\alpha,\lambda}(s)\,\sigma_s dL_s$ does not belong to $ID(\mathbb{R})$ if $r > pq$. Note that if $r > 0$, N and L are not independent. In contrast, when $r = 0$, such random variable is infinitely divisible.

Arising from the discussion above, in presence of stochastic volatility, the stationary distribution of Y cannot be characterized for a general Lévy process. Nevertheless, if we restrict ourselves to the Gaussian case, it is possible to relate the marginal distribution of Y to a subclass of infinitely divisible distributions called type G. Let X be a real-valued random variable. We say that X is *type G* if $X \stackrel{d}{=} V^{1/2}Z$, where Z is a standard normal random variable and V is a non negative infinitely divisible random variable independent of Z. In this case, if $X \sim \mu$ then $\mu \in ID(\mathbb{R})$ and if V has triplet $(\gamma, 0, \nu)$, then X has triplet $\gamma_X = 0$, $b_X = \gamma$ and

$$\nu_X(B) = \mathbb{E}\left[\nu\left(Z^{-1}B\right)\right], \quad B \in \mathcal{B}(\mathbb{R}\backslash\{0\}).$$

See Maejima and Rosiński [17] for more details.

Note that when L is a Brownian motion and σ is independent of L, the random variable $\int_0^\infty \varphi_{\alpha,\lambda}(s)\,\sigma_s dL_s$ conditioned on σ is normally distributed with mean zero and variance $\int_0^\infty \varphi^2_{\alpha,\lambda}(s)\,\sigma_s^2 ds$, thus Y as in (1.1) has marginal distribution of type G if $\int_0^\infty \varphi^2_{\alpha,\lambda}(s)\,\sigma_s^2 ds$ is infinitely divisible.

Before presenting sufficient conditions for this property, we introduce the concept of infinitely divisible processes as in Barndorff-Nielsen et al. [3]. Let $X = (X_t)_{t\in\mathbb{R}}$ be a stochastic processes. We say that X is *infinitely divisible* if for any $n \in \mathbb{N}$, there are $X^{1,n}, \ldots, X^{n,n}$ independent and identically distributed processes, such that

$$X \stackrel{d}{=} X^{1,n} + \cdots + X^{n,n},$$

where by $\stackrel{d}{=}$ we mean equality in law (finite-dimensional distributions). Note that if X is infinitely divisible, then for any $(t_1, \ldots, t_n), (a_1, \ldots, a_n) \in \mathbb{R}^n$ and $n \in \mathbb{N}$, the law of $\sum_{i=1}^{n} a_i X_{t_i}$ belongs to $ID(\mathbb{R})$. The reverse is not true in general. For an extensive discussion and properties for infinitely divisible processes see Barndorff-Nielsen et al. [3]. With this definition in mind we proceed to describe the law of Y in the presence of stochastic volatility.

Proposition 4.1 *Assume that σ is a strongly stationary such that σ^2 is an infinitely divisible process which is continuous in $\mathcal{L}^1(\Omega, \mathcal{F}, \mathbb{P})$ and L is a Brownian motion independent of σ. Then, for $\alpha > -1/2$ the distribution of $\int_0^\infty \varphi_{\alpha,\lambda}(s)\,\sigma_s dL_s$ is type G.*

Proof Firstly, observe that due to the stationarity of σ, the random variable $\int_0^\infty \varphi_{\alpha,\lambda}^2(s)\,\sigma_s^2 ds$ exists in $\mathcal{L}^1(\Omega, \mathcal{F}, \mathbb{P})$. Moreover, since the mapping $t \mapsto \sigma_t^2$ is continuous in $\mathcal{L}^1(\Omega, \mathcal{F}, \mathbb{P})$

$$\int_0^\infty \varphi_{\alpha,\lambda}^2(s)\,\sigma_s^2 ds = \mathcal{L}^1\text{-}\lim_{n\to\infty} \sum_{k\in\mathbb{N}} \sigma_{\frac{k}{2^n}}^2 \int_{\frac{k}{2^n}}^{\frac{k+1}{2^n}} \varphi_{\alpha,\lambda}^2(s)\,ds.$$

By hypothesis σ^2 is infinitely divisible, thus the series on the right-hand side of the previous equation is infinitely divisible, which means that the law of $\int_0^\infty \varphi_{\alpha,\lambda}^2(s)\,\sigma_s^2 ds$ is in $ID(\mathbb{R})$. This proves the result. □

References

1. O.E. Barndorff-Nielsen, F.E. Benth, J. Pedersen, A.E. Veraart, On stochastic integration for volatility modulated Lévy-driven volterra processes. Stoch. Process. Appl. **124**(1), 812–847 (2014)
2. O.E. Barndorff-Nielsen, F.E. Benth, A.E. Veraart, Modelling energy spot prices by volatility modulated Lévy-driven volterra processes. Bernoulli **19**(3), 803–845 (2013)
3. O.E. Barndorff-Nielsen, M. Maejima, K. Sato, Infinite divisibility for stochastic processes and time change. J. Theor. Probab. **19**(2), 411–446 (2006)
4. O.E. Barndorff-Nielsen, M. Maejima, K. Sato, Some classes of multivariate infinitely divisible distributions admitting stochastic integral representations. Bernoulli **12**(1), 1–33 (2006)
5. O.E. Barndorff-Nielsen, J. Rosiński, S. Thorbjørnsen, General Υ-transformations. ALEA Lat. Am. J. Probab. Math. Stat. **4**, 131–165 (2008)
6. O.E. Barndorff-Nielsen, J. Schmiegel, Lévy-based tempo-spatial modelling; with applications to turbulence. Uspekhi Mat. Nauk (159), 65–91 (2003)
7. O.E. Barndorff-Nielsen, J. Schmiegel, Ambit processes; with applications to turbulence and cancer growth, in *Stochastic Analysis and Applications: The Abel Symposium 2005*, Oslo (2007)
8. O.E. Barndorff-Nielsen, J. Schmiegel, Brownian semistationary processes and volatility/intermittency, in *Advanced Financial Modelling*, ed. by H. Albrecher, W. Rungaldier, W. Schachermeyer (de Gruyter, Berlin/New York, 2009), pp. 1–26
9. A. Basse-O'Connor, Some properties of a class of continuous time moving average processes, in *Proceedings of the 18th EYSM*, ed. by N. Suvak (2013)
10. A. Basse-O'Connor, S. Graversen, J. Pedersen, A unified approach to stochastic integration on the real line. Theory Probab. Appl. **58**, 355–380 (2013)
11. F.E. Benth, H. Eyjolfsson, A. Veraart, Approximating Lévy semistationary processes via Fourier methods in the context of power markets. SIAM J. Financ. Math. **5**(1), 71–98 (2014)
12. P.J. Brockwell, V. Ferrazzano, C. Klüoppelberg, High-frequency sampling and kernel estimation for continuous-time moving average processes. J. Time Ser. Anal. **34**(3), 385–404 (2013)
13. J.M. Corcuera, E. Hedevang, M.S. Pakkanen, M. Podolskij, Asymptotic theory for Brownian semi-stationary processes with application to turbulence. Stoch. Process. Appl. **123**(7), 2552–2574 (2013)
14. J. Hoffmann-Jørgensen, *Probability with a View Toward Statistics*. Chapman and Hall Probability Series, vol. I (Chapman & Hall, New York, 1994)
15. M. Jacobsen, T. Mikosch, J. Rosiński, G. Samorodnitsky, Inverse problems for regular variation of linear filters, a cancellation property for σ-finite measures and identification of stable laws. Ann. Appl. Probab. **19**, 210–242 (2009)

16. A. Lindner, K. Sato, Properties of stationary distributions of a sequence of generalized Ornstein-Uhlenbeck processes. Mathematische Nachrichten **284**(17–18), 2225–2248 (2011)
17. M. Maejima, J. Rosiński, Type G distributions on $\mathbb{R}^d$. J. Theor. Probab. **15**(2), 323–341 (2002)
18. J. Pedersen, K. Sato, The class of distributions of periodic Ornstein-Uhlenbeck processes driven by Lévy processes. J. Theor. Probab. **18**(1), 209–235 (2005)
19. B.S. Rajput, J. Rosiński, Spectral representations of infinitely divisible processes. Probab. Theory Relat. Fields **82**(3), 451–487 (1989)
20. K. Sato, *Lévy Processes and Infinitely Divisible Distributions*, 1st edn. (Cambridge University Press, Cambridge, 1999)
21. K. Sato, Additive processes and stochastic integrals. Ill. J. Math. **50**(1–4), 825–851 (2006)
22. K. Sato, Fractional integrals and extensions of selfdecomposability, in *Lévy Matters I*. Lecture Notes in Mathematics, ed. by O.E. Barndorff-Nielsen, J. Bertoin, J. Jacod, C. Klüppelberg (Springer, Berlin/Heidelberg, 2010), pp. 1–91
23. A.E. Veraart, L.A. Veraart, Modelling electricity day-ahead prices by multivariate Lévy semistationary processes, in *Quantitative Energy Finance*, ed. by F.E. Benth, V.A. Kholodnyi, P. Laurence (Springer, New York, 2014), pp. 157–188

Ambit Fields: Survey and New Challenges

Mark Podolskij

Abstract In this paper we present a survey on recent developments in the study of ambit fields and point out some open problems. Ambit fields is a class of spatio-temporal stochastic processes, which by its general structure constitutes a flexible model for dynamical structures in time and/or in space. We will review their basic probabilistic properties, main stochastic integration concepts and recent limit theory for high frequency statistics of ambit fields.

Keywords Ambit fields • High frequency data • Limit theorems • Numerical schemes • Stochastic integration

AMS 2010 Subject Classification: 60F05, 60G22, 60G57, 60H05, 60H07.

1 Introduction

The recent years have witnessed a strongly increasing interest in ambit stochastics. Ambit fields[1] is a class of spatio-temporal stochastic processes that has been originally introduced by Barndorff-Nielsen and Schmiegel in a series of papers [3–5] in the context of turbulence modelling, but which found manifold applications in mathematical finance and biology among other sciences; see e.g. [7, 10].

Ambit processes describe the dynamics in a stochastically developing field, for instance a turbulent wind field, along curves embedded in such a field. An illustration of the various concepts involved is demonstrated by Fig. 1. This shows a curve running with time through space. To each point on the curve is associated

[1]From Latin *ambitus*: a sphere of influence

M. Podolskij (✉)
Department of Mathematics, University of Aarhus, Ny Munkegade 118, 8000 Aarhus C, Denmark
e-mail: mpodolskij@creates.au.dk

R.H. Mena et al. (eds.), *XI Symposium on Probability and Stochastic Processes*, Progress in Probability 69, DOI 10.1007/978-3-319-13984-5_12

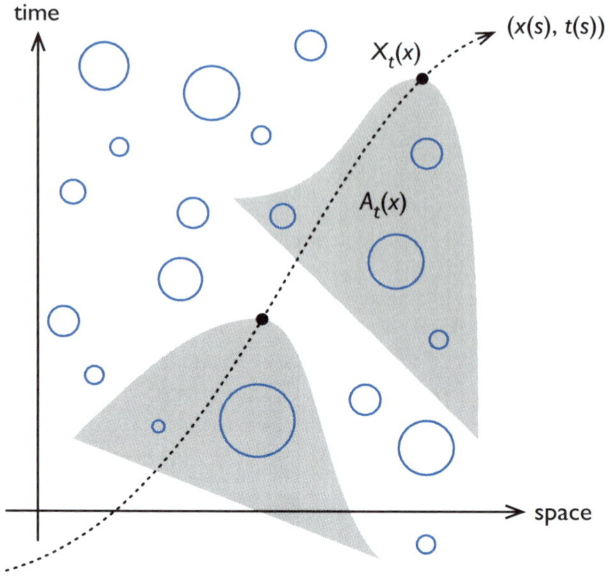

Fig. 1 A spatio-temporal ambit field. The value $X_t(x)$ of the field at the point (x, t) marked by the nearby *dot* is defined through an integral over the corresponding ambit set $A_t(x)$ marked by the *shaded region*. The *blue circles* of varying sizes indicate the stochastic volatility/intermittency. By considering the field along the path $(x(s), t(s))$ in space-time an ambit process is obtained

a random variable X_θ which should be thought of as the value of an observation at that point. This value is assumed to depend only on a region of previous space-time, indicated by the solid line in the figure, called the ambit set associated to the point in question. Further, a key characteristic of the modelling framework, which distinguishes ambit fields from other approaches is that beyond the most basic kind of random input it also specifically incorporates additional, often drastically changing, inputs referred to as *volatility* or *intermittency*. This feature is illustrated by the blue circles whose sizes indicate volatility effects of varying degree.

In terms of mathematical formulae an ambit field is specified via

$$X_t(x) = \mu + \int_{A_t(x)} g(t, s, x, \xi)\sigma_s(\xi)L(\mathrm{d}s, \mathrm{d}\xi) + \int_{D_t(x)} q(t, s, x, \xi)a_s(\xi)\mathrm{d}s\mathrm{d}\xi \qquad (1.1)$$

where t denotes time while x gives the position in space. Further, $A_t(x)$ and $D_t(x)$ are *ambit sets*, g and q are deterministic weight functions, σ represents the volatility or intermittency field, a is a drift field and L denotes a *Lévy basis*. We recall that a Lévy basis $L = \{L(B) : B \in \mathcal{S}\}$, where $\mathcal{S}$ is a δ-ring of an arbitrary non-empty set S such that there exists an increasing sequence of sets $(S_n) \subset \mathcal{S}$ with $\cup_{n\in\mathbb{N}} S_n = S$,

is an independently scattered random measure with Lévy-Khinchin representation

$$\log \mathbb{E}[\exp(iuL(B))] = iuv_1(B) - \frac{1}{2}u^2 v_2(B) + \int_{\mathbb{R}} \big(\exp(iuy) - 1 - iuy1_{[-1,1]}(y)\big)\, v_3(dy, B). \tag{1.2}$$

Here v_1 is a signed measure on $\mathcal{S}$, v_2 is a measure on $\mathcal{S}$ and $v_3(\cdot,\cdot)$ is a generalised Lévy measure on $\mathbb{R} \times \mathcal{S}$ (see e.g. [46] for details). In the turbulence framework the stochastic field $(X_t(x))_{t\geq 0, x\in\mathbb{R}^3}$ describes the velocity of a turbulent flow at time t and position x, while the ambit sets $A_t(x), D_t(x)$ are typically bounded.

Triggered by the wide applicability of ambit fields there is an increasing need in understanding their mathematical properties. The main mathematical topics of interest are: stochastic analysis for ambit fields, modelling of volatility/intermittency, asymptotic theory for ambit fields, numerical approximations of ambit fields, and asymptotic statistics. The aim of this paper is to give an overview of recent studies and to point out the main challenges in future. We will see that most of the mathematical problems already arise in a null spatial setting, and therefore parts of the presented theory will deal with this case only. However, possible extensions to spatial dimension will be discussed. At this stage we would like to mention another review on ambit fields [14], which focuses on integration theory, various examples of ambit fields and their applications in physics and finance. Numerical schemes for ambit processes, which will not be demonstrated in this work, are investigated in [23].

The paper is structured as follows. We introduce the main mathematical setting and discuss some basic probabilistic properties of ambit fields in Sect. 2. Section 3 gives a review on integration concepts with respect to ambit fields. Section 4 is devoted to limit theorems for high frequency statistics of ambit fields and their statistical applications.

2 The Model and Basic Probabilistic Properties

Throughout this paper all stochastic processes are defined on a given probability space $(\Omega, \mathcal{F}, \mathbb{P})$. We first consider a subclass of an ambit field defined via

$$X_t(x) = \mu + \int_{A+(t,x)} g(t-s, x-\xi)\sigma_s(\xi)L(\mathrm{d}s\mathrm{d}\xi) + \int_{D+(t,x)} q(t-s, x-\xi)a_s(\xi)\mathrm{d}s\mathrm{d}\xi, \tag{2.1}$$

where $A, D \subset \mathbb{R} \times \mathbb{R}^d$ are fixed ambit sets. This specific framework allows for modelling stationary random fields, namely when the field (σ, a) is stationary and independent of the driving Lévy basis L, X turns out to be stationary too. Another potentially important features of a random field, such as e.g. isotropy or skewness, can be easily modelled via an appropriate choice of the deterministic kernels g and q, and the stochastic field (σ, a).

Specializing even further, we introduce a *Lévy semi-stationary process* given by

$$Y_t = \mu + \int_{-\infty}^{t} g(t-s)\sigma_s L(\mathrm{d}s) + \int_{-\infty}^{t} q(t-s)a_s \mathrm{d}s, \tag{2.2}$$

where now L is a two-sided one dimensional Lévy motion and $A = D = (-\infty, 0)$, which is a purely temporal analogue of the ambit field at (2.1). It is indeed this special subclass of ambit fields, which will be the focus of our interest, as many mathematical problems have to be solved for Lévy semi-stationary processes first before transferring the principles to more general random fields. In the following we will check the validity and discuss some basic properties of ambit fields. Note that we sometimes use dL_s instead of $L(\mathrm{d}s)$ in the setting of (2.2). We also remark that the meaning of some notations may change from section to section.

2.1 *Definition of the Stochastic Integral*

The first natural question when introducing an ambit field is the definition of the first integral appearing in the decomposition (1.1). Here we recall two classical approaches of Rajput and Rosinski [46] and Walsh [50].

2.1.1 Rajput and Rosinski Theory

The first simplified mathematical problem is the definition of stochastic integral in case of a *deterministic* intermittency field σ. In this situation we may study a more general integral

$$\int_A f dL,$$

where L is a Lévy basis on a δ-ring $\mathcal{S}, f : (S, \sigma(\mathcal{S})) \to (\mathbb{R}, \mathcal{B}(\mathbb{R}))$ a measurable real valued function and $A \in \sigma(\mathcal{S})$, which is the main object of a seminal paper [46]. We will briefly recall the most important results of this work. By definition of a Lévy basis the law of $(L(A_1), \ldots, L(A_d))$, $A_i \in \mathcal{S}$, is infinitely divisible, the random variables $L(A_1), \ldots, L(A_d)$ are independent when the sets $A_1, \ldots, A_d$ are disjoint, and

$$L\left(\cup_{i=1}^{\infty} A_i\right) = \sum_{i=1}^{\infty} L(A_i) \qquad \mathbb{P} - \text{almost surely}$$

for disjoint A_i's with $\cup_{i=1}^{\infty} A_i \in \mathcal{S}$. Recalling the characteristic triplet (v_1, v_2, v_3) from the Lévy-Khinchin representation (1.2), the *control measure* λ of L is defined via

$$\lambda(A) := |v_1|(A) + v_2(A) + \int_{\mathbb{R}} \min(1, x^2) v_3(dx, A), \qquad A \in \mathcal{S},$$

where $|v_1|$ denotes the total variation measure associated with v_1. In this subsection we will use the truncation function

$$\tau(z) := \begin{cases} z : & \|z\| \leq 1 \\ z/\|z\| : & \|z\| > 1 \end{cases}$$

Now, for any *simple function*

$$f(x) = \sum_{i=1}^{d} a_i 1_{A_i}(x), \qquad a_i \in \mathbb{R}, A_i \in \mathcal{S},$$

the stochastic integral is defined as

$$\int_A f dL := \sum_{i=1}^{d} a_i L(A \cap A_i), \qquad A \in \sigma(\mathcal{S}).$$

The extension of this definition is as follows.

Definition 2.1 A measurable function $f : (\mathcal{S}, \sigma(\mathcal{S})) \to (\mathbb{R}, \mathcal{B}(\mathbb{R}))$ is called L-*integrable* if there exists a sequence of simple functions $(f_n)_{n \geq 1}$ such that

(i) $f_n \to f$ λ-almost surely.
(ii) For any $A \in \sigma(\mathcal{S})$ the sequence $(\int_A f_n dL)$ converges in probability.

In this case the stochastic integral is defined by

$$\int_A f dL := \mathbb{P} - \lim_{n \to \infty} \int_A f_n dL.$$

Although this definition is quite intuitive, it does not specify the class of L-integrable functions explicitly. The next theorem, which is one of the main results of [46], gives an explicit condition on the L-integrability of a function f.

Theorem 2.2 ([46, Theorem 2.7]) *Let $f : (S, \sigma(\mathcal{S})) \to (\mathbb{R}, \mathcal{B}(\mathbb{R}))$ be a measurable function. Then f is L-integrable if and only if the following conditions hold:*

$$\int_S U(f(s), s)\lambda(ds) < \infty, \quad \int_S f^2(s) v_2(s)\lambda(ds) < \infty, \quad \int_S V_0(f(s), s)\lambda(ds) < \infty,$$

where

$$U(u,s) := u v_1(s) + \int_{\mathbb{R}} (\tau(xu) - u\tau(x)) v_3(dx,s),$$

$$V_0(u,s) := \int_{\mathbb{R}} \min(1, |xu|^2) v_3(dx,s).$$

Furthermore, the real valued random variable $X = \int_S f dL$ *is infinitely divisible with Lévy-Khinchin representation*

$$\log \mathbb{E}[\exp(iuX)] = iu v_1(f) - \frac{1}{2} u^2 v_2(f) + \int_{\mathbb{R}} \left(\exp(iuy) - 1 - iuy 1_{[-1,1]}(y) \right) v_3^f(dy),$$

where

$$v_1(f) = \int_S U(f(s), s) \lambda(ds),$$

$$v_2(f) = \int_S f^2(s) v_2(s) \lambda(ds),$$

$$v_3^f(B) = v_3 \{(x,s) \in \mathbb{R} \times S : \ xf(s) \in B \setminus \{0\}\}, \qquad B \in \mathcal{B}(\mathbb{R}).$$

Remark 2.3 The most basic example of a null spatial ambit field is the Lévy moving average process given by

$$X_t = \int_{-\infty}^{t} g(t-s) dL_s, \tag{2.3}$$

where L is a two-sided Lévy motion with $\mathbb{E}L_t = 0$. In this situation the control measure λ is just the Lebesgue measure and the sufficient conditions from Theorem 2.2 translate to

$$\int_0^\infty g^2(s) ds < \infty, \qquad \int_0^\infty \left(\int_{\mathbb{R}} \min(|xg(s)|, |xg(s)|^2) v_3(dx) \right) ds < \infty,$$

where v_3 is the Lévy measure of L. □

Theorem 2.2 gives a precise condition for existence of the stochastic integral. However, it does not say anything about the existence of moments. To study this question let us assume for the rest of this subsection that

$$\mathbb{E}[|L(A)|^q] < \infty \qquad \forall A \in \mathcal{S},$$

for some $q > 0$. Clearly, in general we can not expect the existence of moments higher than q for the integral $\int_S f dL$, but we may study the existence of pth moment with $p \le q$. For this purpose we introduce the function $\Phi_p : \mathbb{R} \times S \to \mathbb{R}$ via

$$\Phi_p(u, s) := U^\star(u, s) + u^2 v_2(s) + V_p(u, s),$$

where

$$U^\star(u, s) := \sup_{c \in [-1,1]} |U(cu, s)|,$$

$$V_p(u, s) := \int_{\mathbb{R}} \left(|ux|^p 1_{\{|ux|>1\}} + |ux|^2 1_{\{|ux| \le 1\}} \right) v_3(dx, s),$$

and the function U is introduced in Theorem 2.2. Then

$$\|f\|_{\Phi_p} := \inf \left\{ c > 0 : \int_S \Phi_p(c^{-1} |f(s)|, s) \lambda(ds) \right\}$$

defines a norm and the vector space

$$\mathbb{L}^{\Phi_p} := \left\{ f : \int_S \Phi_p(|f(s)|, s) \lambda(ds) < \infty \right\}$$

equipped with $\|\cdot\|_{\Phi_p}$ is the so called *Musielak-Orlicz space* (see [46] for details). The next result gives a connection between the existence of pth norm of the stochastic integral and the finiteness of $\|f\|_{\Phi_p}$.

Theorem 2.4 ([46, Theorem 3.3]) *Let* $p \in [0, q]$. *Then it holds that*

$$\mathbb{E}\left[\left| \int_S f dL \right|^p \right] < \infty \iff \|f\|_{\Phi_p} < \infty.$$

Remark 2.5 Let X be a Lévy moving average process as defined at (2.3), where the driving Lévy motion has characteristic triplet $(0, 0, \text{const.}|x|^{-1-\beta} dx)$, i.e. L is a symmetric β-stable process ($\beta \in (0, 2)$) without drift. In this case the pth absolute moment of X_t exists whenever $g \in L^\beta(\mathbb{R}_{\ge 0})$ and $p < \beta$. Moreover it holds that

$$C_1 \mathbb{E}\left[\left| \int_0^\infty |g(x)|^\beta dx \right| \right]^{1/\beta} \le \mathbb{E}[|X_t|^p]^{1/p} \le C_2 \mathbb{E}\left[\left| \int_0^\infty |g(x)|^\beta dx \right| \right]^{1/\beta}$$

for some positive constants C_1, C_2. □

2.1.2 Walsh Approach

The integration concept proposed by Walsh [50] is, in some sense, an extension of Itô's theory to spatio-temporal setting. As in the classical Itô calculus the martingale theory and isometry play an essential role.

Here we review the main ideas of Walsh approach. In the following we will concentrate on integration with respect to a Lévy basis on *bounded* domains (the theory can be extended to unbounded domains in a straightforward manner, see [50]). Let $\mathcal{B}_b(\mathbb{R}^d)$ the Borel σ-field generated by bounded Borel sets on $\mathbb{R}^d$ and L a Lévy basis on $[0,T] \times S$ with $S \in \mathcal{B}_b(\mathbb{R}^d)$. For $S \supseteq A \in \mathcal{B}_b(\mathbb{R}^d)$ we introduce the notation

$$L_t(A) := L((0,t] \times A).$$

Unlike in the approach of Rajput and Rosinski, which does not require any moment assumptions (which however deals only with deterministic integrands), the basic assumption of Walsh concept is:

$$\text{For any } A \in \mathcal{B}_b(\mathbb{R}^d) \text{ it holds that } \mathbb{E}[L_t(A)] = 0 \text{ and } L_t(A) \in L^2(\Omega, \mathcal{F}, \mathbb{P}).$$

Now, we introduce a right continuous filtration $(\mathcal{F}_t)_{t \geq 0}$ via

$$\mathcal{F}_t := \cap_{\epsilon>0} \mathcal{F}^0_{t+\epsilon} \quad \text{with} \quad F^0_t = \sigma\{L_s(A) : \ S \supseteq A \in \mathcal{B}_b(\mathbb{R}^d), 0 < s \leq t\} \vee \mathcal{N},$$

where $\mathcal{N}$ denotes the $\mathbb{P}$-null sets of $\mathcal{F}$. Since L is a Lévy basis, the stochastic field $(L_t(A))_{t \geq 0, S \supseteq A \in \mathcal{B}_b(\mathbb{R}^d)}$ is an *orthogonal martingale measure* with respect to $(\mathcal{F}_t)_{t \geq 0}$ in the sense of Walsh, i.e. $(L_t(A))_{t \geq 0}$ is a square integrable martingale with respect to the filtration $(\mathcal{F}_t)_{t \geq 0}$ for any $A \in \mathcal{B}_b(\mathbb{R}^d)$ and the random variables $L_t(A), L_t(B)$ are independent for disjoint sets $A, B \in \mathcal{B}_b(\mathbb{R}^d)$.

In the next step we introduce the *covariance measure* Q via

$$Q([0,t] \times A) := \langle L(A) \rangle_t, \qquad A \in \mathcal{B}_b(\mathbb{R}^d),$$

and the associated L^2-norm

$$\|\psi\|_Q^2 := \mathbb{E}\left[\int_{[0,T]\times S} \psi^2(t,\xi) Q(dt, d\xi)\right]$$

for a random field ψ on $[0,T] \times S$. Now, we are following the Itô's program: We call ψ an *elementary random field* if it has the form

$$\psi(\omega, s, \xi) = X(\omega) 1_{(a,b]}(s) 1_A(\xi),$$

where X is a bounded $\mathcal{F}_a$-measurable random variable and $A \in \mathcal{B}_b(\mathbb{R}^d)$. For such an elementary random field the integral with respect to a Lévy basis is defined by ($t \leq T$)

$$\int_0^t \int_B \psi(s,\xi)L(ds,d\xi) := X(L_{t\wedge b}(A\cap B) - L_{t\wedge a}(A\cap B)), \qquad B \in \mathcal{B}_b(\mathbb{R}^d),\ A, B \subseteq S.$$

By linearity this definition can be extended to the linear span given by elementary random fields. The σ-field $\mathcal{P}$ generated by elementary random fields is called *predictable*, and the space $\mathcal{P}_L = L^2(\Omega \times [0,T] \times S, \mathcal{P}, Q)$ equipped with $\|\cdot\|_Q$ is a Hilbert space. Furthermore, the space of elementary functions is dense in $\mathcal{P}_L$. Thus, for any random field $\psi \in \mathcal{P}_L$, we may find a sequence of elementary random fields ψ_n with $\|\psi_n - \psi\|_Q \to 0$ and

$$\int_0^t \int_B \psi(t,\xi)L(dt,d\xi) := \lim_{n\to\infty} \int_0^t \int_B \psi_n(t,\xi)L(dt,d\xi) \qquad \text{in } L^2(\Omega, \mathcal{F}, \mathbb{P}).$$

Moreover, the Itô type isometry

$$\mathbb{E}\left[\left|\int_0^T \int_S \psi(s,\xi)L(ds,d\xi)\right|^2\right] = \|\psi\|_Q^2$$

is satisfied by construction.

Remark 2.6 The Walsh approach can be used to define ambit fields in (1.1) only when the driving Lévy basis is square integrable, which excludes e.g. β-stable processes ($\beta \in (0,2)$). The recent work of [26] combines the ideas of Walsh [50] and Rajput and Rosinski [46] to propose an integration concept for random integrands and general Lévy bases. It relies on an earlier work [24] by Bichteler and Jacod. A general reference for comparison of various integration concepts is [30].

□

2.2 *Is an Ambit Field a Semimartingale?*

Semimartingales is nowadays a well studied object in probability theory. From theoretical perspective it is important to understand whether a given null spatial subclass of ambit fields is a semimartingale or not, since in this case one may better study its fine structure properties. Furthermore, limit theory for high frequency statistics, which will be the focus of our discussion in Sect. 4, has been investigated in great generality in the framework of semimartingales; we refer to [35] for a comprehensive study. Thus, new asymptotic results are only needed for ambit fields, which do not belong to the class of semimartingales.

For simplicity of exposition we concentrate on the study of *Volterra type* processes

$$X_t = \int_{-\infty}^{t} g(t,s) dL_s, \tag{2.4}$$

where L is a Lévy motion with $L_0 = 0$, which is obviously a purely temporal subclass of ambit fields. The semimartingale property can be studied with respect to the filtration generated by the driving Lévy motion L or with respect to the natural filtration (which is a harder task). For any $t \geq 0$, we define

$$\mathcal{F}_t^L := \sigma(L_s : s \leq t), \qquad \mathcal{F}_t^X := \sigma(X_s : s \leq t).$$

In the following we will review the main studies of the semimartingale property.

2.2.1 Semimartingale Property with Respect to $(\mathcal{F}_t^L)_{t\geq 0}$

Semimartingale property of various subclasses of Volterra type processes $(X_t)_{t\geq 0}$ have been studied in the literature, see e.g. [17, 21, 38] among many others. However, in this subsection we closely follow the recent results of [18, Section 4]. We note that the original work [18] contains a study of more general processes, which are specialized in this paper to models of the form (2.4).

Let L be a Lévy process with characteristic triplet (a, b^2, ν). In the following we consider *stationary increments moving average* model of the type

$$X_t = \int_{\mathbb{R}} [g_1(t-s) - g_0(-s)] dL_s, \tag{2.5}$$

where g_0, g_1 are measurable functions satisfying $g_0(x) = g_1(x) = 0$ for $x < 0$. This obviously constitutes a subclass of (2.4). Integrability conditions can be directly extracted from Theorem 2.2. Notice that by construction the process $(X_t)_{t\geq 0}$ has stationary increments, which explains the aforementioned notion. The main result of this subsection is [18, Theorem 4.2].

Theorem 2.7 *Assume that the process $(X_t)_{t\geq 0}$ is defined as in (2.5) and the following conditions are satisfied:*

$$\int_0^\infty |g_1'(s)|^2 ds < \infty \qquad (\textit{when } b^2 > 0),$$

$$\int_0^\infty \int_{\mathbb{R}} \min(|x g_1'(s)|, |x g_1'(s)|^2) \nu(dx) ds < \infty.$$

Then $(X_t)_{t\geq 0}$ is a semimartingale with respect to the filtration $(\mathcal{F}_t^L)_{t\geq 0}$.

Theorem 2.7 gives sufficient conditions for the semimartingale property of the process $(X_t)_{t\geq 0}$. In certain special cases these conditions are also necessary as the following result from [18, Corollary 4.8] shows.

Theorem 2.8 *Assume that the process $(X_t)_{t\geq 0}$ is defined as in (2.5), L has infinite variation on compact sets and L_1 is either square integrable or has a regular varying distribution at ∞ with index $\beta \in [-2,-1)$. Then $(X_t)_{t\geq 0}$ is a semimartingale with respect to the filtration $(\mathcal{F}_t^L)_{t\geq 0}$ if and only if the following conditions are satisfied:*

$$\int_0^\infty |g_1'(s)|^2 ds < \infty \qquad (when\ b^2 > 0),$$

$$\int_0^\infty \int_{\mathbb{R}} \min(|xg_1'(s)|, |xg_1'(s)|^2)\nu(dx)ds < \infty.$$

In this case it has the decomposition

$$X_t = X_0 + g_1(0)L_t + \int_0^t \left(\int_{\mathbb{R}} g_1'(s-u)dL_u \right) ds.$$

When the driving process L is a symmetric β-stable Lévy motion with $\beta \in (1,2)$, i.e. the characteristic triplet is given via $(0, 0, \text{const.}|x|^{-1-\beta}dx)$, the condition of Theorem 2.8 translates to

$$\int_0^\infty |g_1'(s)|^\beta ds < \infty.$$

Remark 2.9 Another important subclass of stationary increments moving average models, which will be the object of investigation in Sect. 4, is a *fractional Lévy motion*. A fractional Lévy motion is defined as

$$X_t = \int_{-\infty}^t [(t-s)_+^\alpha - (-s)_+^\alpha] dL_s. \tag{2.6}$$

When L is a Brownian motion, this is one of the representation of a fractional Brownian motion with Hurst parameter $H = \alpha + 1/2$ (with $\alpha \in (-1/2, 1/2)$). Assume now that $\alpha > 0$. Then X is a semimartingale with respect to the filtration $(\mathcal{F}_t^L)_{t\geq 0}$ if and only if $b^2 = 0, \alpha \in (0, 1/2)$ and

$$\int_{\mathbb{R}} |x|^{1/(1-\alpha)} \nu(dx) < \infty.$$

We refer to [18, Proposition 4.6] for details. □

Remark 2.10 Sufficient conditions for the semimartingale property for general Lévy semi-stationary processes defined at (2.2) can be deduced from a seminal work [45] on Volterra type equations. □

In physical applications it may appear that an ambit field is observed along a curve in time-space. Let us consider for simplicity an ambit field of the form

$$X_t(x) = \int_{A_t(x)} g(t,s,x,\xi)L(\mathrm{d}s,\mathrm{d}\xi),$$

where L is a Lévy basis on $\mathbb{R}_{\geq 0} \times \mathbb{R}^d$. Let $\theta = (\theta_1,\theta_2) : [0,T] \to \mathbb{R}_{\geq 0} \times \mathbb{R}^d$ be a curve in time-space. Then the observed process is given by

$$Y_t = X_{\theta_1(t)}(\theta_2(t)).$$

The semimartingale property of the process Y is still an open problem.

2.2.2 Semimartingale Property with Respect to $(\mathcal{F}_t^X)_{t\geq 0}$

Showing the semimartingale property with respect to the natural filtration $(\mathcal{F}_t^X)_{t\geq 0}$ is by far a more delicate issue. In this subsection we restrict our attention to Volterra processes of the form

$$\int_{\mathbb{R}} [g_1(t-s) - g_0(-s)]dW_s.$$

where W is a Brownian motion, since, to the best of our knowledge, little is known for general driving Lévy processes. Here g_0, g_1 are measurable functions satisfying the integrability condition $s \mapsto g_1(t-s) - g_0(-s) \in L^2(\mathbb{R})$ for any $t \in \mathbb{R}$ (this insures the existence of the integral). In the case $g_0 = 0$ the paper [37] provides necessary and sufficient conditions for the semimartingale property with respect to the natural filtration $(\mathcal{F}_t^X)_{t\geq 0}$. The work [16] extends the results to include the model defined above. The methodology of proofs is based upon Fourier transforms and Hardy functions.

We need to introduce some notation. For a function $h : \mathbb{R} \to \mathbb{R}$, $\hat{h} : \mathbb{R} \to \mathbb{C}$ denotes its Fourier transform, i.e.

$$\hat{h}(t) := \int_{\mathbb{R}} \exp(itx)h(x)dx.$$

For a function $f : \mathbb{R} \to S^1$, where S^1 denotes the unit circle on a complex plane, satisfying $\overline{f}(\cdot) = f(-\cdot)$ we define the function $\tilde{f} : \mathbb{R} \to \mathbb{R}$ via

$$\tilde{f}(t) := \lim_{a\to\infty} \int_{-a}^{a} \frac{\exp(its) - 1_{[-1,1]}(s)}{is} f(s)ds.$$

The main result of this subsection is [16, Theorem 3.2].

Theorem 2.11 *The Gaussian process $(X_t)_{t\geq 0}$ is a semimartingale with respect to the natural filtration $(\mathcal{F}_t^X)_{t\geq 0}$ if and only if the following conditions are satisfied:*

(i) The function g_1 has the representation

$$g_1(t) = b + a\tilde{f}(t) + \int_0^t \widehat{f\hat{h}}(s)ds \qquad \text{for Lebesgue almost all } t \in \mathbb{R},$$

where $a, b \in \mathbb{R}$, $f : \mathbb{R} \to S^1$ is a measurable function with $\overline{f}(\cdot) = f(-\cdot)$, and $h \in L^2(\mathbb{R})$ is 0 whenever $a \neq 0$.

(ii) Set $\zeta = \widehat{f(\widehat{g_1 - g_0})}$. When $a \neq 0$ then

$$\int_0^r \left(\frac{|\zeta(s)|}{\sqrt{\int_s^\infty \zeta^2(u)du}} \right) ds < \infty \qquad \forall r > 0,$$

where $0/0 := 0$.

The second part of [16, Theorem 3.2] gives the canonical decomposition of $(X_t)_{t\geq 0}$ in case, where the above conditions (i) and (ii) are satisfied. If $X_0 = 0$ one may choose a, b, h and f such that the canonical decomposition $X_t = M_t + A_t$ is given via

$$M_t = a \int_{\mathbb{R}} (\tilde{f}(t-s) - \tilde{f}(-s))dW_s, \qquad A_t = \int_0^t \left(\int_{\mathbb{R}} \widehat{f\hat{h}}(s-u)dW_u \right) ds.$$

In this decomposition the martingale part M is a Wiener process with scaling parameter $(2\pi a)^2$. Finally, let us remark that in the special case $g_1 = g_0$, as e.g. in the fractional Brownian motion setting, the condition (ii) of Theorem 2.11 is trivially satisfied.

2.3 Fine Properties of Lévy Semi-stationary Processes

In Sect. 4 we will study the limit theory for *power variation* of Lévy semi-stationary processes

$$\sum_{i=1}^{[t/\Delta_n]} |Y_{i\Delta_n} - Y_{(i-1)\Delta_n}|^p \qquad \text{with } \Delta_n \to 0,$$

where the process Y is defined at (2.2). A first step towards understanding the first and second order asymptotics for this class of statistics is to determine the fine scale behaviour of Lévy semi-stationary processes. For simplicity of exposition we start our discussion with Lévy moving average processes defined at (2.3), i.e.

$$X_t = \int_{-\infty}^{t} g(t-s)dL_s,$$

which constitute a subclass of Lévy semi-stationary processes with constant intermittency and zero drift. We restrict ourselves to symmetric β-stable Lévy processes L with $\beta \in (0,2]$ and zero drift (thus, the Brownian motion is included).

The most interesting class of weight functions g in physical applications is given via

$$g(x) = x^{\alpha} f(x) \qquad \text{for } x > 0, \tag{2.7}$$

and $g(x) = 0$ for $x \le 0$, where $f : \mathbb{R}_{\ge 0} \to \mathbb{R}$ is a smooth function with $f(0) \neq 0$ decaying fast enough at infinity to ensure the existence of the integral (see Remark 2.3). When $\beta \in (0,2)$, i.e. L is a pure jump process, we further always assume that $\alpha > 0$, since $\alpha < 0$ leads to explosive behaviour of X near jump times of L. Hence, for any $\beta \in (0,2]$, the Lévy moving average process X is continuous since $g(0) = 0$ when $\beta < 2$. A formal differentiation leads to the identity

$$dX_t = g(0+)dL_t + \left(\int_{-\infty}^{t} g'(t-u)dL_u \right) dt.$$

According to Theorem 2.8 this identity indeed makes sense when (a) $\beta = 2$, $g(0+) < \infty$ and $g' \in L^2(\mathbb{R}_{\ge 0})$, or (b) $\beta \in (0,2)$ and

$$\int_0^{\infty} |g'(x)|^{\beta} dx < \infty.$$

In this case the Lévy moving average process X is an Itô semimartingale and the law of large numbers for its power variation is well understood (see e.g. the monograph [35]). In Sect. 4 we will be specifically interested in situations where X is not a semimartingale. It is particularly the case under the conditions

$$\beta = 2 \quad \text{and} \quad \alpha \in (-1/2, 1/2) \quad \text{or}$$
$$\beta \in (0,2) \quad \text{and} \quad \alpha \in (0, 1-1/\beta),$$

since the above integrability condition for the function g' is not satisfied near 0. Under these conditions the derivative of the function g explodes at 0. For a small $\Delta > 0$, we intuitively deduce the following approximation for the increments of X:

$$\begin{aligned} X_{t+\Delta} - X_t &= \int_{\mathbb{R}} [g(t+\Delta-s) - g(t-s)]dL_s \\ &\approx \int_{t+\Delta-\epsilon}^{t+\Delta} [g(t+\Delta-s) - g(t-s)]dL_s \end{aligned}$$

$$\approx f(0) \int_{t+\Delta-\epsilon}^{t+\Delta} [(t+\Delta-s)_+^\alpha - (t-s)_+^\alpha] dL_s$$

$$\approx f(0) \int_{\mathbb{R}} [(t+\Delta-s)_+^\alpha - (t-s)_+^\alpha] dL_s = \tilde{X}_{t+\Delta} - \tilde{X}_t,$$

where

$$\tilde{X}_t = f(0) \int_{\mathbb{R}} [(t-s)_+^\alpha - (-s)_+^\alpha] dL_s, \tag{2.8}$$

and $\epsilon > 0$ is an arbitrary small real number with $\epsilon \gg \Delta$. The formal proof of this first order approximation relies on the fact that the weight $g(t+\Delta-s)-g(t-s)$ attains asymptotically highest values when $s \approx t$, since g' explodes at 0. The stochastic process $\tilde{X}$ is called a *fractional β-stable Lévy motion* and its properties have been studied in several papers, see e.g. [20, 39] among others (for $\beta = 2$ it is just an ordinary scaled fractional Brownian motion with Hurst parameter $H = \alpha + 1/2$). In particular, $\tilde{X}$ has stationary increments, it is $(\alpha + 1/\beta)$-self similar with symmetric β-stable marginals.

The key fact to learn from this approximation is that, under above assumptions on g, the fine structure of a Lévy moving average process X with symmetric β-stable driver L is similar to the fine structure of a fractional β-stable Lévy motion $\tilde{X}$. Thus, under certain conditions, one may transfer the asymptotic theory for power variation of $\tilde{X}$ to the corresponding results for power variation of X. The limit theory for power variation of $\tilde{X}$ is sometimes easier to handle than the original statistic due to stationarity of increments of $\tilde{X}$ and their self similarity property, which allows to transform the original triangular observation scheme into a usual one when studying distributional properties (for the latter, ergodic limit theory might apply). Indeed, it is one method of proofs of laws of large numbers presented in Theorem 4.7(ii) of Sect. 4.

Remark 2.12 As mentioned above the fractional Lévy motion defined at (2.8) is, up to a scaling factor, a fractional Brownian motion B^H with Hurst parameter $H = \alpha + 1/2 \in (0, 1)$ when $\beta = 2$. Due to self similarity property of B^H it is sufficient to study the asymptotic behaviour of the statistic

$$\sum_{i=1}^{[t/\Delta_n]} |B_i^H - B_{i-1}^H|^p, \qquad p > 0,$$

to investigate the limit theory for power variation of B^H. Below we review some classical results of [25, 49]. First of all, we obtain the convergence

$$\Delta_n \sum_{i=1}^{[t/\Delta_n]} |B_i^H - B_{i-1}^H|^p \stackrel{\text{u.c.p.}}{\Longrightarrow} m_p t, \qquad m_p := \mathbb{E}[|\mathcal{N}(0,1)|^p], \tag{2.9}$$

where $Z^n \overset{\text{u.c.p.}}{\Longrightarrow} Z$ stands for uniform convergence in probability on compact intervals, i.e. $\sup_{t\in[0,T]} |Z_t^n - Z_t| \overset{\mathbb{P}}{\longrightarrow} 0$. The associated weak limit theory depends on the correlation kernel of the fractional Brownian noise and the *Hermite rank* of the function $h(x) = |x|^p - m_p$. Recall that the correlation kernel of the fractional Brownian noise is given via

$$\rho(j) := \text{corr}(B_1^H - B_0^H, B_{j+1}^H - B_j^H) = \frac{1}{2}\Big(|j+1|^{2H} - 2|j|^{2H} + |j-1|^{2H}\Big).$$

The Hermite expansion of the function h is defined as

$$h(x) = |x|^p - m_p = \sum_{l=2}^{\infty} a_l H_l(x),$$

where $(H_l)_{l\geq 0}$ are Hermite polynomials, i.e.

$$H_0 = 1 \quad \text{and} \quad H_l = (-1)^l \exp(x^2/2) \frac{d}{dx^l}\{-\exp(x^2/2)\} \quad \text{for } l \geq 1.$$

The Hermite rank of h is the smallest index l with $a_l \neq 0$, which is 2 in our case. The condition for the validity of a central limit theorem associated with (2.9) is then

$$\sum_{j=1}^{\infty} \rho^2(j) < \infty,$$

where the power 2 indicates the Hermite rank of h. This condition holds if and only if $H \in (0, 3/4)$. For $H > 3/4$ the limiting process is non-central. More precisely, the following functional limit theorems hold:

$$0 < H < 3/4: \qquad \Delta_n^{-1/2}\left(\Delta_n \sum_{i=1}^{[t/\Delta_n]} |B_i^H - B_{i-1}^H|^p - m_p t\right) \overset{d}{\longrightarrow} v_p W_t',$$

$$H = 3/4: \qquad (\Delta_n \log \Delta_n^{-1})^{-1/2}\left(\Delta_n \sum_{i=1}^{[t/\Delta_n]} |B_i^H - B_{i-1}^H|^p - m_p t\right) \overset{d}{\longrightarrow} \tilde{v}_p W_t',$$

$$3/4 < H < 1: \qquad \Delta_n^{2H-2}\left(\Delta_n \sum_{i=1}^{[t/\Delta_n]} |B_i^H - B_{i-1}^H|^p - m_p t\right) \overset{d}{\longrightarrow} Z_t,$$

where the weak convergence takes place on $\mathbb{D}([0,T])$ equipped with the uniform topology, W' denotes a Brownian motion and Z is a Rosenblatt process (see e.g.

[49]). Finally, the constants v_p and $\tilde{v}_p$ are given by

$$v_p := \sum_{l=2}^{\infty} l! a_l^2 \Big(1 + 2\sum_{j=1}^{\infty} \rho^l(j)\Big),$$

$$\tilde{v}_p := 4a_2 \lim_{n\to\infty} \frac{1}{\log n} \sum_{j=1}^{n-1} \frac{n-k}{n} \rho^2(j).$$

□

Remark 2.13 When $\beta \in (0, 2)$ and $\alpha \in (0, 1 - 1/\beta)$ we deduce via the $(\alpha + 1/\beta)$-self similarity property and the strong ergodicity of the fractional β-stable Lévy process $\tilde{X}$

$$\Delta_n^{1-p(\alpha+1/\beta)} \sum_{i=1}^{[t/\Delta_n]} |\tilde{X}_{i\Delta_n} - \tilde{X}_{(i-1)\Delta_n}|^p \overset{\text{u.c.p.}}{\Longrightarrow} c_p t, \qquad c_p := \mathbb{E}[|\tilde{X}_1 - \tilde{X}_0|^p]$$

whenever $p < \beta$. For $p > \beta$ the constant c_p is infinite and non-ergodic limits appear, see Theorem 4.7 in Sect. 4. □

Remark 2.14 Once we have proved the law of large numbers for power variation of the basic process $\tilde{X}$ as in Remarks 2.12 and 2.13 (or for the Lévy moving average process X defined at (2.3)), the main principles of the proof usually transfer to the integral process

$$I_t = \int_0^t \sigma_s d\tilde{X}_t,$$

whenever the latter is well defined. As it was shown in e.g. [8, 27] the Bernstein's blocking technique can be applied to deduce the law of large numbers for power variation of the process I. For instance, when $\tilde{X} = B^H$ is a fractional Brownian motion with Hurst parameter H, it holds that

$$\Delta_n^{1-pH} \sum_{i=1}^{[t/\Delta_n]} |I_{i\Delta_n} - I_{(i-1)\Delta_n}|^p \overset{\text{u.c.p.}}{\Longrightarrow} m_p \int_0^t |\sigma_s|^p ds,$$

where σ is a stochastic process, which has finite q-variation with $q < 1/(1-H)$ (see [27]). Quite often the same asymptotic result holds also for a subclass of ambit processes given by

$$Y_t = \int_{-\infty}^t g(t-s)\sigma_s dW_s,$$

where W is a Brownian motion (cf. Sect. 4). The reason is again a similar fine structure of the processes I and Y (see e.g. [28, Section 2.2.3] for a detailed exposition).

Transferring a central limit theorem for the power variation of the driver B^H to the integral process I (and also in case of the process Y) is a more delicate issue. Apart from further assumptions on the integrand σ a more technical and precise treatment of the Bernstein's blocking technique is required. We refer to a recent work [29] for a detailed description of such a method, which relies on fractional calculus. □

3 Integration with Respect to Ambit Fields

In this section we will discuss the integration concepts with respect to ambit processes of the type

$$X_t = \int_0^t g(t,s)\sigma_s dL_s,$$

where σ is a stochastic intermittency process and L is a Lévy motion. Our presentation is mainly based upon the recent work [12], where Malliavin calculus is applied to define the stochastic integral. The introduction of the integral

$$\int_0^t Z_s dX_s, \tag{3.1}$$

for a stochastic integrand Z, strongly depends on whether the driving process L is a Brownian motion or a pure jump Lévy motion, since the main notions of Malliavin calculus differ in those two cases. An alternative way of defining the stochastic integral at (3.1) without imposing L^2-structure of the integrand Z is proposed in [13]. The authors apply white noise analysis to construct the integral in the situation, where X is driven by a Brownian motion.

3.1 Integration with Respect to Ambit Processes Driven by Brownian Motion

Before we present the definition of the integral at (3.1) for $L = W$, we start by introducing the main notions of Malliavin calculus on Gaussian spaces. The reader is referred to the monograph [40] for any unexplained definition or result.

Let $\mathbb{H}$ be a real separable Hilbert space. We denote by $B = \{B(h) : h \in \mathbb{H}\}$ an *isonormal Gaussian process* over $\mathbb{H}$, i.e. B is a centered Gaussian family indexed by the elements of $\mathbb{H}$ and such that, for every $h_1, h_2 \in \mathbb{H}$,

$$\mathbb{E}\big[B(h_1)B(h_2)\big] = \langle h_1, h_2 \rangle_{\mathbb{H}}. \tag{3.2}$$

In what follows, we shall use the notation $L^2(B) = L^2(\Omega, \sigma(B), \mathbb{P})$. For every $q \geq 1$, we write $\mathbb{H}^{\otimes q}$ to indicate the qth tensor power of $\mathbb{H}$; the symbol $\mathbb{H}^{\odot q}$ indicates the qth *symmetric* tensor power of $\mathbb{H}$, equipped with the norm $\sqrt{q!}\|\cdot\|_{\mathbb{H}^{\otimes q}}$. We denote by I_q the isometry between $\mathbb{H}^{\odot q}$ and the qth Wiener chaos of X, which is a linear map satisfying the property

$$I_q(h^{\otimes q}) := H_q(B(h)), \qquad h^{\otimes q} := h \otimes \cdots \otimes h \in \mathbb{H}^{\otimes q} \quad \text{with } \|h\|_{\mathbb{H}} = 1,$$

where H_q is the qth Hermite polynomial defined in Remark 2.12. It is well-known (see [40, Chapter 1]) that any random variable $F \in L^2(B)$ admits an orthogonal *chaotic expansion*:

$$F = \sum_{q=0}^{\infty} I_q(f_q), \tag{3.3}$$

where $I_0(f_0) = \mathbb{E}[F]$, the series converges in L^2 and the kernels $f_q \in \mathbb{H}^{\odot q}$, $q \geq 1$, are uniquely determined by F. In the particular case where $\mathbb{H} = L^2(A, \mathcal{A}, \mu)$, with $(A, \mathcal{A})$ a measurable space and μ a σ-finite and non-atomic measure, one has that $\mathbb{H}^{\odot q} = L^2_s(A^q, \mathcal{A}^{\otimes q}, \mu^{\otimes q})$ is the space of symmetric and square integrable functions on A^q. Moreover, for every $f \in \mathbb{H}^{\odot q}$, $I_q(f)$ coincides with the multiple Wiener-Itô integral (of order q) of f with respect to B (see [40, Chapter 1]).

Now, we introduce the Malliavin derivative. Let $\mathcal{S}$ be the set of all smooth cylindrical random variables of the form

$$F = f\big(B(h_1), \ldots, B(h_n)\big),$$

where $n \geq 1, f : \mathbb{R}^n \to \mathbb{R}$ is a smooth function with compact support and $h_i \in \mathbb{H}$. The Malliavin derivative of F is the element of $L^2(\Omega, \mathbb{H})$ defined as

$$DF := \sum_{i=1}^{n} \frac{\partial f}{\partial x_i}\big(B(h_1), \ldots, B(h_n)\big) h_i.$$

For instance, $DW(h) = h$ for every $h \in \mathbb{H}$. We denote by $\mathbb{D}^{1,2}$ the closure of $\mathcal{S}$ with respect to the norm $\|\cdot\|_{1,2}$, defined by the relation

$$\|F\|_{1,2}^2 = \mathbb{E}[F^2] + \mathbb{E}[\|DF\|_{\mathbb{H}}^2].$$

Note that, if F is equal to a finite sum of multiple Wiener-Itô integrals, then $F \in \mathbb{D}^{1,2}$. The Malliavin derivative D verifies the following *chain rule*: if $\varphi : \mathbb{R}^n \to \mathbb{R}$ is in C_b^1 (that is, the collection of continuously differentiable functions with bounded partial derivatives) and if $\{F_i\}_{i=1,\dots,n}$ is a vector of elements of $\mathbb{D}^{1,2}$, then $\varphi(F_1, \dots, F_n) \in \mathbb{D}^{1,2}$ and

$$D\varphi(F_1, \dots, F_n) = \sum_{i=1}^n \frac{\partial \varphi}{\partial x_i}(F_1, \dots, F_n) DF_i.$$

We denote by δ the adjoint of the unbounded operator D, also called the *divergence operator*. A random element $u \in L^2(\Omega, \mathbb{H})$ belongs to the domain of δ, noted $\mathrm{Dom}\delta$, if and only if it verifies

$$|\mathbb{E}\langle DF, u\rangle_{\mathbb{H}}| \le c_u \, \|F\|_{L^2} \quad \text{for any } F \in \mathcal{S},$$

where c_u is a constant depending only on u. If $u \in \mathrm{Dom}\delta$, then the random variable $\delta(u)$ is defined by the duality relationship (sometimes called 'integration by parts formula'):

$$\mathbb{E}[F\delta(u)] = \mathbb{E}\langle DF, u\rangle_{\mathbb{H}}, \tag{3.4}$$

which holds for every $F \in \mathbb{D}^{1,2}$. An immediate consequence of (3.4) is the following identity

$$\delta(Fu) = F\delta(u) - \langle DF, u\rangle_{\mathbb{H}}, \tag{3.5}$$

which holds for all $F \in \mathbb{D}^{1,2}$ and $u \in \mathrm{Dom}\delta$ such that $Fu \in \mathrm{Dom}\delta$.

Remark 3.1 When $\mathbb{H} = L^2([0, T], dx)$, which is the most basic example, then the isonormal Gaussian process B is a standard Brownian motion on $[0, T]$. In this case we have

$$\delta(h) = \int_0^T h_s dB_s, \qquad \forall h \in L^2([0, T], dx).$$

The divergence operator δ is often called Skorohod integral. One can show that for a stochastic process $u \in \mathrm{Dom}\delta$, the Skorohod integral $\delta(u)$ and the Itô integral $\int_0^T u_s dB_s$ coincide whenever the latter is well defined.

In case $\mathbb{H} = L^2([0, T], dx)$ the Malliavin derivative D and the divergence operator δ can be computed directly using chaos expansion. Indeed, the derivative of a random variable F as in (3.3) can be identified with the element of $L^2([0, T] \times \Omega)$ given by

$$D_a F = \sum_{q=1}^\infty q I_{q-1}\big(f_q(\cdot, a)\big), \quad a \in [0, T]. \tag{3.6}$$

On the other hand, for any $u \in L^2([0,T] \times \Omega)$ there exists a chaos decomposition

$$u_s = \sum_{q=0}^{\infty} I_q(f_q(\cdot, s)), \qquad f_q(\cdot, s) \in L_s^2([0,T]^q \times \Omega).$$

Let $\tilde{f}_q \in L_s^2([0,T]^{q+1} \times \Omega)$ denote the symmetrization of $f_q(\cdot,\cdot)$. Then the element $\delta(u)$ can be written in terms of chaotic decomposition as

$$\delta(u) = \sum_{q=0}^{\infty} I_{q+1}(\tilde{f}_q).$$

□

Now, we start introducing the definition of the integral $\int_0^t Z_s dX_s$, where the ambit process X is driven by a Brownian motion W. The following exposition is related to a seminal work [2], where the integration with respect to Gaussian processes has been investigated. Throughout this subsection we assume that

$$\mathbb{E}\left[\int_0^t g^2(t,s)\sigma_s^2 ds\right] < \infty \tag{3.7}$$

and $\mathcal{F} = \sigma(W_t : t \in [0,T])$. The following definition is due to [12, Section 4].

Definition 3.2 Assume that, for any $s \geq 0$, the function $t \mapsto g(t,s)$ has bounded variation on the interval $[t,v]$ for all $0 \leq s < t < v < \infty$. We say that the process Z belongs to the class $I^X(0,T)$ when the following conditions are satisfied:

(i) For any $s \in [0,T]$ the process $(Z_u - Z_s)_{u \in (s,T]}$ is integrable with respect to $g(du,s)$.
(ii) Define the operator

$$\mathcal{K}_g(h)(t,s) := h(s)g(t,s) + \int_s^t (h(u) - h(s))g(du,s).$$

The process $s \mapsto \mathcal{K}_g(Z)(T,s)\sigma_s 1_{[0,T]}(s)$ belongs to Domδ.
(iii) $\mathcal{K}_g(Z)(T,s)$ is Malliavin differentiable with respect to D_s with $s \in [0,T]$, such that the mapping $s \mapsto D_s[\mathcal{K}_g(Z)(T,s)]\sigma_s$ is Lebesgue integrable on $[0,T]$.

When $Z \in I^X(0,T)$ we define

$$\int_0^T Z_s dX_s := \delta\left(\mathcal{K}_g(Z)(T,s)\sigma_s 1_{[0,T]}(s)\right) + \int_0^T D_s[\mathcal{K}_g(Z)(T,s)]\sigma_s ds.$$

We remark that the proposed definition is linear in the integrand. It also holds that

$$\int_0^T YZ_s dX_s = Y \int_0^T Z_s dX_s$$

for any bounded random variable Y such that $Z, YZ \in I^X(0,T)$. We refer to [12] for further properties and applications.

The operator $\mathcal{K}$ has been introduced in [2]. The intuition behind Definition 3.2 is explained by the following heuristic derivation. Using classical integration by parts formula and (3.5) we conclude that

$$\int_0^T Z_s dX_s = Z_T X_T - \int_0^T \frac{dZ_u}{du}\left(\int_0^u g(u,s)\sigma_s dW_s\right) du$$

$$= Z_T X_T - \int_0^T \delta\left(\frac{dZ_u}{du} g(u,s)\sigma_s 1_{[0,T]}(s)\right) du$$

$$- \int_0^T \int_0^u D_s\left[\frac{dZ_u}{du}\right] g(u,s)\sigma_s ds du.$$

Next, the stochastic Fubini theorem applied to the last two quantities implies the identity

$$\int_0^T Z_s dX_s = Z_T X_T - \delta\left(\sigma_s \int_s^T g(u,s)\frac{dZ_u}{du} du 1_{[0,T]}(s)\right)$$

$$- \int_0^T D_s\left[\int_s^T g(u,s)\frac{dZ_u}{du} du\right] \sigma_s ds.$$

Similarly, we deduce that

$$Z_T X_T = \delta\left(Z_T g(T,s)\sigma_s 1_{[0,T]}(s)\right) + \int_0^T D_s\left[Z_T\right] g(T,s)\sigma_s ds.$$

Thus, putting things together and applying the classical integration by parts formula once again we obtain the heuristic formula

$$\int_0^T Z_s dX_s = \delta\left(\sigma_s \left[Z_T g(T,s) - \int_s^T g(u,s)\frac{dZ_u}{du} du\right] 1_{[0,T]}(s)\right)$$

$$- \int_0^T D_s\left[Z_T g(T,s) - \int_s^T g(u,s)\frac{dZ_u}{du} du\right] \sigma_s ds$$

$$= \delta\left(\mathcal{K}_g(Z)(T,s)\sigma_s 1_{[0,T]}(s)\right) + \int_0^T D_s[\mathcal{K}_g(Z)(T,s)]\sigma_s ds.$$

This explains the intuition behind Definition 3.2.

Remark 3.3 In a recent work [22] the integration concept has been extended to the class of Hilbert-valued processes. □

3.2 Integration with Respect to Ambit Processes Driven by Pure Jump Lévy Motion

In this subsection we will introduce the definition of the integral

$$\int_0^t Z_s dX_s,$$

where the ambit process X is driven by a square integrable pure jump Lévy motion L with characteristic triplet $(0, 0, \nu)$. We assume that the condition (3.7) holds and $\mathcal{F} = \sigma(L_t : t \in [0, T])$.

The definition of the stochastic integral proposed in [12] relies again on Malliavin calculus. However, in contrast to the Gaussian space, there exist different variations of Malliavin calculus for Poisson random measures. Here we follow an approach described in [31]. We deal with the Hilbert space $\mathbb{H} = L^2([0, T], dx)$. Let $N(dt, dz)$ denote the Poisson random measure on $[0, T] \times (\mathbb{R} \setminus \{0\})$ associated with L and $\tilde{N}(dt, dz) = N(dt, dz) - dt\nu(dz)$ the compensated Poisson random measure. We have

$$L_t = \int_0^t \int_{\mathbb{R}\setminus\{0\}} z\tilde{N}(dt, dz).$$

As in the Gaussian case there exists an orthogonal chaos decomposition of the type (3.3) in terms of multiple integrals. For any $f \in L_s^2(([0, T] \times (\mathbb{R} \setminus \{0\}))^q)$, we introduce the qth order multiple integral of f with respect to $\tilde{N}(dt, dz)$ via

$$I_q(f) = q! \int_0^T \int_{\mathbb{R}\setminus\{0\}} \cdots \int_0^{t_2-} \int_{\mathbb{R}\setminus\{0\}} f(t_1, z_1, \ldots, t_q, z_q)\tilde{N}(dt_1, dz_1) \ldots N(dt_q, dz_q).$$

Then, for any random variable $F \in L^2(\Omega, \mathcal{F}, \mathbb{P})$, there exists a unique sequence of symmetric functions $(f_q)_{q\geq 0}$ with $f_q \in L_s^2(([0, T] \times (\mathbb{R} \setminus \{0\}))^q)$ such that

$$F = \sum_{q=0}^{\infty} I_q(f_q), \tag{3.8}$$

which is obviously an analogue of (3.3). Furthermore, it holds that

$$\mathbb{E}[F^2] = \sum_{q=0}^{\infty} q! \|f_q\|_{q,\nu}^2,$$

where the norm $\|f_q\|_{q,\nu}^2$ is defined by

$$\|f_q\|_{q,\nu}^2 := \int_{([0,T]\times(\mathbb{R}\setminus\{0\}))^q} f^2(t_1, z_1, \ldots, t_q, z_q)dt_1\nu(dz_1)\ldots dt_q\nu(dz_q).$$

Similarly to the exposition of Remark 3.1, the Malliavin derivative D and the divergence operator δ are introduced using the above chaos representation. We say that a random variable F with chaos decomposition (3.8) belongs to the space $\mathbb{D}^{1,2}$ whenever the condition

$$\sum_{q=0}^{\infty} qq!\|f_q\|_{q,\nu}^2 < \infty$$

holds. Whenever $F \in \mathbb{D}^{1,2}$ we define

$$D_{t,z}F := \sum_{q=1}^{\infty} qI_{q-1}(f_q(\cdot, t, z)).$$

Now, we say that a random field $u \in L^2([0,T] \times \mathbb{R} \setminus \{0\} \times \Omega)$ belongs to the domain of the divergence operator δ (Domδ) when

$$\left|\mathbb{E}\left[\int_0^T \int_{\mathbb{R}\setminus\{0\}} u(t,z)D_{t,z}F\nu(dz)dt\right]\right| \le c_u\|F\|_{L^2}$$

for all $F \in \mathbb{D}^{1,2}$. Whenever $u \in$ Domδ the element $\delta(u)$ is uniquely characterized via the identity

$$\mathbb{E}[F\delta(u)] = \mathbb{E}\left[\int_0^T \int_{\mathbb{R}\setminus\{0\}} u(t,z)D_{t,z}F\nu(dz)dt\right] \qquad \forall F \in \mathbb{D}^{1,2},$$

which is an integration by parts formula (cf. (3.4)). An immediate consequence of the integration by parts formula is the following equation:

$$F\delta(u) = \delta(u(F + DF)) + \int_0^T \int_{\mathbb{R}\setminus\{0\}} u(t,z)D_{t,z}F\nu(dz)dt, \tag{3.9}$$

which holds for any $F \in \mathbb{D}^{1,2}$, $u \in$ Domδ such that $u(F + DF) \in$ Domδ. Notice the appearance of the term uDF on the right hand side that is absent in the Gaussian case (cf. (3.5)).

Now, we proceed with the introduction of the stochastic integral. Its definition in the pure jump Lévy framework is essentially analogous to the Gaussian case. We refer again to [12, Section 4] for a more detailed exposition and the intuition behind this definition.

Definition 3.4 Assume that, for any $s \geq 0$, the function $t \mapsto g(t,s)$ has bounded variation on the interval $[t,T]$. We say that the process Z belongs to the class $I^X(0,T)$ when the following conditions are satisfied:

(i) For any $s \in [0,T]$ the process $(Z_u - Z_s)_{u\in(s,T]}$ is integrable with respect to $g(du,s)$.
(ii) The process $(s,z) \mapsto z(\mathcal{K}_g(Z)(T,s) + D_{s,z}[\mathcal{K}_g(Z)(T,s)])\sigma_s 1_{[0,T]}(s)$ belongs to Domδ.
(iii) $\mathcal{K}_g(Z)(T,s)$ is Malliavin differentiable with respect to $D_{s,z}$ with $(s,z) \in [0,T]\times \mathbb{R}$, such that the mapping $(s,z) \mapsto zD_{s,z}[\mathcal{K}_g(Z)(T,s)]\sigma_s$ is $\nu(dz)dt$-integrable.

When $Z \in I^X(0,T)$ we define

$$\int_0^T Z_s dX_s := \delta\left(z(\mathcal{K}_g(Z)(T,s) + D_{s,z}[\mathcal{K}_g(Z)(T,s)])\sigma_s 1_{[0,T]}(s)\right)$$

$$+ \int_0^T \int_{\mathbb{R}} zD_{s,z}[\mathcal{K}_g(Z)(T,s)]\sigma_s \nu(dz)ds.$$

4 Limit Theory for High Frequency Observations of Ambit Fields

In this section we will review the asymptotic results for power variation of Lévy semi-stationary processes (*LSS*) without drift, i.e.

$$Y_t = \mu + \int_{-\infty}^t g(t-s)\sigma_s L(\mathrm{d}s).$$

We will see that the limit theory heavily depends on whether the driving Lévy motion is a Brownian motion or a pure jump process. Furthermore, the structure of the weight function g plays an important role as we will see below. More precisely, the *singularity points* of g determine the type of the limit.

In what follows we assume that the underlying observations of Lévy semi-stationary process Y are

$$Y_0, Y_{\Delta_n}, Y_{2\Delta_n}, \ldots, Y_{\Delta_n[t/\Delta_n]}$$

with $\Delta_n \to 0$ and t fixed. In other words, we are in the infill asymptotics setting. For statistical purposes we introduce kth order differences $\Delta_{i,k}^n Y$ of Y defined via

$$\Delta_{i,k}^n Y := \sum_{j=0}^k (-1)^j \binom{k}{j} Y_{(i-j)\Delta_n}. \tag{4.1}$$

For instance,

$$\Delta_{i,1}^n Y = Y_{i\Delta_n} - Y_{(i-1)\Delta_n} \quad \text{and} \quad \Delta_{i,2}^n Y = Y_{i\Delta_n} - 2Y_{(i-1)\Delta_n} + Y_{(i-2)\Delta_n}.$$

The power variation of kth order differences of Y is given by the statistic

$$V(Y,p,k;\Delta_n)_t := \sum_{i=k}^{[t/\Delta_n]} |\Delta_{i,k}^n Y|^p. \tag{4.2}$$

In the following we will study the asymptotic behaviour of the functional $V(Y,p,k;\Delta_n)_t$.

4.1 LSS Processes Driven by Brownian Motion

In this section we consider the case of *Brownian semi-stationary processes* given via

$$Y_t = \mu + \int_{-\infty}^t g(t-s)\sigma_s W(\mathrm{d}s),$$

defined on a filtered probability space $(\Omega, \mathcal{F}, (\mathcal{F}_t)_{t\in\mathbb{R}}, \mathbb{P})$. The following asymptotic results have been investigated in a series of papers [9, 11, 28, 32]. We also refer to a related work [8, 27], where the power variation of integral processes as defined in Remark 2.14 has been studied.

In the model introduced above W is an $(\mathcal{F}_t)_{t\in\mathbb{R}}$-adapted white noise on $\mathbb{R}$, $g : \mathbb{R} \to \mathbb{R}$ is a deterministic weight function satisfying $g(t) = 0$ for $t \le 0$ and $g \in \mathbb{L}^2(\mathbb{R})$. The intermittency process σ is assumed to be an $(\mathcal{F}_t)_{t\in\mathbb{R}}$-adapted càdlàg process. The finiteness of the process Y is guaranteed by the condition

$$\int_{-\infty}^t g^2(t-s)\sigma_s^2 ds < \infty \quad \text{almost surely}, \tag{4.3}$$

for any $t \in \mathbb{R}$, which we assume from now on. As pointed out in [9, 11] and also briefly discussed in Remark 2.14, the *Gaussian core* G is crucial for understanding the fine structure of Y. The process $G = (G_t)_{t\in\mathbb{R}}$ is a zero-mean stationary Gaussian process given by

$$G_t := \int_{-\infty}^t g(t-s)W(ds), \qquad t \in \mathbb{R}. \tag{4.4}$$

We remark that $G_t < \infty$ since $g \in \mathbb{L}^2(\mathbb{R})$. A straightforward computation shows that the correlation kernel r of G has the form

$$r(t) = \frac{\int_0^\infty g(u)g(u+t)du}{\|g\|^2_{\mathbb{L}^2(\mathbb{R})}}, \qquad t \geq 0.$$

Another important quantity for the asymptotic theory is the variogram R, i.e.

$$R(t) := \mathbb{E}[(G_{t+s} - G_s)^2] = 2\|g\|^2_{\mathbb{L}^2(\mathbb{R})}(1 - r(t)), \qquad \tau_k(\Delta_n) := \sqrt{\mathbb{E}[(\Delta^n_{i,k} G)^2]}. \tag{4.5}$$

The quantity $\tau_k(\Delta_n)$ will appear as a proper scaling in the law of large numbers for the statistic $V(Y, p, k; \Delta_n)$ introduced at (4.2).

As mentioned above the set of singularity points $0 = \theta_0 < \theta_1 < \cdots < \theta_l < \infty$ of g will determine the limit theory for the power variation $V(Y, p, k; \Delta_n)$. Let $\alpha_0, \ldots, \alpha_l \in (-1/2, 0) \cup (0, 1/2)$ be given real numbers. For any function $h \in C^m(\mathbb{R})$, $h^{(m)}$ denotes the m-th derivative of h. Recall that $k \geq 1$ stands for the order of the filter defined in (4.1). We introduce the following set of assumptions.

(A): For $\delta < \frac{1}{2}\min_{1 \leq i \leq l}(\theta_i - \theta_{i-1})$ it holds that

- (i) $g(x) = x^{\alpha_0} f_0(x)$ for $x \in (0, \delta)$ and $g(x) = |x - \theta_l|^{\alpha_l} f_l(x)$ for $x \in (\theta_l - \delta, \theta_l) \cup (\theta_l, \infty)$.
- (ii) $g(x) = |x - \theta_i|^{\alpha_i} f_i(x)$ for $x \in (\theta_i - \delta, \theta_i) \cup (\theta_i, \theta_i + \delta)$, $i = 1, \ldots, l-1$.
- (iii) $g(\theta_i) = 0, f_i \in C^k((\theta_i - \delta, \theta_i + \delta))$ and $f_i(\theta_i) \neq 0$ for $i = 0, \ldots, l$.
- (iv) $g \in C^k(\mathbb{R} \setminus \{\theta_0, \ldots, \theta_l\})$ and $g^{(k)} \in \mathbb{L}^2\left(\mathbb{R} \setminus \cup_{i=0}^l (\theta_i - \delta, \theta_i + \delta)\right)$.
- (v) For any $t > 0$

$$F_t = \int_{\theta_l + 1}^\infty g^{(k)}(s)^2 \sigma^2_{t-s} ds < \infty. \tag{4.6}$$

We also set

$$\alpha := \min\{\alpha_0, \ldots, \alpha_l\}, \qquad \mathcal{A} := \{0 \leq i \leq l : \alpha_i = \alpha\}. \tag{4.7}$$

The points $\theta_0, \ldots, \theta_l$ are singularities of g in the sense that $g^{(k)}$ is not square integrable around these points, because $\alpha_0, \ldots, \alpha_l \in (-1/2, 0) \cup (0, 1/2)$ and conditions (A)(i)–(iii) hold. Condition (A)(iv) indicates that g exhibits no further singularities.

Remark 4.1 According to the discussion of Sect. 2.3, the Brownian semi-stationary processes Y (or even the Gaussian core G) is not a semimartingale, since $g' \notin L^2(\mathbb{R}_{\geq 0})$ due to the presence of the singularity points $\theta_0, \ldots, \theta_l$. For this reason we can not rely on limit theory for power variations of continuous semimartingales investigated in e.g. [6, 34]. Although some of the asymptotic results look similar to the semimartingale case, the methodology behind the proof is completely different. The main steps of the proof are based on methods of Malliavin calculus developed in e.g. [41, 44] and on Bernstein's blocking technique. □

The limit theory for the power variation is quite different according to whether we have a single singularity at, say, $\theta_0 = 0$ (i.e. $l = 0$) or multiple singularity points. Hence, we will treat the corresponding results separately.

4.1.1 The Case $l = 0$

The theory presented in this section is mainly investigated in [9, 11]. Below we will intensively use the concept of stable convergence, which is originally due to Rényi [47]. We say that a sequence of processes X^n converges stably in law to a process X, where X is defined on an extension $(\Omega', \mathcal{F}', \mathbb{P}')$ of the original probability $(\Omega, \mathcal{F}, \mathbb{P})$, in the space $\mathbb{D}([0, T])$ equipped with the uniform topology ($X^n \xrightarrow{d_{st}} X$) if and only if

$$\lim_{n\to\infty} \mathbb{E}[f(X^n)Z] = \mathbb{E}'[f(X)Z]$$

for any bounded and continuous function $f : \mathbb{D}([0, T]) \to \mathbb{R}$ and any bounded $\mathcal{F}$-measurable random variable Z. We refer to [1, 36] or [47] for a detailed study of stable convergence. Note that stable convergence is a stronger mode of convergence than weak convergence, but it is weaker that u.c.p. convergence.

The following theorem has been shown in [11, Theorems 1 and 2].

Theorem 4.2 *Assume that condition (A) holds.*

(i) We obtain that

$$\Delta_n \tau_k(\Delta_n)^{-p} V(Y, p, k; \Delta_n)_t \xrightarrow{u.c.p.} V(Y, p)_t := m_p \int_0^t |\sigma_s|^p ds, \tag{4.8}$$

where the power variation $V(Y, p, k; \Delta_n)_t$ is defined at (4.2) and the constant m_p is given by (2.9).

(ii) Assume that the intermittency process σ is Hölder continuous of order $\gamma \in (0, 1)$ and $\gamma(p \wedge 1) > 1/2$. When $k = 1$ we further assume that $\alpha \in (-1/2, 0)$. Then we obtain the stable convergence

$$\Delta_n^{-1/2}\Big(\Delta_n \tau_k(\Delta_n)^{-p} V(Y, p, k; \Delta_n)_t - V(Y, p)_t\Big) \xrightarrow{d_{st}} \lambda \int_0^t |\sigma_s|^p \, dB_s \tag{4.9}$$

on $\mathbb{D}([0, T])$ equipped with the uniform topology, where B is a Brownian motion that is defined on an extension of the original probability space $(\Omega, \mathcal{F}, \mathbb{P})$ and *is independent of $\mathcal{F}$, and the constant λ is given by*

$$\lambda^2 = \lim_{n\to\infty} \Delta_n^{-1}\mathrm{var}\Big(\Delta_n^{1-pH} V(B^H, p, k; \Delta_n)_1\Big), \tag{4.10}$$

with B^H being a fractional Brownian motion with Hurst parameter $H = \alpha + 1/2$.

Remark 4.3 The appearance of the fractional Brownian motion in the definition of the constant λ^2 is not surprising given the discussion of fine properties of the Gaussian core G in Sect. 2.3. In case $k = 1$ the factor λ^2 coincides with the quantity v_p defined in Remark 2.12. We also remark that the validity region of the central limit theorem in (4.9) in the case $k = 1$ ($\alpha \in (-1/2, 0)$) is smaller than the region $H = \alpha + 1/2 \in (0, 3/4)$ described in Remark 2.12. This is due to a bias problem, which appears in the context of Brownian semi-stationary processes. □

Notice that the asymptotic result at (4.8) and (4.9) are not feasible from the statistical point of view, since the scaling $\tau_k(\Delta_n)$ depends on the unknown weight function g. Nevertheless, Theorem 4.2 is useful for statistical applications. Our first example is the estimation of the *smoothness parameter* α. Under mild conditions on the intermittency process σ, the Brownian semi-stationary process Y (as well as its Gaussian core G) has Hölder continuous paths of any order smaller than $H = \alpha + 1/2 \in (0, 1)$. In turbulence the smoothness parameter α is related to the so called *Kolmogorov's* 2/3-*law*, which predicts that

$$\mathbb{E}[(X_{t+\Delta} - X_t)^2] \propto \Delta^{2/3},$$

or in other words $\alpha \approx -1/6$, which holds for a certain range of frequencies Δ. Hence, estimation of the parameter α is extremely important.

A typical model for the weight function g is the Gamma kernel given via

$$g(x) = x^{\alpha} \exp(-cx), \qquad c > 0, \alpha \in (-1/2, 0) \cup (0, 1/2),$$

which obviously satisfies the assumption (A)(i)-(iv) with $l = 0$. An application of the law of large numbers at (4.8) for a fixed $t > 0$ gives

$$S_n := \frac{V(Y, p, k; 2\Delta_n)_t}{V(Y, p, k; \Delta_n)_t} \xrightarrow{\mathbb{P}} 2^{\frac{(2\alpha+1)p}{2}},$$

since $\tau_k(2\Delta_n)^2/\tau_k(\Delta_n)^2 \to 2^{2\alpha+1}$. The latter is due to $\tau_k(\Delta_n)^2 \sim \Delta_n^{2\alpha+1}$, which follows from the fact that the Gaussian core G and the fractional Brownian motion B^H with Hurst parameter $H = \alpha + 1/2$ have the same small scale behaviour. Thus, a consistent estimator of α is given via

$$\hat{\alpha}_n = \frac{1}{2}\left(\frac{2\log_2 S_n}{p} - 1\right) \xrightarrow{\mathbb{P}} \alpha, \tag{4.11}$$

where $\log_2$ denotes the logarithm at basis 2. Note that the estimator $\hat{\alpha}_n$ is feasible, i.e. it does not depend on the unknown scaling $\tau_k(\Delta_n)$. One may also deduce a standard feasible central limit theorem for $\hat{\alpha}_n$ as it was shown in [9, 11, 28], and thus obtain asymptotic confidence regions for the smoothness parameter α. We also refer to [28] for empirical implementation of this estimation method to turbulence data.

Another useful application of Theorem 4.2 is the estimation of the *relative intermittency*, which is defined as

$$RI_t := \frac{\int_0^t \sigma_s^2 ds}{\int_0^T \sigma_s^2 ds}, \qquad t \le T,$$

where $T > 0$ is a fixed time. While the intermittency process σ is not identifiable when no structural assumption on g are imposed, the relative intermittency RI_t is easy to estimate. Indeed, the convergence in (4.8) immediately implies that

$$\widehat{RI}_t^n := \frac{V(Y, p, k; \Delta_n)_t}{V(Y, p, k; \Delta_n)_T} \xrightarrow{\mathbb{P}} RI_t.$$

We refer to [15] for the limit theory and physical applications of the statistic $\widehat{RI}_t^n$.

4.1.2 The Case $l \ge 1$

The limit theory for the case $l \ge 1$ appears to be more complex. The asymptotic results presented below have been investigated in [32].

First of all, we need to introduce some notations. Recall that $k \in \mathbb{N}$ denotes the order of increments. The k-th order filter associated with g is introduced via

$$\Delta_k^n g(x) := \sum_{j=0}^k (-1)^j \binom{k}{j} g(x - j\Delta_n), \qquad x \in \mathbb{R}. \tag{4.12}$$

There is a straightforward relationship between the scaling quantity $\tau_k(\Delta_n)$ defined at (4.5) and the function $\Delta_k^{n,} g$, namely

$$\tau_k(\Delta_n)^2 = \|\Delta_k^n g\|_{\mathbb{L}^2(\mathbb{R})}^2.$$

Now, we define the concentration measure associated with $\Delta_k^n g$:

$$\pi_{n,k}(A) := \frac{\int_A (\Delta_k^n g(x))^2 dx}{\|\Delta_k^n g\|_{\mathbb{L}^2(\mathbb{R})}^2}, \qquad A \in \mathcal{B}(\mathbb{R}). \tag{4.13}$$

Observe that $\pi_{n,k}$ is a probability measure. Its asymptotic behaviour determines the law of large numbers for the power variation. In order to identify the limit of $\pi_{n,k}$, we define the following functions

$$h_0(x) := f_0(\theta_0) \sum_{j=0}^{k} (-1)^j \binom{k}{j} (x-j)_+^{\alpha_0}, \tag{4.14}$$

$$h_i(x) := f_i(\theta_i) \sum_{j=0}^{k} (-1)^j \binom{k}{j} |x-j|^{\alpha_i}, \qquad i = 1, \ldots, l,$$

where $x_+ := \max\{x, 0\}$. The following results determines the asymptotic behaviour of the power variation $V(Y, p, k; \Delta_n)_t$ for $p = 2$ and $l \geq 1$. We refer to [32, Proposition 3.1, Theorems 3.2 and 3.3] for further details.

Theorem 4.4 *Assume that the condition (A) holds.*

(i) It holds that

$$\pi_{n,k} \xrightarrow{d} \pi_k,$$

for any $k \geq 1$, where the probability measure π_k is given as

$$supp(\pi_k) = \{\theta_i\}_{i \in \mathcal{A}}, \qquad \pi_k(\theta_i) = \frac{\|h_i\|^2_{\mathbb{L}^2(\mathbb{R})} 1_{i \in \mathcal{A}}}{\sum_{j=0}^{l} \|h_j\|^2_{\mathbb{L}^2(\mathbb{R})} 1_{j \in \mathcal{A}}}, \tag{4.15}$$

where the set $\mathcal{A}$ has been defined at (4.7).

(ii) We obtain the convergence

$$\frac{\Delta_n}{\tau_k(\Delta_n)^2} V(Y, 2, k; \Delta_n)_t \overset{u.c.p.}{\Longrightarrow} QV(Y, k)_t := \int_0^\infty \left(\int_{-\theta}^{t-\theta} \sigma_s^2 ds \right) \pi_k(d\theta). \tag{4.16}$$

(iii) Assume that the intermittency process σ is Hölder continuous of order $\gamma > 1/2$. When $k = 1$ we further assume that $\alpha_j \in (-\frac{1}{2}, 0)$ for all $0 \leq j \leq l$. Then, under condition

$$\alpha_i - \alpha > 1/4 \qquad \textit{for all } i \notin \mathcal{A}, \tag{4.17}$$

we obtain the stable convergence

$$\Delta_n^{-1/2} \left(\frac{\Delta_n}{\tau_k(\Delta_n)^2} V(Y, 2, k; \Delta_n)_t - QV(X, k)_t \right) \xrightarrow{d_{st}} \int_0^t v_s^{1/2} dB_s \tag{4.18}$$

on $\mathbb{D}([0, \min_{1 \leq j \leq l}(\theta_j - \theta_{j-1})])$ equipped with the uniform topology, where B is a Brownian motion, independent of $\mathcal{F}$, defined on an extension of the original *probability space* $(\Omega, \mathcal{F}, \mathbb{P})$*. The stochastic process* v *is given by*

$$v_s = \Lambda_k \left(\int_0^\infty \sigma_{s-\theta}^2 \pi_k(d\theta) \right)^2, \tag{4.19}$$

where Λ_k *is defined by*

$$\Lambda_k = \lim_{n\to\infty} \Delta_n^{-1} \mathrm{var}\Big(\frac{\Delta_n}{\hat{\tau}_k(\Delta_n)^2} V(B^H, 2, k; \Delta_n)_1 \Big)$$

with B^H *being a fractional Brownian motion with Hurst parameter* $H = \alpha + 1/2$ *and* $\hat{\tau}_k(\Delta_n)^2 := \mathbb{E}[(\Delta_{i,k}^n B^H)^2]$*.*

In order to explain the mathematical intuition behind the results of Theorem 4.4 we present some remarks.

Remark 4.5 We notice that $\mathrm{supp}(\pi_k) = \{\theta_i\}_{i\in\mathcal{A}}$, which means that only those singularity points contribute to the limit, which correspond to the minimal index α. This fact is not surprising from the statistical point of view, since a process with the roughest path always dominates when considering a power variation.

When $l = 0$ it holds that $\pi_{n,k} \stackrel{d}{\longrightarrow} \delta_{\{0\}}$, hence the convergence in (4.8) is a particular case of (4.16). Otherwise, the limiting measure π_k is a discrete probability measure. It is an open problem whether the result of (4.16) can be deduced for a continuous probability measure π_k. □

Remark 4.6 Although the singularity points θ_i with $i \notin \mathcal{A}$ do not contribute to the limit at (4.16), they cause a certain bias, which might explode in the central limit theorem. Condition (4.17) ensures that it does not happen. We also remark that the functional stable convergence at (4.18) does not hold on any interval $[0, T]$, but just on $[0, \min_{1\le j\le l}(\theta_j - \theta_{j-1})]$. One may still show a stable central limit theorem with an $\mathcal{F}$-conditional Gaussian process as the limit on a larger interval, but only when $\theta_j - \theta_{j-1} \in \mathbb{N}$ for all j, since otherwise the covariance structure of the original statistic does not converge. □

Notice that the minimal parameter α defined at (4.7) still determines the Hölder continuity of the process Y (and the Gaussian core G). The estimator $\hat{\alpha}_n$ defined at (4.11) remains consistent, i.e.

$$\hat{\alpha}_n = \frac{1}{2}\left(\log_2 \frac{V(Y, 2, k; 2\Delta_n)_t}{V(Y, 2, k; \Delta_n)_t} - 1 \right) \stackrel{\mathbb{P}}{\longrightarrow} \alpha.$$

One may also construct a standardized version of the statistic $\hat{\alpha}_n$, which satisfies a standard central limit theorem (see [32, Section 4] for a detailed exposition). But in this case the time $t < \min_{1\le j\le l}(\theta_j - \theta_{j-1})$ must be used, which requires the knowledge of singularity points θ_i.

For potential applications in turbulence the asymptotic results need to be extended to ambit fields X driven by a Gaussian random measure, i.e.

$$X_t(x) = \mu + \int_{A_t(x)} g(t,s,x,\xi)\sigma_s(\xi)W(\mathrm{d}s,\mathrm{d}\xi), \qquad t \geq 0, x \in \mathbb{R}^3,$$

where W is a Gaussian random measure. This type of high frequency limit theory has not been yet investigated neither for observations of X on a grid in time-space nor for observations of X along a curve. In the multiparameter setting there exist a related work on generalized variation of fractional Brownian sheet (see e.g. [43]) and integral processes (see e.g. [42, 48]).

4.2 LSS Processes Driven by a Pure Jump Lévy Motion

In this section we will mainly study the power variation of a Lévy moving average process defined via

$$Y_t = \mu + \int_{-\infty}^{t} g(t-s)dL_s,$$

where L is a symmetric pure jump Lévy motion with Lévy measure ν and drift zero. Notice that this is a subclass of *LSS* processes with $\sigma = 1$, and hence it plays a similar role as the Gaussian core G defined at (4.4). The asymptotic theory for power variation of Y is likely to transfer to limit theory for general *LSS* processes as it was indicated in Remark 2.14. In contrast to Gaussian moving averages, little is known about power variation of Lévy moving average processes driven by a pure jump Lévy motion. Below we present some recent results from [19], which completely determine the first order structure of power variation of Y. We need to impose a somewhat similar set of assumptions as presented in (A) for case $l = 0$ (in particular, they insure the existence of Y_t, cf. Remark 2.9).

(A′): It holds that

(i) $g(x) = x^{\alpha}f(x)$ for $x \geq 0$ and $g(x) = 0$ for $x < 0$ with $\alpha > 0$.

(ii) For some $\theta > 0$ it holds that

$$\limsup_{t\to\infty} t^{\theta}\nu\{x : |x| > t\} < \infty$$

(iii) $g \in C^k(\mathbb{R}_{\geq 0})$,

$$|g^{(j)}(x)| \leq K|x|^{\alpha-j}, \qquad x \in (0,\delta)$$

and $g^{(j)} \in L^{\theta}((\delta,\infty))$ for some $\delta > 0$. Moreover, $|g^{(j)}|$ is decreasing on (δ,∞).

(A′-log): In addition to assumption (A′) suppose that for all $j = 1, \dots, k$ we have $\int_\delta^\infty |g^{(j)}(s)|^\theta \log(1/|g^{(j)}(s)|)\, ds < \infty$.

Another important parameter in our limit theory is the *Blumenthal-Getoor index* of the driving Lévy motion L, which is defined by

$$\beta := \inf\left\{ r \geq 0 : \sum_{s \in [0,1]} |\Delta L_s|^r < \infty \right\}, \qquad \Delta L_s = L_s - L_{s-}. \tag{4.20}$$

Obviously, finite activity Lévy processes have Blumenthal-Getoor index $\beta = 0$ while Lévy processes with finite variation satisfy $\beta \leq 1$. In general, the Blumenthal-Getoor index is a non-random number $\beta \in [0, 2]$ and it can characterized by the Lévy measure ν of L as follows:

$$\beta = \inf\left\{ r \geq 0 : \int_{-1}^{1} |x|^r \nu(dx) < \infty \right\}.$$

The latter implies that β-stable Lévy processes with $\beta \in (0, 2)$ have Blumenthal-Getoor index β.

We recall the definition of kth order increments $\Delta_{i,k}^n Y$ introduced at (4.1) and consider the power variation

$$V(Y, p, k; \Delta_n)_t = \sum_{i=k}^{[t/\Delta_n]} |\Delta_{i,k}^n Y|^p.$$

The following theorem from [19] determines the first order structure of the statistic $V(Y, p, k; \Delta_n)_t$. Notice that the results below are stated for a fixed $t > 0$.

Theorem 4.7 *Assume that condition (A′) holds and fix $t > 0$.*

(i) Suppose that $p \geq 1$ and assume (A′-log) when $\theta = 1$. If $\alpha \in (0, k - 1/p)$ and $p > \beta$, we obtain the stable convergence

$$\Delta_n^{-\alpha p} V(Y, p, k; \Delta_n)_t \xrightarrow{d_{st}} |f(0)|^p \sum_{m : T_m \in [0,t]} |\Delta L_{T_m}|^p \left(\sum_{l=0}^{\infty} |h(l + U_m)|^p \right), \tag{4.21}$$

where $(U_m)_{m \geq 1}$ are i.i.d. $\mathcal{U}([0, 1])$-distributed random variables independent of the original σ-algebra $\mathcal{F}$ and the function $h = h_0$ is defined at (4.14).

(ii) Assume that L is a symmetric β-stable Lévy process with $\beta \in (0, 2)$. If $\alpha \in (0, k - 1/\beta)$ and $p < \beta$ then it holds

$$\Delta_n^{1 - p(\alpha + 1/\beta)} V(Y, p, k; \Delta_n)_t \xrightarrow{\mathbb{P}} t c_p, \qquad c_p := \mathbb{E}[|L_1(k)|^p] < \infty, \tag{4.22}$$

where the process $L_t(k)$ *is defined as*

$$L_t(k) := \int_{\mathbb{R}} h(s) dL_s$$

and the function $h = h_0$ *is defined at* (4.14).

(iii) *Suppose that* $p \geq 1$ *and assume* (A'-log) *when* $\theta = 1$. *When* $\alpha > k - 1/p$, $p > \beta$ *or* $\beta > k - 1/\beta$, $p < \beta$, *we deduce*

$$\Delta_n^{1-pk} V(Y, p, k; \Delta_n)_t \xrightarrow{\mathbb{P}} \int_0^t |F_k(u)|^p du, \qquad F_k(u) := \int_{-\infty}^u g^{(k)}(u-s) dL_s. \tag{4.23}$$

We remark that the result of Theorem 4.7(i) is sharp in a sense that the conditions $\alpha \in (0, k - 1/p)$ and $p > \beta$ are sufficient and (essentially) necessary to conclude (4.21). Indeed, since $|h(l + U_m)| \leq \text{const.} l^{\alpha - k}$ for $l \geq 1$, we obtain that

$$\sum_{l=0}^{\infty} |h(l + U_m)|^p \leq \text{const} < \infty$$

when $\alpha \in (0, k - 1/p)$, and on the other hand $\sum_{m: T_m \in [0,t]} |\Delta L_{T_m}|^p < \infty$ for $p > \beta$, which follows from the definition of the Blumenthal-Getoor index.

The idea behind the proof of Theorem 4.7(ii) has been described in Sect. 2.3. Note that for $k = 1$ the random variable $L_1(1)$ coincides with the increments $\tilde{X}_1 - \tilde{X}_0$ of the fractional β-stable Lévy motion introduced in (2.8). Following the mathematical intuition of the aforementioned discussion, the limit in (4.22) is not really surprising. Also notice that the limit is indeed finite since $p < \beta$.

We remark that for values of α close to $k - 1/p$ or $k - 1/\beta$ in Theorem 4.7(iii), the function $g^{(k)}$ explodes at 0. This leads to unboundedness of the process F_k defined at (4.23). Nevertheless, under conditions of Theorem 4.7(iii), the limiting process is still finite.

Remark 4.8 Notice that the critical cases, i.e. $\alpha = k - 1/p, p > \beta$ and $\alpha = k - 1/, p < \beta$, are not described in Theorem 4.7. In this cases an additional log factor appears, and for $\alpha = k - 1/p, p > \beta$ the mode of convergence changes from stable convergence to convergence in probability (clearly the limits change too). We refer to [19] for a detailed discussion of critical cases. □

Remark 4.9 The asymptotic results of Theorem 4.7 uniquely identify the parameters α and β. First of all, note that the convergence rates in (4.21)–(4.23) are all different under the corresponding conditions. Indeed, it holds that

$$p(\alpha + 1/\beta) - 1 < \alpha p < pk - 1,$$

since in case (i) we have $\alpha < k - 1/p$ and in case (ii) we have $p < \beta$. Hence, computing the statistic $V(Y, p, k; \Delta_n)_t$ at log scale for all $p \in [0, 2]$ identifies the parameters α and β. □

Remark 4.10 A related study of the asymptotic theory is presented in [20], who investigated the fine structure of Lévy moving average processes driven by a truncated β-stable Lévy motion. The authors showed the result of Theorem 4.7(ii) (see Theorem 5.1 therein), whose prove was however incorrect, since it was based on the computation of the variance that diverges to infinity.

In a recent work [33] extended the law of large numbers in Theorem 4.7(ii) to integral processes driven by fractional Lévy motion. The main idea relies on the mathematical intuition described in Remark 2.14. □

The next theorem demonstrates a central limit theorem associated with Theorem 4.7(ii) (see [19]).

Theorem 4.11 *Assume that condition* (A′) *holds and fix* $t > 0$. *Let* L *be a symmetric* β*-stable Lévy process with characteristic triplet* $(0, 0, c|x|^{-1-\beta}dx)$ *and* $\beta \in (0, 2)$. *When* $k \geq 2$, $\alpha \in (0, k - 2/\beta)$ *and* $p < \beta/2$ *then it holds*

$$\Delta_n^{-1/2}\left(\Delta_n^{1-p(\alpha+1/\beta)}V(Y, p, k; \Delta_n)_t - tc_p\right) \xrightarrow{d} \mathcal{N}(0, t\eta^2), \tag{4.24}$$

where the quantity η^2 *is defined via*

$$\eta^2 = \theta(0) + 2\sum_{i=1}^{\infty}\theta(i),$$

$$\theta(i) = a_p^{-2}\int_{\mathbb{R}^2}\frac{1}{|s_1 s_2|^{1+p}}\psi_i(s_1, s_2)ds_1 ds_2,$$

$$\psi_i(s_1, s_2) = \exp\left(-c|f(0)|^\beta\int_{\mathbb{R}}|s_1 h(x) - s_2 h(x+i)|^\beta dx\right),$$

$$- \exp\left(-c|f(0)|^\beta\int_{\mathbb{R}}|s_1 h(x)|^\beta + |s_2 h(x+i)|^\beta dx\right),$$

where the function h *is defined at* (4.14) *and* $a_p := \int_{\mathbb{R}}(\exp(iu) - 1)|u|^{-1-p}du$.

Let us explain the various conditions of Theorem 4.11. The assumption $p < \beta/2$ ensures the existence of variance of the statistic $V(Y, p, k; \Delta_n)_t$. The validity range of the central limit theorem ($\alpha \in (0, k - 2/\beta)$) is smaller than the validity range of the law of large numbers in Theorem 4.7(ii) ($\alpha \in (0, k - 1/\beta)$). It is not clear which limit distribution appears in case $\alpha \in (k - 2/\beta, k - 1/\beta)$. A more severe assumption is $k \geq 2$, which excludes the first order increments. The limit theory in this case is also unknown.

Remark 4.12 Let us explain a somewhat complex form of the variance η^2. A major problem of proving Theorems 4.7(ii) and 4.11 is that neither the expectation of $|\Delta_{i,k}^n Y|^p$ nor its variance can be computed directly. However, the identity

$$|x|^p = a_p^{-1} \int_{\mathbb{R}} (\exp(iux) - 1)|u|^{-1-p} du \qquad \text{for } p \in (0, 1),$$

which can be shown by substitution $y = ux$, turns out to be a useful instrument. Indeed, for any deterministic function $\varphi : \mathbb{R} \to \mathbb{R}$ satisfying the conditions of Remark 2.3, it holds that

$$\mathbb{E}\left[\exp\left(iu \int_{\mathbb{R}} \varphi_s dL_s\right)\right] = \exp\left(-c|u|^\beta \int_{\mathbb{R}} |\varphi_s|^\beta ds\right).$$

This two identities are used to compute the variance of the statistic $V(Y, p, k; \Delta_n)_t$ and they are both reflected in the formula for the quantity $\theta(i)$.

Remark 4.13 As in the case of a Gaussian driver, Theorem 4.7(ii) might be useful for statistical applications. Indeed, for any fixed $t > 0$, it holds that

$$S_n = \frac{V(Y, p, k; 2\Delta_n)_t}{V(Y, p, k; \Delta_n)_t} \xrightarrow{\mathbb{P}} 2^{p(\alpha+1/\beta)},$$

under conditions of Theorem 4.7(ii). Thus, a consistent estimator of α (resp. β) can be constructed given the knowledge of β (resp. α) and the validity of conditions $\alpha \in (0, k - 1/\beta)$ and $p < \beta$. A bivariate version of the central limit theorem (4.24) for frequencies Δ_n and $2\Delta_n$ would give a possibility to construct feasible confidence regions. □

Obviously, the presented limiting results still need to be extended to spatio-temporal setting. Asymptotic theory for ambit fields observed on a grid in time-space or along a curve would be extremely useful for statistical analysis of turbulent flows.

References

1. D.J. Aldous, G.K. Eagleson, On mixing and stability of limit theorems. Ann. Probab. **6**(2), 325–331 (1978)
2. E. Alos, O. Mazet, D. Nualart, Stochastic calculus with respect to Gaussian processes. Ann. Probab. **29**(2), 766–801 (2001)
3. O.E. Barndorff-Nielsen, J. Schmiegel, Ambit processes; with applications to turbulence and cancer growth, in *Stochastic Analysis and Applications: The Abel Symposium 2005*, ed. by F.E. Benth, G.D. Nunno, T. Linstrøm, B. Øksendal, T. Zhang (Springer, Heidelberg, 2007), pp. 93–124
4. O.E. Barndorff-Nielsen, J. Schmiegel, Time change, volatility and turbulence, in *Proceedings of the Workshop on Mathematical Control Theory and Finance*, Lisbon 2007, ed. by A. Sarychev, A. Shiryaev, M. Guerra, M.d.R. Grossinho (Springer, Berlin, 2008), pp. 29–53

5. O.E. Barndorff-Nielsen, J. Schmiegel, Brownian semistationary processes and volatility/intermittency, in *Advanced Financial Modelling*, ed. by H. Albrecher, W. Runggaldier, W. Schachermayer (Walter de Gruyter, Berlin, 2009), pp. 1–26
6. O.E. Barndorff-Nielsen, S.E. Graversen, J. Jacod, M. Podolskij, N. Shephard, A central limit theorem for realised power and bipower variations of continuous semimartingales, in *From Stochastic Calculus to Mathematical Finance. Festschrift in Honour of A.N. Shiryaev*, ed. by Yu. Kabanov, R. Liptser, J. Stoyanov (Springer, Heidelberg, 2006), pp. 33–68
7. O.E. Barndorff-Nielsen, E.B.V. Jensen, K.Y. Jónsdóttir, J. Schmiegel, Spatio-temporal modelling – with a view to biological growth, in *Statistical Methods for Spatio-Temporal Systems*, ed. by B. Finkenstädt, L. Held, V. Isham (Chapman and Hall/CRC, London, 2007), pp. 47–75
8. O.E. Barndorff-Nielsen, J.M. Corcuera, M. Podolskij, Power variation for Gaussian processes with stationary increments. Stoch. Process. Their Appl. **119**, 1845–865 (2009)
9. O.E. Barndorff-Nielsen, J.M. Corcuera, M. Podolskij, Multipower variation for Brownian semistationary processes. *Bernoulli* **17**(4), 1159–1194 (2011)
10. O.E. Barndorff-Nielsen, F.E. Benth, A. Veraart, Modelling energy spot prices by volatility modulated Lévy-driven Volterra processes. Bernoulli **19**(3), 803–845 (2013)
11. O.E. Barndorff-Nielsen, J.M. Corcuera, M. Podolskij, Limit theorems for functionals of higher order differences of Brownian semi-stationary processes, in *Prokhorov and Contemporary Probability Theory: In Honor of Yuri V. Prokhorov*, ed. by A.N. Shiryaev, S.R.S. Varadhan, E.L. Presman (Springer, Berlin/New York, 2013)
12. O.E. Barndorff-Nielsen, F.E. Benth, J. Pedersen, A. Veraart, On stochastic integration for volatility modulated Lévydriven Volterra processes. Stoch. Process. Their Appl. **124**(1), 812–847 (2014)
13. O.E. Barndorff-Nielsen, F.E. Benth, B. Szozda, On stochastic integration for volatility modulated Brownian-driven Volterra processes via white noise analysis. Infin. Dimens. Anal. Quantum Probab. Relat. Top. **17**(2), 1450011 (2014)
14. O.E. Barndorff-Nielsen, F.E. Benth, A. Veraart, Recent advances in ambit stochastics with a view towards tempo-spatial stochastic volatility/intermittency. Banach Center Publ. **104**, 25–60 (2015)
15. O.E. Barndorff-Nielsen, M. Pakkanen, J. Schmiegel, Assessing relative volatility/intermittency/energy dissipation. Electron. J. Stat. **8**(2), 1996–2021 (2014)
16. A. Basse, Gaussian moving averages and semimartingales. Electron. J. Probab. **13**(39), 1140–1165 (2008)
17. A. Basse, J. Pedersen, Lévy driven moving averages and semimartingales. Stoch. Process. Their Appl. **119**(9), 2970–2991 (2009)
18. A. Basse-O'Connor, J. Rosinski, On infinitely divisible semimartingales. Probab. Theory Relat. Fields (2012, to appear)
19. A. Basse-O'Connor, R. Lachieze-Rey, M. Podolskij, Limit theorems for Lévy moving average processes. Working paper (2014)
20. A. Benassi, S. Cohen, J. Istas, On roughness indices for fractional fields. Bernoulli **10**(2), 357–373 (2004)
21. C. Bender, A. Lindner, M. Schicks, Finite variation of fractional Levy processes. J. Theor. Probab. **25**(2), 594–612 (2012)
22. F.E. Benth, A. Süß, Integration theory for infinite dimensional volatility modulated Volterra processes. Working paper (2013). Available at arXiv:1303.7143
23. F.E. Benth, H. Eyjolfsson, A.E.D. Veraart, Approximating Lévy semistationary processes via Fourier methods in the context of power markets. SIAM J. Financ. Math. **5**(1), 71–98 (2014)
24. K. Bichteler, J. Jacod, Random measures and stochastic integration, in *Theory and Application of Random Fields*. Lecture Notes in Control and Information Sciences, vol. 49, (Springer, Berlin/Heidelberg, 1983), pp. 1–18
25. P. Breuer, P. Major, Central limit theorems for nonlinear functionals of Gaussian fields. J. Multivar. Anal. **13**(3), 425–441 (1983)
26. C. Chong, C. Klüppelberg, Integrability conditions for space-time stochastic integrals: theory and applications. Bernoulli (2014, to appear)

27. J.M. Corcuera, D. Nualart, J.H.C. Woerner, Power variation of some integral fractional processes. Bernoulli **12**(4), 713–735 (2006)
28. J.M. Corcuera, E. Hedevang, M. Pakkanen, M. Podolskij, Asymptotic theory for Brownian semi-stationary processes with application to turbulence. Stoch. Process. Their Appl. **123**, 2552–2574 (2013)
29. J.M. Corcuera, D. Nualart, M. Podolskij, Asymptotics of weighted random sums. Commun. Appl. Ind. Math. (2014, to appear)
30. R.C. Dalang, L. Quer-Sardanyons, Stochastic integrals for spde's: a comparison. Expos. Math. **12**(1), 67–109 (2011)
31. G. Di Nunno, B. Oksendal, F. Proske, *Malliavin Calculus for Lévy Processes with Applications to Finance* (Springer, Berlin, 2009)
32. K. Gärtner, M. Podolskij, On non-standard limits of Brownian semi-stationary processes. Stoch. Process. Their Appl. **125**(2), 653–677 (2014)
33. S. Glaser, A law of large numbers for the power variation of fractional Lévy processes. Stoch. Anal. Appl. **33**(1), 1–20 (2015)
34. J. Jacod, Asymptotic properties of realized power variations and related functionals of semimartingales. Stoch. Process. Appl. **118**, 517–559 (2008)
35. J. Jacod, P.E. Protter, *Discretization of Processes* (Springer, Berlin/Heidelberg/New York, 2012)
36. J. Jacod, A.N. Shiryaev, *Limit Theorems for Stochastic Processes*, 2d edn. (Springer, Berlin. 2002)
37. T. Jeulin, M. Yor, Moyennes mobiles et semimartingales, in *Séminaire de Probabilités XXVII*, vol. 1557 (Springer, Berlin, 1993), pp. 53–77
38. F.B. Knight, *Foundations of the Prediction Process*. Volume 1 of Oxford Studies in Probability (The Clarendon Press/Oxford University Press/Oxford Science Publications, New York, 1992)
39. T. Marquardt, Fractional Lévy processes with an application to long memory moving average processes. Bernoulli **12**, 1090–1126 (2006)
40. D. Nualart, *The Malliavin Calculus and Related Topics*, 2nd edn. (Springer, Berlin, 2006)
41. D. Nualart, G. Peccati, Central limit theorems for multiple stochastic integrals. Ann. Probab. **33**(1), 177–193 (2005)
42. M. Pakkanen, Limit theorems for power variations of ambit fields driven by white noise. Stoch. Process. Their Appl. **124**(5), 1942–1973 (2014)
43. M. Pakkanen, A. Réveillac, Functional limit theorems for generalized variations of the fractional Brownian sheet. Bernoulli (2015, to appear)
44. G. Peccati, C.A. Tudor, Gaussian limits for vector-values multiple stochastic integrals, in *Séminaire de Probabilités XXXVIII*. Lecture Notes in Mathematics, vol. 1857 (Springer, Berlin, 2005), pp. 247–193
45. P. Protter, Volterra equations driven by semimartingales. Ann. Probab. **13**(2), 519–530 (1985)
46. B. Rajput, J. Rosinski, Spectral representation of infinitely divisible distributions. Probab. Theory Relat. Fields **82**, 451–487 (1989)
47. A. Rényi, On stable sequences of events. Sankhyā Ser. A **25**, 293–302 (1963)
48. A. Réveillac, Estimation of quadratic variation for two-parameter diffusions. Stoch. Process. Their Appl. **119**(5), 1652–1672 (2009)
49. M. Taqqu, Convergence of integrated processes of arbitrary Hermite rank. Z. Wahrsch. Verw. Gebiete **50**(1), 53–83 (1979)
50. J. Walsh, An introduction to stochastic partial differential equations, in *École d'Eté de Prob. de St. Flour XIV*. Lecture Notes in Mathematics, vol. 1180 (Springer, Berlin/Heidelberg, 1986), pp. 265–439

Progress in Probability (PP)

This series is designed for the publication of workshops, seminars and conference proceedings on all aspects of probability theory and stochastic processes, as well as their connections with and applications to other areas such as mathematical statistics and statistical physics.

Edited by

Davar Khoshnevisan, The University of Utah, Salt Lake City, USA
Andreas Kyprianou, University of Bath, UK
Sidney I. Resnick, Cornell University, Ithaca, USA

■ **Vol. 68. Dalang, R.C., Dozzi, M., Flandoli, F., Russo, F. (Eds.)**
Stochastic Analysis: A Series of Lectures (2015).
ISBN 978-3-0348-0908-5
This book presents an up-to-date picture of research on stochastic analysis and highlights recent developments in the form of lecture notes contributed by leading international experts. The topics include the influence of stochastic perturbations on differential and partial differential equations, the modeling of stochastic perturbations by Lévy processes and the numerical aspects of the approximation of such equations, stochastic integration and stochastic calculus in abstract function spaces, stochastic partial differential equations arising from measure-valued diffusions, and reflection problems for multidimensional Brownian motion.
Based on a course given at the Ecole Polytechnique Fédérale's Bernoulli Center in Lausanne, Switzerland, from January to June 2012, these lecture notes offer a valuable resource not only for specialists, but also for other researchers and PhD students in the fields of probability, analysis and mathematical physics.

■ **Vol. 67. Dalang, R.; Dozzi, M.; Russo, F. (Eds.)**
Seminar on Stochastic Analysis, Random Fields and Applications VII (2013).
ISBN 978-3-0348-0544-5
This volume contains refereed research or review articles presented at the 7th Seminar on Stochastic Analysis, Random Fields and Applications which took place at the Centro Stefano Franscini (Monte Verità) in Ascona , Switzerland, in May 2011. The seminar focused mainly on: stochastic (partial) differential equations, especially with jump processes, construction of solutions and approximations; Malliavin calculus and Stein methods, and other techniques in stochastic analysis, especially chaos representations and convergence, and applications to models of interacting particle systems; stochastic methods in financial models, especially models for power markets or for risk analysis, empirical estimation and approximation, stochastic control and optimal pricing. The book will be a valuable resource for researchers in stochastic analysis and for professionals interested in stochastic methods in finance.

■ **Vol. 66. Houdré, C.; Mason, D.M.; Rosiński, J.; Wellner, J.A. (Eds.)**
High Dimensional Probability VI. The Banff Volume (2013).
ISBN 978-3-0348-0489-9
This is a collection of papers by participants at the High Dimensional Probability VI Meeting held from October 9–14, 2011 at the Banff International Research Station in Banff, Alberta, Canada. High Dimensional Probability (HDP) has resulted in the creation of powerful new tools and perspectives, whose range of application has led to interactions with other areas of mathematics, statistics, and computer science. The papers in this volume show that HDP theory continues to develop new tools, methods, techniques and perspectives to analyze the random phenomena. Both researchers and advanced students will find this book of great use for learning about new avenues of research.

■ **Vol. 65. Kohatsu-Higa, A.; Privault, N.; Sheu, S.-J. (Eds.)**
Stochastic Analysis with Financial Applications (2011).
ISBN 978-3-0348-0096-9
This book presents a broad overview of the range of applications of stochastic analysis and some of its recent theoretical developments. This includes numerical simulation, error analysis, parameter estimation, as well as control and robustness properties for stochastic equations. The book also covers the areas of backward stochastic differential equations via the (non-linear) G-Brownian motion and the case of jump processes. Concerning the applications to finance, many of the articles deal with the valuation and hedging of credit risk in various forms, and include recent results on markets with transaction costs.

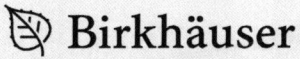

Printed by Printforce, the Netherlands